Methods in Microarray Normalization

Drug Discovery Series

Series Editor

Andrew A. Carmen
illumina, Inc.
San Diego, California, U.S.A.

1. Virtual Screening in Drug Discovery, *edited by Juan Alvarez and Brian Shoichet*
2. Industrialization of Drug Discovery: From Target Selection Through Lead Optimization, *edited by Jeffrey S. Handen, Ph.D.*
3. Phage Display in Biotechnology and Drug Discovery, *edited by Sachdev S. Sidhu*
4. G Protein-Coupled Receptors in Drug Discovery, *edited by Kenneth H. Lundstrom and Mark L. Chiu*
5. Handbook of Assay Development in Drug Discovery, *edited by Lisa K. Minor*
6. In Silico Technologies in Drug Target Identification and Validation, *edited by Darryl León and Scott Markel*
7. Biochips as Pathways to Drug Discovery, *edited by Andrew Carmen and Gary Hardiman*
8. Functional Protein Microarrays in Drug Discovery, *edited by Paul F. Predki*
9. Functional Informatics in Drug Discovery, *edited by Sergey Ilyin*
10. Methods in Microarray Normalization, *edited by Phillip Stafford*

Drug Discovery Series/10

Methods in Microarray Normalization

Edited by
Phillip Stafford

CRC Press
Taylor & Francis Group
Boca Raton London New York

CRC Press is an imprint of the
Taylor & Francis Group, an **informa** business

CRC Press
Taylor & Francis Group
6000 Broken Sound Parkway NW, Suite 300
Boca Raton, FL 33487-2742

© 2008 by Taylor & Francis Group, LLC
CRC Press is an imprint of Taylor & Francis Group, an Informa business

No claim to original U.S. Government works
Printed in the United States of America on acid-free paper
10 9 8 7 6 5 4 3 2 1

International Standard Book Number-13: 978-1-4200-5278-7 (Hardcover)

This book contains information obtained from authentic and highly regarded sources. Reprinted material is quoted with permission, and sources are indicated. A wide variety of references are listed. Reasonable efforts have been made to publish reliable data and information, but the author and the publisher cannot assume responsibility for the validity of all materials or for the consequences of their use.

Except as permitted under U.S. Copyright Law, no part of this book may be reprinted, reproduced, transmitted, or utilized in any form by any electronic, mechanical, or other means, now known or hereafter invented, including photocopying, microfilming, and recording, or in any information storage or retrieval system, without written permission from the publishers.

For permission to photocopy or use material electronically from this work, please access www.copyright.com (http://www.copyright.com/) or contact the Copyright Clearance Center, Inc. (CCC) 222 Rosewood Drive, Danvers, MA 01923, 978-750-8400. CCC is a not-for-profit organization that provides licenses and registration for a variety of users. For organizations that have been granted a photocopy license by the CCC, a separate system of payment has been arranged.

Trademark Notice: Product or corporate names may be trademarks or registered trademarks, and are used only for identification and explanation without intent to infringe.

Library of Congress Cataloging-in-Publication Data

Methods in microarray normalization / editor, Phillip Stafford.
  p. ; cm. -- (Drug discovery series ; 10)
  Includes bibliographical references and index.
  ISBN 978-1-4200-5278-7 (hardback : alk. paper)
  1. DNA microarrays. 2. Gene expression. I. Stafford, Phillip. II. Series.
  [DNLM: 1. Microarray Analysis--methods. 2. Gene Expression Profiling--methods. QZ 52 M592 2007]

QP624.5.D726M47 2007
572.8'636--dc22   2007046114

Visit the Taylor & Francis Web site at
http://www.taylorandfrancis.com

and the CRC Press Web site at
http://www.crcpress.com

Contents

Preface vii
The Editor ix
Contributors xi
Summary of Chapters xiii

Chapter 1 A Comprehensive Analysis of the Effect of Microarray Data Preprocessing Methods on Differentially Expressed Transcript Selection 1

Monika Ray, Johannes Freudenberg, and Weixiong Zhang

Chapter 2 Differentiation Detection in Microarray Normalization 19

Lei M. Li and Chao Cheng

Chapter 3 Preprocessing and Normalization for Affymetrix GeneChip Expression Microarrays 41

Ben Bolstad

Chapter 4 Spatial Detrending and Normalization Methods for Two-Channel DNA and Protein Microarray Data 61

Adi Laurentiu Tarca, Sorin Draghici, Roberto Romero, and Michael Tainsky

Chapter 5 A Survey of cDNA Microarray Normalization and a Comparison by *k*-NN Classification 81

Wei Wu and Eric P. Xing

Chapter 6 Technical Variation in Modeling the Joint Expression of Several Genes 121

Walter Liggett

Chapter 7 Biological Interpretation for Microarray Normalization Selection 151

Phillip Stafford and Younghee Tak

Chapter 8 Methodology of Functional Analysis for Omics Data Normalization 173

Ally Perlina, Tatiana Nikolskaya, and Yuri Nikolsky

Chapter 9 Exon Array Analysis for the Detection of Alternative Splicing 205

Paul Gardina and Yaron Turpaz

Chapter 10 Normalization of Array CGH Data 233

Bo Curry, Jayati Ghosh, and Charles Troup

Chapter 11 SNP Array-Based Analysis for Detection of Chromosomal Aberrations and Copy Number Variations 245

Ben Bolstad, Srinka Ghosh, and Yaron Turpaz

Index 267

Preface

Gene expression profiling using microarrays has been increasingly brought to task for the simple goal of ensuring that data from a particular tissue and a particular platform match the same tissue run on another platform in another lab. Expression arrays are becoming extremely valuable in the clinic and, as such, are being closely monitored by the Food and Drug Administration (FDA). MammaPrint uses the Agilent microarray system, which has been FDA approved. The National Institute for Standards has created the Microarray Quality Control Consortium in order to identify those factors that tend to throw lab-to-lab and cross-platform concordance off. They have made remarkable progress, and papers and new insight continue to pour out of their initial investment. The External RNA Consortium is purposed to create standards and spike-ins to allow proper measurements of array performance no matter what the location or environmental conditions are.

These efforts are moving the expression microarray from the experimental lab into the position of a trusted and widespread technological commodity. At that point it becomes a tool, but before that happens, the concept of data preprocessing and cross-platform concordance must be thoroughly addressed. Both topics have received much attention, but a line in the sand must be drawn beyond which we should assume that proper normalization will be used to minimize noise, bias, and cross-platform discordance, and to improve accuracy of biological interpretation.

Normalization and array preprocessing encompass a variety of data manipulations, but in this book we will look at those mathematical formulae that remove known and intrinsic biases in order to level the playing field—leaving expression data to stand on their own without accommodating the platform manufacturer, the fluorescent dye used, the varying quality of RNA, the laboratory, scanner, or any of a number of potential nuisance factors. In this book, many of the most respected authors in the field present their work for the reader to use and judge. In addition, several scientists have chosen to present their novel techniques for next-generation technologies such as the single nucleotide polymorphism (SNP) chip, the exon array, the comparative genomic hybridization (CGH) array, and the protein array.

Some useful references for the reader to engage in his or her own application of normalization include:

DNMAD (Diagnosis and Normalization for Microarray Data): http://dnmad.bioinfo.cnio.es/
Global normalization of two-color microarray data with loess: http://nbc11.biologie.uni-kl.de/framed/left/menu/auto/right/microarray/loess.shtml
SNOMAD: http://pevsnerlab.kennedykrieger.org/snomadinput.html
Terry Speed's "Normalization of Microarray Data—How to Do It!": http://www.maths.lth.se/bioinformatics/calendar/20010409/

Two-color DNA microarray normalization tool: http://128.122.133.135/cgi-bin/rodrigo/normalization.cgi

University of Pittsburgh Department of Bioinformatics: http://bioinformatics.upmc.edu/index.html

The Editor

Mathematics and biology have always gone together for me. As an undergraduate in physics and chemistry, I was interested in coercing computers (then the mighty Apple IIe) to compute as many simulations of my models as possible. At the time, I had to be content with sampling hundreds of times; a few decades later, I could have sampled billions of times in the same amount of computer time for less money. When I followed my older brother's career into the burgeoning field of biotechnology, I kept my mathematical bent. My brother pursued bioreactor engineering at Genetech and was one of the early developers of large-scale, continuous-flow fermentors for recombinant bacteria. I subsequently graduated with degrees in microbiology and botany, with particular emphasis on physics and chemistry, but in graduate school I jumped into the molecular biology of human pathogenic fungi. My professor at the time, Dr. Douglas Rhoads, besides being a molecular wizard, was also a bit of a computer geek and allowed me to pursue what would eventually become the field of bioinformatics.

As the genome of *Saccharomyces cervisiae* was completed, I became the de facto expert at sequence alignments and comparative genomics (which had not actually been invented). Gopher, Veronica, Mosaic, GCG, and GenBank on a single CD were signs of the time. As a post-doc at Dartmouth College, I studied cancer genomics, again in a yeast model. Yeast was the darling of biochemistry because so many functions can be reconstituted *in vitro*. My career in mathematical genomics soon became computational biochemistry of protein–protein interactions. Drug companies, academics, and everyone in between became suddenly interested in the Brent/Golemis two-hybrid screens. The bind database in Canada was just an idea, and Curagen had just completed a massive two-hybrid screen of all baits against all prey in yeast. Although some interesting findings emerged, more questions than answers were uncovered. Somewhere along the way, yeast lost favor for more sophisticated models, drug companies merged in order to stay solvent, and the Internet paid homage to a few dot com millionaires and even more dot failures.

However, this was also the time when the human and mouse genomes were accessible online, and programming and computer skills were merging with biological skills to create a new breed of scientist: the bioinformatician. Now, of course, there are dozens of subspecialties, but at the time if you could write a Perl script and knew the difference between the nucleus and a nucleotide you could write your own ticket. I was content to study the functional biochemistry of molecular motors in a wonderful model organism: *Loligo pealii*, the longfin squid, found offshore near the Woods Hole Biological Institute. Several summers under the watchful eye of Dr. George Langford provided me with an enormous respect for the extravagant levels of care that biological molecules require in order to function properly in an extracellular environment. As well, I learned to cook calamari. With this in mind, I returned to New Hampshire with the goal of integrating protein biochemistry, genetics, and mathematics. We had collected thousands of observations of vesicles moving along axons on actin and microtubules. This was a compelling data set, especially when examined using the appropriate statistics. I wished to extract insight using highly repetitive observations. This was the realm of molecular (nonclinical trials-based) biostatistics.

It was at this time that Motorola Life Sciences was growing their CodeLink microarray business. I joined their group and took advantage of Motorola University's Blackbelt Program to apply Six Sigma statistical methods, which were then strictly associated with semiconductor manufacturing. Mixing Six Sigma ideas with biological data was revolutionary. The first few meetings of Cambridge Healthtech's microarray data analysis conferences were filled with statisticians arguing about confidence, false positives, and how best to normalize cDNA data. Affymetrix was just making their early U95 chips; Agilent had split from Hewlett Packard and was optimizing the ink-jet printing technology to spot cDNA molecules. As commercial expression arrays grew in demand, more emphasis was placed on quality control, replication, and experimental design. I enjoyed the business of creating and analyzing new microarray technology, but I was drawn to a new field: translational genomics. Here was a field that was using microarrays to their fullest extent; TGen, the Translational Genomics Research Institute, and Jeff Trent were leading the charge.

Imagine a clinical trial where you could measure improvement in health between patients whose doctor did or did not have access to information from an expression microarray. This was the type of clinical research that led to Agendia's array-based breast cancer detection system, MammaPrint. Something dramatic happened: Normalization of expression data could have a direct effect on an individual's health. No longer was mathematics abstract; a mistake in data processing could ruin the outcome of a cancer treatment. This is arguably the current state of events, and it was at this point I began seeking the next big, highly parallel biomedical technology. Antibody arrays, glycoarrays, protein and peptide chips, methylation arrays, aptomer, CGH, exon, and SNP chips are emerging quickly. With them come more mathematical challenges. The Biodesign Institute at Arizona State University prides itself on tackling these challenges on the way to creating disruptive paradigms in medicine and health care. This is where I pursue my research; it is here that new bioinformatic solutions are being created to interpret highly parallel measurements of biomolecules.

To those who are interested in pursuing this field, I would leave some advice: Learn mathematics while you learn biology. Develop models that predict biological behavior. Think about modular software design even as you think about biological behavior. The newest biomedical technologies are only as good as the computational support behind them. These are the new integrative sciences that will propel science into the next millennium, and today's bioinformatics students will be creating tomorrow's medical advances.

Phillip Stafford
Tempe, Arizona

Contributors

Ben Bolstad
Affymetrix, Inc.
Santa Clara, California

Chao Cheng
Computational Biology
University of Southern California
Los Angeles, California

Bo Curry
Agilent Technologies, Inc.
Santa Clara, California

Sorin Draghici
Bioinformatics Core
Karmanos Cancer Institute
Detroit, Michigan

Department of Computer Science
Wayne State University
Detroit, Michigan

Johannes Freudenberg
Biomedical Informatics
Cincinnati Children's Hospital Medical Center
Cincinnati, Ohio

Paul Gardina
Affymetrix, Inc.
Santa Clara, California

Jayati Ghosh
Agilent Technologies, Inc.
Santa Clara, California

Srinka Ghosh
Affymetrix, Inc.
Santa Clara, California

Lei M. Li
Computational Biology and Mathematics
University of Southern California
Los Angeles, California

Walter Liggett
Statistical Engineering Division
National Institute of Standards and Technology
Gaithersburg, MD

Tatiana Nikolskaya
GeneGo, Inc.
St. Joseph, Michigan

Yuri Nikolsky
GeneGo, Inc.
St. Joseph, Michigan

Ally Perlina
GeneGo, Inc.
St. Joseph, Michigan

Monika Ray
Department of Computer Science and Engineering
Washington University in Saint Louis
Saint Louis, Missouri

Roberto Romero
Perinatology Research Branch
NICHD/NIH/DHHS
Bethesda, Maryland and
Detroit, Michigan

Phillip Stafford
Arizona State University
Tempe, Arizona

Michael Tainsky
Program in Molecular Biology and
 Human Genetics
Karmanos Cancer Institute
Wayne State University
Detroit, Michigan

Younghee Tak
Arizona State University
Tempe, Arizona

Adi Laurentiu Tarca
Perinatology Research Branch
NICHD/NIH/DHHS
Bethesda, Maryland and
Detroit, Michigan

Bioinformatics Core
Karmanos Cancer Institute
Detroit, Michigan

Department of Computer Science
Wayne State University
Detroit, Michigan

Charles Troup
Agilent Technologies, Inc.
Santa Clara, California

Yaron Turpaz
Affymetrix, Inc.
Santa Clara, California

Wei Wu
Division of Pulmonary, Allergy, and
 Critical Care Medicine
School of Medicine
University of Pittsburgh
Pittsburgh, Pennsylvania

Eric P. Xing
School of Computer Science
Carnegie Mellon University
Pittsburgh, Pennsylvania

Weixiong Zhang
Department of Genetics
Department of Computer Science and
 Engineering
Washington University in Saint Louis
Saint Louis, Missouri

Summary of Chapters

Chapter 1: A Comprehensive Analysis of the Effect of Microarray Data Preprocessing Methods on Differentially Expressed Transcript Selection
Normalization techniques have a profound influence on microarray expression data. Recently, in an attempt to create a robust consensus across multiple expression platforms, several well-meaning groups have proposed methods that strongly weight cross-platform concordance. However, when normalization techniques affect downstream analysis on the same platform—in fact, when they are applied to the same data—are we getting ahead of ourselves by trying to achieve cross-platform concordance without satisfactorily addressing this problem first? The authors have investigated the effects of normalization techniques on the same platform and same data set. They present a thorough report on arguably the most important output of expression analysis: the list of genes shown to be differentially expressed across conditions, which is a usual step following expression data generation. This comprehensive approach provides the reader with a nonbiased view of the effect of normalization on detection of differential expression.

Chapter 2: Differentiation Detection in Microarray Normalization
Microarray data are both complex and rich in information content. The complexity of microarray expression data creates a heavy dependence on normalization methods that minimize nuisance factors that arise from uncontrollable factors. One important factor that arises is nonrandom patterns of differential expression between control and treatment samples. It is important to distinguish the difference between outliers and true differences on a per-gene basis. The authors address several important nuisance factors by utilizing least trimmed squares (LTS) and a sampling technique that ameliorates spatial artifacts from either manufacturing or hybridization effects.

Chapter 3: Preprocessing and Normalization for Affymetrix GeneChip Expression Microarrays
Affymetrix is one of the leaders in microarray fabrication and sales and arguably at the forefront of the drive to increase the density of probes. In order to estimate the quality for these arrays, one must utilize probes that report hybridization efficiency, mismatch thermodynamics, and general kinetics of the hybridization reaction. Affymetrix relies on perfect and single-base-pair mismatch probes to gather these important indicators of quality. Several Affymetrix normalization methods also make use of these probes. This chapter introduces the reader to the preprocessing steps needed for Affymetrix arrays, details about hybridization, and a description of RMA, GCRMA, and MAS5 normalization methods.

Chapter 4: Spatial Detrending and Normalization Methods for Two-Channel DNA and Protein Microarray Data
Although many commercial expression microarrays are now available with extremely high-quality, low spatial signal bias, many specialized and custom genomes are

printed with robotic microarray printers. Additionally, new technologies such as protein and peptide arrays, antibody arrays, and glycoarrays must be printed using contact (quill) printing simply due to the complexity and heterogeneity of the probe solution. Spatial artifacts can occur via a number of physical limitations of the printing process, hybridization conditions, and wash steps. This chapter introduces the reader to spatial bias problems and illustrates several useful solutions using freely available software packages from the Bioconductor project (www.bioconductor.org) implemented in the statistical environment R.

Chapter 5: A Survey of cDNA Microarray Normalization and a Comparison by k-NN Classification
Complementary DNA (cDNA) expression microarrays continue to be prevalent as a useful platform for high-throughput analysis of gene expressions in biological subjects. They can be flexibly custom made to target genes of special interest and thus are particularly useful for nonroutine exploratory investigations. One of the most obvious impacts of the two-color cDNA arrays is the effect of fluorescent dyes on the resulting data. Factors such as dye size and incorporation efficiency and hybridization efficiency can cause variance in the outcome. The authors present a comprehensive survey of normalization methods and an empirical assessment, based on *k*-NN leave-one-out cross-validation of the effectiveness of each normalization scheme in removing noise from the array data.

Chapter 6: Technical Variation in Modeling the Joint Expression of Several Genes
Recently, interest has increased about a long discussed topic: cross-laboratory expression array concordance. With several companies already selling an expression-based clinical product (Agendia's MammaPrint being the first to receive clearance by the FDA), the need to standardize conditions and analyses becomes critical if one hopes to see consistent results. The MicroArray Quality Control (MAQC) project collected interlaboratory data on several microarray platforms. Analysis of these data by the National Institute of Standards and Technology (NIST) shows how comparable data can be used to diminish sources of measurement noise. The author presents a useful concept in modeling joint expression of genes that includes assay-to-assay normalization, microarray response linearity, variance stabilization, variance of individual genes and covariance between pairs of genes, and site-to-site variation in microarray response.

Chapter 7: Biological Interpretation for Microarray Normalization Selection
Many normalization methods exist for most microarray platforms, but more exist for Affymetrix GeneChip arrays than for any other single platform other than the accumulation of methods for cDNA arrays. Normalization affects not only chip-to-chip comparisons, but also within-chip comparisons. The authors propose a method of looking at patterns of similar functional groups and metabolic pathways that result from gene lists created from identical data normalized with seven different normalization methods.

Chapter 8: Methodology of Functional Analysis for Omics Data Normalization
The authors present a novel method of assessing data integrity and normalization effects on the analytical outcome of the expression data. Although correlation, t-tests,

fold-changes, p-values, or log ratios are useful for determining variance sources and trends between data, the most important assessment of data quality is how closely the normalized data correlate to meaningful changes at the transcriptional level. Relationships between housekeeping genes have been used as a proxy for internal consistency and bias, but the authors present a more robust method involving many more genes: pathways analysis of microarray data. GeneGo's MetaCore is an advanced pathway analysis suite containing an extensive database of gene–protein–RNA–metabolite–compound relationships from a combination of manual and automatically text-mined public literature. Regulatory pathways can be skewed by gene lists that change due to normalization effects. The authors present comparative analysis of highly significant genes among regular tissue expression data, filtered by the same statistical means, but via different normalization methods and from different microarray platforms, to provide a more "biologically informed" assessment of expression data.

Chapter 9: Exon Array Analysis for the Detection of Alternative Splicing
The human genome consists of approximately 20,000 genes—a remarkably low number considering human complexity. However, a rich and complex mixture of mRNAs can be generated by the combinatorial assembly of modular coding sections called exons, which can subtly or dramatically alter the function of proteins depending on whether they are included or skipped in a particular gene product. This efficient method of enhancing genetic flexibility in a compact genome comes at a cost: misspliced variants may escape cellular proofreading and potentiate some genetic diseases (cystic fibrosis transmembrane conductance regulator and cystic fibrosis) as well as several types of cancer (CCK-B and colon carcinoma). The authors present normalization and signal estimation methods focusing on the detection of differential usage of the estimated 100,000–200,000 exons distributed within human genes.

Chapter 10: Normalization of Array CGH Data
Underlying the phenotypic changes observed in cancer and in many heritable disorders are genetic changes. During tumor development, chromosomal aberrations accumulate, including amplifications, deletions, and rearrangements. Some aberrations, such as amplification of genes associated with cell proliferation and deletion or partial deletion of genes regulating cell growth and apoptosis, are selected for and play a direct role in oncogenesis. Other aberrations are apparently accidental side effects of increasing genetic instability. Comparative genomic hybridization to DNA microarrays (array CGH) is a technique capable of measuring the copy number of hundreds of thousands of genomic regions in a single experiment. This chapter describes normalization of array CGH data to generate robust copy number measurements.

Chapter 11: SNP Array-Based Analysis for Detection of Chromosomal Aberrations and Copy Number Variations
Single nucleotide polymorphism microarrays (aka SNP chips) are increasingly being used, not only to map SNPs, but also to detect chromosomal instability. Microsatellites have long been the signposts used by geneticists to search for nearby genetic aberrations responsible for a variety of diseases. Now, scientists have a much

higher density tool: SNP arrays that enable them to find rare alleles and multigenic loci, as well as to track polymorphisms that change frequency as the world becomes less geographically isolated. Losses and gains of chromosome segments are one of the most important ways that normal cells become immortalized and, therefore, cancerous. Analysis of SNP data is challenging, but the authors present several useful methods for processing and interpreting data from these high-density SNP arrays.

1 A Comprehensive Analysis of the Effect of Microarray Data Preprocessing Methods on Differentially Expressed Transcript Selection

Monika Ray, Johannes Freudenberg, and Weixiong Zhang

CONTENTS

Motivation 1
Results 2
Introduction 2
Materials and Methods 5
    Data 5
    Data Preprocessing 5
        Background Correction 6
        Normalization 6
        Perfect Match Correction 7
        Summarization 7
    Differentially Expressed Gene Identification 8
Results 9
Discussion 13
Acknowledgment 15
References 15

MOTIVATION

Data preprocessing is performed prior to differentially expressed (DE) gene selection. Numerous preprocessing methods have been proposed with little consensus as to which is the most suitable. Due to the poor concordance among results from cross-platform analyses, protocols are being developed to enable cross-platform reproducibility. However, the effect of data analysis on a single platform is still unknown. Our objective is twofold: to check for consistency in the results from a single platform and to investigate the effect of preprocessing on DE gene selection.

RESULTS

Differentially expressed gene lists are variable and dependent on the preprocessing method used. The data set characteristics greatly affect the outcome of the analysis. There is a lot of variability in the results, despite using a single platform. These issues have been overlooked in past reports. This is the first comprehensive analysis using multiple data sets to assess the effect of data preprocessing on downstream analysis.

INTRODUCTION

Since its invention in 1995 [1], microarray technology has become a principal tool for high throughput gene expression detection and analysis in basic science research and clinical studies. A simple search on PubMed will return the titles of over 10,000 papers on microarray analysis published within the last 2 years. A plethora of gene expression data on model systems and human diseases has been collected and organized at the National Center for Biotechnology Information (NCBI) and many other microarray data repositories. Large microarray data have been produced by many different microarray platforms, including the widely used Affymetrix GeneChip.

Concomitant with the production of massive amounts of data, there has been a proliferation of computational methods for quantifying the level of expression on a DNA chip for various microarray platforms. Even with the same microarray platform, the quality of microarray gene expression can also be significantly impacted by the data preprocessing methods adopted. Several studies have called into question the validity of microarray results, mainly due to the disparities between results obtained by different groups analyzing similar samples [2–4]. Therefore, efforts have been directed to understanding the causes of discrepancies across platforms and the development of protocols that would allow for cross-platform comparisons and reproducibility [5–7]. Unfortunately, the effect of data preprocessing methods, even for the same microarray platform, has not been well understood or adequately addressed so far.

An important aspect of our investigation is that, although we used multiple data sets, we focused on data generated from a single platform. Hence, one of the main objectives of this chapter is to provide the first comprehensive analysis of data preprocessing methods on the *same microarray platform*. By focusing on one microarray platform, we are able to avoid complications from cross-platform gene expression detection and minimize the variability in the results. Furthermore, if there is a huge discrepancy despite using a single platform, then there will be even greater disparities in a cross-platform study.

Along with the increased use of microarray technology, preprocessing methodology for Affymetrix GeneChip—a market leader—is a quickly expanding research subject. In order to compare preprocessing methods, the affycomp webtool [8] was established in 2003, and over 40 microarray preprocessing methods have been submitted by users [9]. In our study, we focused on the Affymetrix GeneChip platform because it is widely used and a large number of data preprocessing methods have been developed for it. Gene expression measurements are confounded by different sources of systematic variation. Sources of noise include differences in dye efficiencies, scanner malfunction, uneven hybridization, array design, and experimenter and other extraneous factors.

In order to compare the mRNA expression levels of each probe on a microarray chip as well as compare microarrays from different experiments for the purposes of performing higher order analyses such as clustering, prediction, and building regulatory networks, it is necessary to remove unwanted variations while retaining intrinsic biological variations. Thus, data preprocessing is a critical step that affects accuracy and validity of downstream analyses. The four main preprocessing steps for Affymetrix GeneChip are background correction, normalization, perfect match (PM) probe correction, and summarization. New processing algorithms that claim to handle the characteristics of microarrays well are continuously being developed. However, there still is no consensus on which processing method is the most efficient [10,11] or how a preprocessing method should be chosen.

Differentially expressed gene ranking or selection, more often than not, is the next course of action after data generation. The use of microarrays to determine bona fide changes in gene expression between experimental paradigms is confounded by noise due to variability in measurement. Since scientists are eager to use microarrays to identify marker genes and understand the pathogenesis of diseases, there should be an acute awareness of what procedures are being used to identify the molecular signature of a phenotype. Therefore, the second objective of our analysis is to assess the variability in measurement by investigating the effect of different background correction, normalization, PM correction, and summarization techniques on downstream analysis—in particular, DE gene selection.

Variance stabilization is achieved through data preprocessing. Most literature on comparison of microarray processing methods usually analyzes either background correction or normalization. Furthermore, not only do these studies focus on just one data set, but they also use tightly controlled calibration data derived from spike-in or dilution studies [9,12,13] or data from very simple organisms [14]. However, to date, we have not found any literature that has thoroughly examined the influence of each of these preprocessing stages on DE gene selection on multiple, diverse, real-life data sets. If the DE gene selection stage is highly influenced by these preprocessing mechanisms, all subsequent stages will also be affected.

Based on biological assumptions and microarray chip design, many types of preprocessing algorithms have been developed to date [15,16]. Background correction is the process of correcting probe intensities on an array using information only from that array. Normalization is the process of removing nonbiological variability across arrays. Summarization is the process of combining the preprocessed PM probes to compute an expression measure for each probe set on the array. In this chapter, we focus on the two most widely used preprocessing pipelines: Microarray Suite 5.0 (MAS5.0, now GeneChip operating system—GCOS) developed by Affymetrix [17] and robust multiarray analysis (RMA) [13].

We first used RMA and MAS5.0 for background correction. However, GeneChip (GC)–robust multichip average (GCRMA) [18] is a popular method for background correction and many consider it as one of the best current preprocessing methods [9,19]. Hence, we decided also to apply GCRMA for background correction. For the normalization stage, we applied the constant normalization and quantile normalization [13] methods. Finally, MAS5.0 and RMA were applied for PM correction while MAS5.0 (Tukey biweight) and RMA (medianpolish) were employed for the summarization stage. The gene selection tools investigated in this study were significance analysis of microarrays (SAM) [20] and RankGene [21]. These two packages were chosen mainly due to their widespread usage. Finally, we wanted to verify whether the choice of a preprocessing method has a greater effect on gene selection than the choice of gene selection tool, which was the conclusion reached by Hoffmann et al. [22].

The focus of this chapter is on the effect of microarray data preprocessing and *not* on how one can manipulate gene selection criteria to overcome the effects of data preprocessing. Following is a brief overview of the study. Four diverse, non-time-series, human data sets generated from an Affymetrix DNA chip were used in this study: Alzheimer's disease (AD), papillary thyroid cancer (PTC), polycystic ovary syndrome (PCOS), and prostate cancer. These particular data sets were chosen since they capture intrinsic gene expression variations underlying these diverse human diseases and have as large a number of samples as one can expect from clinical experiments. Because data quality comes into play while analyzing methods or techniques, we calculated the quality of each of these four data sets using the R package "affyQCReport" [23]. We omitted the prostate cancer data set as it was of questionable quality. On each of the three different data sets we applied 16 different preprocessing strategies resulting from the combining all possible combinations—two for each of the four stages of data preprocessing. In addition, four strategies with GCRMA at the background correction stage were also applied (see Tables 1.1 and 1.2).

A set of processing methods for each stage—for example, RMA (R) for background correction, constant (C) for normalization, MAS5.0 (M) for PM correction,

TABLE 1.1
Processing Methods and Stages in Which They Were Applied

Processing Method	Background Correction (Stage 1)	Normalization (Stage 2)	PM Correction (Stage 3)	Summarization (Stage 4)
RMA (R)	x		x	x
MAS5.0 (M)	x		x	x
GCRMA (G)	x			
Constant (C)		x		
Quantiles (Q)		x		

Note: "x" indicates the stage in which the method was applied. Letters in parentheses refer to the abbreviation of that method.

and RMA (R) for summarization—refers to a "combination." The order of data preprocessing is background correction first, then normalization, followed by PM correction, and, finally, summarization. The four letters in each combination refer to the methods used in the four preprocessing stages in order. "R" refers to RMA, "C" refers to constant, "Q" refers to quantile, "M" refers to MAS5.0, and "G" refers to GCRMA. Differentially expressed gene selection using SAM as well as Rank-Gene was performed on each combination (see Table 1.1). In order to determine which method had the greatest effect on DE gene selection, comparisons were made between combinations that differed in a method for a single stage—for example, between RCMR and RCRR (method for PM correction differs). Number of common transcripts was used as the measure of similarity between methods. If the methods are highly dissimilar, then the number of overlapping transcripts would be small. The results have been presented as bar graphs for RankGene and via hierarchical clustering for SAM.

MATERIALS AND METHODS

DATA

For the purpose of generalizing results, we use three data sets, all of which were generated from an Affymetrix DNA chip. The first is AD data, generated on Affymetrix GeneChip (HG-U133A) by Blaylock et al. [24]. The second data set is on PTC and attempts to identify potential molecular markers for PTC. It is generated from Affymetrix GeneChip (HG-U133 Plus 2.0 Array). The third data set is on PCOS, generated from Affymetrix GeneChip (HG-U133A). The data were generated to study the differences in gene expression in the adipose tissue of women affected by PCOS. All three data sets are available to the public on PubMed. Alzheimer's disease data can be directly downloaded from ftp.ncbi.nih.gov/pub/geo/DATA/supplementary/series/GSE1297/, PTC data from www.ncbi.nlm.nih.gov/geo/gds/gdsbrowse.cgi, and PCOS data from www.ncbi.nlm.nih.gov/geo/query/acc.cgi.

DATA PREPROCESSING

Computation of gene expression measure is a four-step procedure: background correction, normalization, PM probe correction, and, finally, summarization. In most cases, PM and summarization are combined into one step.

Let X be the raw probe intensities across all arrays and E be the final probe set expression measures. Then, if B is the background correction operation on probes on each array, N is the operation that normalizes across arrays, and S is the operation that combines probes to compute an expression measure, the gene expression measure can be formulated as [25]

$$E = S(N(B(X))) \qquad (1.1)$$

Twenty different combinations of background correction methods, normalization methods, PM correction methods, and summarization methods were considered to determine which method has the greatest effect on gene selection.

Background Correction

For the purposes of our analysis, we used the RMA, GCRMA, and MAS5.0 background correction methods.

MAS 5.0 background correction: This method was developed by Affymetrix [17]. Briefly, each chip is split up into 16 zones of equal size. For each zone k, the background, b_k, and the noise, n_k, are defined as the mean and standard deviation of the lowest 2% of zone k's probe intensities, respectively. The probe-specific background value $b(x,y)$ and noise value $n(x,y)$ are defined as a weighted sum over all b_k and n_k, respectively, where x and y denote the position of a probe on the chip. The corrected signal is the raw signal reduced by $b(x,y)$ where physically possible; otherwise, it is $n(x,y)$. Please refer to Affymetrix [17] for more details.

RMA background correction: Robust multiarray average (RMA) is a widely used alternative preprocessing strategy for Affymetrix GeneChips [13]. The procedure is based on the assumption that the observed probe signal, O, consists of a normally distributed background component, N, and an exponentially distributed signal component, S, such that

$$O = N + S, N \sim N(\mu, \sigma^2), S \sim \exp(\alpha) \quad (1.2)$$

The parameters α, μ, and σ^2 are estimated from the data, and the raw intensities are replaced with the estimated expected value $E'(S|O = o)$ given the observed value o. The parameter of the exponential distribution function is α.

GCRMA background correction: GCRMA is a further development of RMA that takes the probe-specific hybridization affinities into account [18]. It also acknowledges the use of mismatch (MM) probes (see later discussion). The following model is fitted (where PM refers to "perfect match probe"):

$$PM = O_{PM} + N_{PM} + SMM = O_{MM} + N_{MM} + \phi S \quad (1.3)$$

where O_{PM} and O_{MM} represent optical noise, N_{PM} and N_{MM} represent nonspecific binding noise, and S is the actual signal of interest. The ϕ is a value between zero and one and accounts for the fact that MM probes tend to be less sensitive than their corresponding PM probes, but often still measure a specific signal. The parameters N_{PM} and N_{MM} are assumed to be a function of the probe-specific hybridization affinity, which is precomputed based on either a calibration data set or the experimental data; in our study we chose the latter. Equation (1.3) is then fitted using an empirical Bayes approach. Please refer to Wu et al. [18] for more details.

Normalization

For the purposes of our analysis, we used constant or global normalization and quantiles normalization. The former is similar to the one used by Affymetrix; the latter is part of the RMA preprocessing strategy.

Global normalization: The term "global normalization" refers to a family of methods where each probe intensity is scaled by a chip-specific factor such that a given summary statistic m_j' is the same for all chips after scaling. That is, $m_j' = m$,

Microarray Data Preprocessing Methods

for $j = 1, 2, \ldots, J$, where J is the number of samples. Commonly used summary statistics include sum, median, and mean. In global normalization, the mean of the first sample is used (without loss of generality), $m = m_1$. Thus, the chip-specific scaling factor is computed as follows:

$$f_j = m_1/m_j, j = 1, 2, \ldots, J \tag{1.4}$$

where m_j is the mean of all probe intensities in sample j.

Quantiles normalization: Quantiles normalization assumes that the intensities of each chip originate from the same underlying distribution. This implies that the quantile for each chip is the same. However, biases in the signal generating process result in chip-specific distributions. The goal of quantiles normalization is to remove these biases by transforming the data such that each quantile is the same across all chips. Based on this rationale, the following algorithm was proposed [13]:

- Sort probe intensities X_j for each sample j.
- For each quantile (or rank) i, compute the mean $m_i = 1/J \sum_j x_{(i)j}$.
- Replace $x_{(i)j}$ by m_i for each sample j.
- Restore the original order of X_j for each sample j.

Perfect Match Correction

Each PM probe on an Affymetrix GeneChip is complemented by an MM probe that has a different base as its 13th nucleotide. Hence, a mismatch probe is identical to the perfect match sequence except for a single incorrect base in the middle of the oligomer. The rationale for including MM probes in the chip design is to provide a tool for measuring the unspecific binding contribution of the signal. In our study, we applied the MAS5.0 and RMA PM correction methods. The MAS5.0 PM correction is performed as follows. First, compute the ideal mismatch IM_{ij} for each probe pair i in each sample j and then subtract IM_{ij} from the PM probe intensity PM_{ij}, where the ideal mismatch is equal to the mismatch probe intensity MM_{ij}, if $MM_{ij} < PM_{ij}$. Otherwise, IM_{ij} is computed by downscaling the corresponding PM signal. (Please refer to Affymetrix [17] for more details.) In RMA preprocessing, MM probes are simply ignored because these probe signals also have probe-specific signal contributions, which would dilute the PM corrected signal.

Summarization

Affymetrix DNA microarrays contain multiple probes for each target transcript. In order to establish a single expression value for each probe set (i.e., a set of probes corresponding to the same transcript), individual probe values must be summarized. Affymetrix provides chip definition files (CDFs) that can be used to determine probe set assignments. However, recent studies suggest that the manufacturer's probe design and annotation may be outdated and erroneous [26]. Therefore, redefined CDFs (version 8, reference database: RefSeq) were obtained from Brainarray [27] and used instead of the manufacturer's CDFs. For the purposes of our analysis, we used the MAS5 (Tukey biweight) and RMA (medianpolish) summarization methods.

Tukey biweight: This algorithm is described in Affymetrix [17] in detail. Briefly, probe values are log transformed after PM correction. Next, for each probe set,

1. For each probe j, the deviation u_j from the probe set median, which is weighted by the median deviation from the probe set median, is computed.
2. For each probe j, the weight $w_j = (1 - u^2)^2$ is computed.
3. The weighted mean, *TBI*, which is the final summary value, is computed as

$$TBI = \Sigma_j w_j x_j / \Sigma_j w_j \qquad (1.5)$$

Medianpolish: This algorithm is used as part of the RMA framework [13]. Briefly, a two-way analysis of variance (ANOVA)-like model is fitted:

$$\log_2 (y_{ij}) = \alpha_i + \mu_j + \varepsilon_{ij} \qquad (1.6)$$

where
  y_{ij} is the observed signal intensity of probe i on chip j
  α_i is the probe affinity effect of probe i
  $\Sigma_i \alpha_i = 0$
  μ_j represents the expression level for array j
  ε_{ij} is an independent identically distributed error term with zero mean

The estimate μ_j' is the expression value for the respective probe set on array j. The model parameters are estimated using a robust procedure that iteratively estimates the error matrix ε_{ij}' by repeatedly subtracting row medians and column medians in an alternating fashion until reaching convergence.

DIFFERENTIALLY EXPRESSED GENE IDENTIFICATION

Significant gene identification or selection of DE genes has been a subject of active research for many years, resulting in a multitude of algorithms. Significant gene extraction falls into two broad categories: *wrapper* methods and *filter* methods. In wrapper gene selection methods, the DE gene identification phase is integrated with the classification phase. In filter methods, the DE gene extraction phase is independent of the classification phase. Our study focused on filter methods only. Most of the previous studies on preprocessing techniques used filter methods, such as ANOVA and the ratio of expression levels, to obtain a list of DE genes. In this study, we used two packages for the identification of DE genes: RankGene [21] and SAM [20].

RankGene is a C++ program for analyzing gene expression data, feature selection, and ranking genes based on the predictive power of each gene to classify samples into functional or disease categories [21]. It supports eight measures for quantifying a gene's ability to distinguish between classes. Since we are not evaluating different gene selection methods or criteria but rather investigating the effect of preprocessing on DE gene selection strategies, the gene selection technique used is of little concern in this study. In our analysis, we used information-gain measure of predictability to identify DE genes. Information gain is a statistical impurity measure that quantifies the best possible class predictability that can be obtained by dividing the full range of expression of a given gene into two disjoint intervals (e.g., class 1 and class 2). All samples (e.g., patients) in one interval belong to one class (e.g., normal) and all samples in the other interval belong to the other class (e.g., affected). The difference among the various measures is used to quantify the error in prediction. For a given choice of measure, RankGene minimizes the error over all possible thresholds that partition the gene into two intervals.

SAM is an open-source software program that identifies DE genes based on the change in gene expression relative to the standard deviation of repeated measurements [20]. SAM uses the false discovery rate (FDR) and q-value method presented in Storey [28]. For our analysis we aimed for a list of genes at an FDR of 30%. We did not want a very low FDR as the list of genes obtained would be low per preprocessing combination and it would be hard to compute the amount of overlap between them.

While SAM is a gene-selection as well as gene-ranking tool, RankGene is only a ranking tool. Ranking of genes is a less stringent criterion than selecting genes. Hence, by choosing RankGene and SAM, we analyzed two different protocols most commonly used after data preprocessing. For bar graph visualization purposes, we chose to analyze the top 500 transcripts from the complete ranked list. Therefore, we implemented a stringent selection criterion (i.e., FDR with SAM) and a less stringent criterion (i.e., top 500 genes with RankGene) and compared the results.

RESULTS

RankGene results are presented as bar graphs and SAM results are presented as clustering diagrams. In the case of the bar graphs, preprocessing combinations were divided into seven groups/sets of four comparisons each—one comparison for each of the four processing stages. The measure of similarity is the number of overlapping transcripts. Each bar represents one comparison between two combinations. All the graphs have seven sets of four bars (excluding the last group on the far right of each graph). In any particular set, in each of the comparisons, only one of the four preprocessing methods is changed. In any graph, the heights of the bars should be compared within the set of four comparisons (i.e., one group) and not across the whole graph. The greater the effect of the method is, the smaller will be the number of common transcripts between the two preprocessing combinations, implying greater dissimilarity between the combinations.

Figures 1.1, 1.2a, and 1.2b show the amount of overlap between any two combinations in each of the three data sets via bar graphs and heat map. In RankGene, for each combination, 500 transcripts were ranked according to the information-gain criterion. In SAM, the number of significant transcripts was chosen based on the FDR value, which was chosen to be in the range of 28–31%. Hierarchical clustering (the hclust() function of the statistical package R [29]) of the comparisons was generated on each data set with the SAM selected transcripts (Figures 1.2a and 1.2b) to determine if there was any discernable pattern. The clustering was then combined with a heat map using the R function heatmap() such that red indicates high similarity and white indicates no similarity.

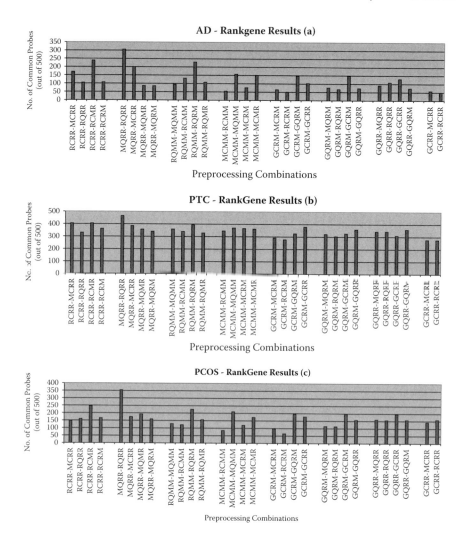

FIGURE 1.1 RankGene comparison of preprocessing combinations by estimating the amount of overlap (number of common probes) between combinations: (a) Alzheimer's disease; (b) papillary thyroid cancer; (c) polycystic ovary syndrome.

Table 1.2 shows the number of significant transcripts obtained using SAM on the PCOS data set. Results indicate that preprocessing technique has a greater effect on gene selection than the kind of gene selection methods used. For example, if the numbers of significant transcripts in RCRR and RQRR are compared, the huge difference in the number of transcripts can easily be observed, despite using the same gene selection method and a change in only one preprocessing stage (i.e., normalization). Similar results were obtained for the other data sets as well.

The relative difference in the heights of the bars in the graphs is phenomenal across data sets for the same gene selection method. From Figure 1.1, one can observe that the preprocessing methods have the greatest similarity when applied to the PTC

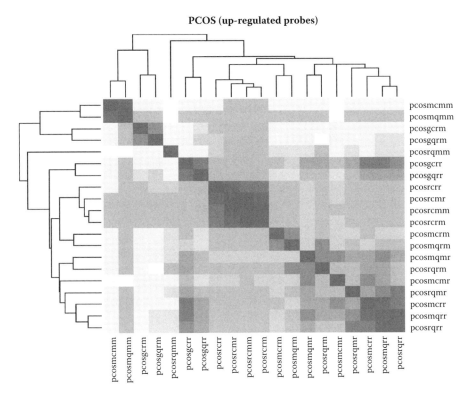

FIGURE 1.2a (See color insert following page 80.) SAM clustering of the preprocessing combinations on polycystic ovary syndrome. Red indicates high similarity, and white indicates no similarity.

data set. However, there is a lot of disparity among the preprocessing combinations when applied to the AD data set. This clearly indicates that the characteristics of data sets—that is, the amount of variability in gene expression—have an effect on the preprocessing method. If any one comparison is analyzed across all data sets for any one gene selection method—for example, if the height of the MCMM–RCMM bar is observed across the different data sets—the bar height varies significantly across data sets. This indicates that the effect of a preprocessing technique also depends on the kind of data set. However, the same comparison across different gene selection methods on the same data set does not produce a drastic difference in the result. These observations along with the results in Table 1.2 indicate that the preprocessing techniques play a greater role on DE gene selection than the gene selection method applied, whether it is RankGene or SAM. This confirms the conclusion attained by Hoffmann et al. [22].

On any data set, the method that has the maximum effect varies across different gene selection techniques. However, since the preprocessing stage depends on the kind of technique used in that stage, we cannot reach any definitive conclusion about whether it is the stage or the technique that has the greater effect. Analysis of this is left for future research.

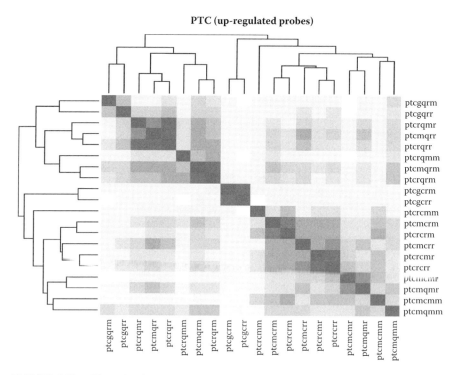

FIGURE 1.2b (See color insert following page 80.) SAM clustering of the preprocessing combinations on papillary thyroid cancer. Red indicates high similarity, and white indicates no similarity.

Hierarchical clustering was performed only for SAM results since the disparity in the number of DE genes selected per combination was so great (as seen in Table 1.2) that a bar graph representation would prove useless. Furthermore, clustering would enable visualization of comparisons between all pairs of preprocessing combinations; hence, pattern detection would be easier. However, as can be seen from Figures 1.2a and 1.2b, the hierarchical clustering does not show any definite pattern with respect to the preprocessing techniques.

Initially, we had only applied RMA and MAS5.0 for background correction. When no conclusive results surfaced, we applied the GCRMA method. Because GCRMA has grown in popularity, we presumed that it would eliminate the discrepancies observed in the results thus far in our study. However, as can be observed from Figures 1.1, 1.2a, and 1.2b, GCRMA did not help in resolving matters. GCRMA and RMA belong to the same family of processing methods. However, if GCRM and RCRM or GQRM and RQRM are compared, the number of overlapping transcripts is sometimes high, as in the case of the PTC data set, and sometimes low, as in the case of the AD data set. This is also true for all the other comparisons involving GCRMA. In all the bar graphs, the far right end consists of comparisons among GCRMA, RMA, and MAS5.0. Again, there does not seem to be any consistency in the results obtained after applying GCRMA. Results tend to be variable depending on the data analyzed and the DE gene selection method employed.

TABLE 1.2
Number of DE Probes Obtained Using SAM on PCOS Data[a]

Preprocessing Combination	No. of DE Probes
GCRM	64
GCRR	74
GQRM	105
GQRR	182
MCMM	2
MCMR	62
MCRM	79
MCRR	54
MQMM	1
MQMR	54
MQRM	52
MQRR	80
RCMM	8861
RCMR	8791
RCRM	7946
RCRR	5047
RQMM	5
RQMR	101
RQRM	26
RQRR	98

[a] FDR = 30%.

DISCUSSION

Our study is the first systematic and comprehensive analysis of the effect of *all* the data preprocessing stages on DE gene selection analyzed on multiple human data sets. There are three key conclusions from this study. The first is that different preprocessing techniques have varying degrees of effect on downstream analysis based on the kinds of data analyzed. Preprocessing techniques are not robust enough to handle the amount of variability present in microarray data. Second, the choice of gene selection methods used to evaluate the effect of preprocessing also affects the outcome. Third, even using a single platform, the discrepancies between results are substantial.

It is a cause for great concern if the results from the same data set, using the same gene selection tool on the same platform, are not more in concordance. This implies that preprocessing methods have a great impact on statistical analyses and this in turn affects the reproducibility of results. If there is little or no consensus among results from the same platform, cross-platform discrepancies cannot be effectively surmounted. Therefore, efforts should be directed towards methods that would increase consistency among results generated from the same platform before extending them to cross-platform analysis. Only statistical analysis techniques and interlaboratory experiment protocols are not the causes of cross-platform disparities, as mentioned by Larkin et al. [5] and Irizarry et al. [6]. Along with standardizing research protocols and statistical analysis tools, attention should be paid to data preprocessing methods as well.

Some reports state that background correction or normalization has the greatest effect [9,22,25] on DE gene selection. As there are no standard evaluation methods, different reports can provide conflicting results. Investigators evaluating processing methods should also address data quality and assessment tools used. Recently, there has been a deluge of research articles, editorials, and commentaries on the reliability of microarray results [30–33]. Ioannidis [30] goes so far as to say that, given information on a gene and 200 patients, he can show that this gene affects survival ($p < .05$) even if it does not! Due to poor accuracy, sensitivity, specificity, and reproducibility, microarrays have not yet passed Food and Drug Administration (FDA) regulations for routine use in clinical settings.

Miklos and Maleszka [2] state that different analyses of the same microarray data lead to different gene prioritizations. Another group performed a multiple random validation strategy to reanalyze data from seven of the largest microarray studies to identify a molecular signature of cancer; they discovered that molecular signatures are unstable [34]. Disparities between genes selected from microarray analysis and biologically relevant genes (i.e., those that are truly relevant) are great even on simple organisms such as yeast [35]. Hence, the incongruity cannot be a characteristic of complex multicellular organisms. On the other hand, many studies claim that their list of genes is significant based on p values, which are typically less than .05. However, statistical significance does not imply biological significance. Due to this emotional dependence on p values, the majority of the published results may be irrelevant if not false [36,37].

This study, along with results in other reports, indicates that microarray analysis is subjective and highly unstable. So how can these problems be solved? Just developing new algorithms for data preprocessing, without addressing the reasons why the current methods fail, will not solve the problems related to microarray analysis. No matter how sophisticated higher analyses become, one cannot ignore the fact that microarray chip design also needs to be improved. The correct design of probes on a DNA chip is essential. Additionally, due to cross-platform and interlaboratory discrepancies, one should be wary of generating or accepting results derived from meta-analysis data. Furthermore, evaluation of bioinformatics protocols should be done on real-life data sets in addition to the well-controlled calibration data sets. Most of the previous work on the analysis of processing algorithms was performed on the dilution or spike-in data from Affymetrix [9,12,13]. However, the variance of the average chip intensity in such data is much lower than those measured in real-life, not well-controlled data sets. Furthermore, the use of a limited number of spike-in genes is not sufficient for a comprehensive analysis of both microarray technology and data analysis methods.

Such issues with spike-in data will cast doubts on the applicability of such data for the development of analytical tools for use in diverse gene expression profiles.

Large sample size is crucial. Although it is expensive and time consuming to gather a large number of samples, it is the price one has to pay for accuracy and reproducibility. It is also essential to understand the limitations of the technology being used. One thing the current microarray technology is still unable to deal with is the reliable detection of transcripts with low expression. Therefore, attempting to do this without improved technology is inadvisable. Characteristics of high-throughput data need to be well understood and addressed accordingly in the design and execution of experiments. Miron and Nadon [32] term this as "inferential literacy." Previous studies have indicated that the high rates of irreproducible results are due to research findings being defended only by statistical significance. Hence, the use and development of other methods of analysis, such as the one suggested by Wacholder et al. [38], should be encouraged. Avoidance of bias in the manipulation of the analyses as well as having clear study designs will help scientists who plan on embarking into the field of microarray analysis and prevent them from performing experiments that would most likely be erroneous.

In conclusion, efforts should be directed to (1) improved chip design, (2) "inferential literacy," (3) large sample sizes, (4) random sampling of data for validation, (5) evaluation of tools on real-life as well as calibration data sets, (6) multiple and completely independent validation of results, (7) standardization of evaluation techniques and laboratory protocols, and (8) application of better methods of significance analysis. The good news is that a fraction of the scientific community is realizing the problems surrounding microarray analysis and appropriately addressing these issues. However, it will be a while before everyone understands and utilizes the fruits of this labor.

Microarray technology is indeed a valuable tool. However, it is important to minimize and separate noise from meaningful information in the data by employing proper experiment and data analysis protocols. Processing of noise masquerading as knowledge or eliminating valuable information while removing noise will lead to faulty or overoptimistic conclusions. In the absence of proper evaluation tools and standardized experiment procedures, microarray analysis has become more of an art than a science, more like an act of faith than scientific inference. There is an elephant in the room.

ACKNOWLEDGMENT

This research was funded in part by NSF grants EIA-0113618 and IIS-0535257, and Monsanto Co.

REFERENCES

1. Schena M, Shalon D, Davis RW, Brown PO: Quantitative monitoring of gene expression patterns with a complementary DNA microarray. *Science* 1995, 270(5235):467–470.
2. Miklos GLG, Maleszka R: Microarray reality checks in the context of a complex disease. *Nature Biotechnology* 2004, 22(5):615–621.
3. Yauk CL, Berndt ML, Williams A, Douglas GR: Comprehensive comparison of six microarray technologies. *Nucleic Acids Research* 2004, 32(15):e124.

4. Tan PK, Downey TJ, Spitznagel ELJ, Xu P, Fu D, Dimitrov DS, Lempicki RA, et al.: Evaluation of gene expression measurements from commercial microarray platforms. *Nucleic Acids Research* 2003, 31(19):5676–5684.
5. Larkin JE, Frank BC, Gavras H, Sultana R, Quackenbush J: Independence and reproducibility across microarray platforms. *Nature Methods* 2005, 2(5):337–344.
6. Irizarry RA, Warren D, Spencer F, Kim IF, Biswal S, Frank BC, Gabrielson E, et al.: Multiple-laboratory comparison of microarray platforms. *Nature Methods* 2005, 2(5):345–350.
7. Wang Y, Barbacioru C, Hyland F, Xiao W, Hunkapiller KL, Blake J, Chan F, et al.: Large-scale real-time PCR validation on gene expression measurements from two commercial long-oligonucleotide microarrays. *BMC Genomics* 2006, 7(59).
8. Irizarry RA: Affycomp II, A benchmark for Affymetrix GeneChip expression measures [http://affycomp. biostat.jhsph.edu/].
9. Irizarry RA, Wu Z, Jaffee HA: Comparison of Affymetrix GeneChip expression measures. *Bioinformatics* 2006, 22(7):789–794.
10. Qin L, Beyer RP, Hudson FN, Linford NJ, Morris DE, Kerr KF: Evaluation of methods for oligonucleotide array data via quantitative real-time PCR. *BMC Bioinformatics* 2006, 7(23).
11. Harr B, Schlotterer C: Comparison of algorithms for the analysis of Affymetrix microarray data as evaluated by co-expression of genes in known operons. *Nucleic Acids Research* 2006, 34(2).
12. Bolstad BM, Irizarry RA, Astrand M, Speed TP: A comparison of normalization methods for high-density oligonucleotide array data based on variance and bias. *Bioinformatics* 2003, 19(2):185–193.
13. Irizarry RA, Hobbs B, Collin F, Beazer-Barclay YD, Antonellis KJ, Scherf U, Speed TP: Exploration, normalization, and summaries of high-density oligonucleotide array probe level data. *Biostatistics* 2003, 4(2).
14. Ding Y, Wilkins D: The effect of normalization on microarray data analysis. *DNA and Cell Biology* 2004, 23:635–642.
15. Quackenbush J: Microarray data normalization and transformation. *Nature Genetics* 2002, 32:496–501.
16. Smyth GK, Speed T: Normalization of cDNA microarray data. *Methods* 2003, 31:265–273.
17. Affymetrix: Statistical algorithms description document [http://www.affymetrix.com/support/technical/whitepapers/sadd% whitepaper.pdf].
18. Wu Z, Irizarry RA, Gentleman R, Murillo FM, Spencer F: A model-based background adjustment for oligonucleotide expression arrays. *Journal of the American Statistical Association* 2004, 99:909–917.
19. University of Rochester Medical Center: Microarray gene expression analysis tools [http://fgc.urmc.rochester.edu/data analysis.html].
20. Tusher V, Tibshirani R, Chu G: Significance analysis of microarrays applied to the ionizing radiation response. *Proceedings of the National Academy of Science USA* 2001, 98:5116–5121.
21. Su Y, Murali TM, Pavlovic V, Schaffer M, Kasif S: RankGene: Identification of diagnostic genes based on expression data. *Bioinformatics* 2003, 19(12):1578–1579.
22. Hoffmann R, Seidl T, Dugas M: Profound effect of normalization on detection of differentially expressed genes in oligonucleotide microarray data analysis. *Genome Biology* 2002, 22(7):789–794.
23. Parman C, Halling C: affyQCReport: A package to generate QC reports for Affymetrix array data [www.bioconductor.org] 2006.

24. Blalock EM, Geddes JW, Chen KC, Porter NM, Markesbery WR, Landfield PW: Incipient Alzheimer's disease: Microarray correlation analyses reveal major transcriptional and tumor suppressor responses. *Proceedings National Academy of Science USA* 2004, 101:2174–2178.
25. Bolstad BM: Comparing the effects of background, normalization, and summarization on gene expression estimates [http://www.stat.berkeley.edu/users/bolstad/stuff/components.%pdf].
26. Dai M, Wang P, Boyd AD, Kostov G, Athey B, Jones EG, Bunney WE, et al.: Evolving gene/transcript definitions significantly alter the interpretation of GeneChip data. *Nucleic Acids Research* 2005, 33(20).
27. Molecular, Behavioral Neuroscience Institute University of Michigan: Brainarray [http://brainarray.mbni.med.umich.edu/Brainarray/Database/CustomCDF/CDF_download.asp].
28. Storey JD: A direct approach to false discovery rates. *Journal of the Royal Statistical Society Series B* 2002, 64(3):479–498.
29. R Development Core Team: *R: A language and environment for statistical computing.* R Foundation for Statistical Computing, Vienna, Austria 2006, [http://www.R-project.org]. [ISBN 3-900051-07-0].
30. Ioannidis JPA: Microarrays and molecular research: Noise discovery? *The Lancet* 2005, 365:454–455.
31. Shields R: MIAME, we have a problem. *Trends in Genetics* 2006, 22(2):65–66.
32. Miron M, Nadon R: Inferential literacy for experimental high-throughput biology. *Trends in Genetics* 2006, 22(2):84–89.
33. Draghici S, Khatri P, Eklund AC, Szallasi Z: Reliability and reproducibility issues in DNA microarray measurements. *Trends in Genetics* 2006, 22(2):101–109.
34. Michiels S, Koscielny S, Hill C: Prediction of cancer outcome with microarrays: A multiple random validation strategy. *The Lancet* 2005, 365:488–492.
35. Birrell GW, Brown JA, Wu HI, Giaeverdagger G, Chudagger AM, Davisdagger RW, Brown JM: Transcriptional response of *Saccharomyces cerevisiae* to DNA-damaging agents does not identify the genes that protect against these agents. *Proceedings of the National Academy of Science USA* 2002, 99(13):8778–8783.
36. Halloran PF, Reeve J, Kaplan B: Lies, damn lies, and statistics: The perils of the *p* value. *American Journal of Transplantation* 2006, 6:10–11.
37. Ioannidis JPA: Why most published research findings are false. *PLOS Medicine* 2005, 2(8):e124.
38. Wacholder S, Chanock S, Garcia-Closas M, Elghormli L, Rothman N: Assessing the probability that a positive report is false: An approach for molecular epidemiology studies. *Journal of the National Cancer Institute* 2004, 96(6):434–442.

2 Differentiation Detection in Microarray Normalization

Lei M. Li and Chao Cheng

CONTENTS

Abstract 20
Introduction 20
Methods 21
    Statistical Principle of Normalization 21
    Little Differentiation Versus Substantial Differentiation 23
    Differentiation Fraction and Undifferentiated Probe Set 24
    Spatial Pattern and Subarrays 24
    Linear Approximation and Its Applicability 24
    High Breakdown Estimation and Simple Least Trimmed Squares 25
    Symmetric and Trimmed Solutions of Simple Linear Regression 26
    Multiple Arrays and Reference 26
    Statistical Breakdown of Normalization 26
Examples and Results 27
    Implementation and SUB-SUB Normalization 27
    Parameter Selection 27
    Variation Reduction by Subarray Normalization 27
    Usage of Mismatch Probes 28
    Affymetrix Spike-in Data Set 29
    Perturbed Spike-in Data Set 32
    Comparative Functional Genomics and Primate Brain Expression
        Data Set 32
    Lung Cancer Data 35
Discussion 36
    External Controls 36
    Differentiation Fraction and Breakdown 36
    Nonlinear Array Transformation Versus Linear Subarray
        Transformation 37
    Transformation 37
    Diagnosis 37
Acknowledgments 38
References 38

ABSTRACT

From measurements of expression microarrays, we seek differentiation of mRNA levels among different biological samples. However, each array has a "block effect" due to uncontrollable variation. The statistical treatment of reducing the block effect is usually referred to as normalization. We formulate normalization as a blind inversion problem. The idea is to find a monotone transformation that matches the distributions of hybridization levels of those probes corresponding to undifferentiated genes between arrays. Two assumptions about the distributions of the reference and target arrays are discussed. The former leads to quantile normalization and the latter leads to regression normalization. We address two important issues. First, array-specific spatial patterns exist due to uneven hybridization and measurement process. Second, in some cases a substantially large portion of genes are differentially expressed, possibly in an asymmetric way, between a target and a reference array. For the purpose of normalization we need to identify a subset that excludes those probes corresponding to differentially expressed genes and abnormal probes due to experimental variation. One approach is least trimmed squares (LTS) and its symmetric variants. Substantial differentiation is protected in LTS by setting an appropriate trimming fraction. To take into account any spatial pattern of hybridization, we divide each array into subarrays and normalize probe intensities within each subarray. We also discuss the issue of statistical breakdown in normalization. The problem and solution are illustrated by various Affymetrix microarray data sets.

INTRODUCTION

Microarray is a key technique in the studies of genomic sciences. It measures abundance of mRNAs or presence of specific DNA polymorphisms by hybridization to appropriate probes on a glass chip. The current technique can hold hundreds of thousands of probes on a single chip. This allows us to have snapshots of expression profiles of a living cell. In this chapter, we mainly consider high-density oligonucleotide expression arrays. The Affymetrix GeneChip® uses 11–20 probes, which have 25 oligonucleotides bases, to represent one gene, and as a whole they are called a probe set [1,16]. Associated with each perfect match (PM) probe, a mismatch (MM) probe that differs only in the middle (13th) base is included in most Affymetrix expression arrays. Mismatch probes are designed to remove the effects of nonspecific binding, cross-hybridization, and electronic noise. Ideally, probes are arranged on a chip in a random fashion. But this is not always true in customized arrays and SNP (single nucleotide polymorphism) arrays.

From measurements of expression microarrays, we seek differentiation of mRNA levels among different cells. However, each array has a "block effect" due to variation in RNA extraction, labeling, fluorescent detection, etc. Without statistical treatment, this block effect is confounded with real expression differentiation. Our definition of normalization is the reduction of the block effect by justifiable statistical methods. It is usually done at the probe level. Several normalization methods for oligonucleotide arrays have been proposed and practiced. One approach uses *lowess* [6] to correct for noncentral and nonlinear bias observed in M-A plots [27]. Another class of approaches corrects for the nonlinear bias seen in Q-Q plots [3,21,24]. As discussed in Workman et al. [24] and Zien et al. [28], several assumptions must hold in the methods using quantiles. First, most genes are not differentially regulated. Second, the number of up-regulated genes roughly equals the number of down-regulated genes. Third, the preceding two assumptions hold across the signal-intensity range. In this chapter, we consider normalization that is more resistant to violations of these assumptions.

We formulate normalization as a blind inversion problem [14]. The basic idea is to find a monotone transformation for the target array so that the joint or marginal distribution of hybridization levels of the target and reference array matches a nominal one. Two different ideas exist to achieve this goal. First, quantiles allow us to compare distributions and the Q-Q plot is the standard graphical tool for the purpose. The normalization proposed in references [3, 21, and 24] aims to match the marginal distribution of hybridization levels of the target with that of the reference. Although a slight and subtle difference exists between the matching joint distribution and matching marginal principles, quantile methods work well for arrays with little differentiation. The second idea is regression, either linear or nonlinear [18,27]. In this chapter we mainly consider the regression method, for it is easy to deal with the situations in which the assumptions mentioned earlier are violated. The mathematical accounts of the quantile and regression normalization will be provided in the methods section.

When we compare two arrays in which a substantially large portion of genes are differentially expressed, we need to identify an undifferentiated subset for the purpose of normalization [12,19,22]. This subset should exclude those probes corresponding to differentially expressed genes and abnormal probes due to experimental variation. We use least trimmed squares (LTS) [17] and its symmetric variants [5] to identify the base for normalization and to estimate the transformation in a simultaneous fashion. Substantial differentiation is protected in LTS by setting an appropriate trimming fraction. The exact LTS solution is computed by a fast and stable algorithm we developed recently [5,15].

Array-specific spatial patterns may exist due to uneven hybridization and other uncontrollable factors in the measurement process. For example, reagent flow during the washing procedure after hybridization may be uneven; scanning may be nonuniform. We have observed different spatial patterns from one array to anther. To take this into account, we divide each array into subarrays that consist of a few hundred or a few thousand probes, and normalize probe intensities within each subarray. Other spatial normalization, such as that in Workman et al. [24], only considers the spatial effect in background. In comparison, we try to adjust for spatial difference in both background and scale. We show that match of distribution at the array level can be achieved by normalization at the subarray level to a great extent. Local subgrid normalization has been proposed in cDNA arrays [23].

METHODS

STATISTICAL PRINCIPLE OF NORMALIZATION

Suppose we have two arrays: one reference and one target. Denote the measured fluorescence intensities from the target and reference arrays by $\{U_i, V_i\}$ and true concentrations of specific binding molecules by $(\tilde{U}_i, \tilde{V}_i)$. Ideally, we expect $(U_i, V_i) = (\tilde{U}_i, \tilde{V}_i)$. In practice, measurement bias exists due to uncontrollable factors and we need a normalization procedure to adjust measurement. In what follows, we have another look at normalization.

Consider a system with $(\tilde{U}_i, \tilde{V}_i)$ as input and (U_i, V_i) as output. Let $\mathbf{h} = (h_1, h_2)$ be the system function that accounts for all uncontrollable biological and instrumental bias; namely,

$$\begin{cases} U_i &= h_1(\tilde{U}_i), \\ V_i &= h_2(\tilde{V}_i). \end{cases}$$

The goal is to reconstruct the input variables $(\tilde{U}_i, \tilde{V}_i)$ based on the output variables (U_i, V_i). It is a blind inversion problem [14], in which both input values and the effective system are unknown. The general scheme of blind inversion, referred to as BIND, has three steps: first, identify the distributions of input and output; second, estimate the system function using the distributional information of input and output; third, invert the system and reconstruct each individual input value.

The BIND scheme leads us to the question: What is the distribution of true concentrations $(\tilde{U}_i, \tilde{V}_i)$? First, let us assume that the target and reference array are biologically undifferentiated. In other words, the two arrays measure expressions of identical or very similar biological samples. The differences between the target and reference are purely caused by random variation and uncontrollable factors. If the two experiments are subject to the same mechanism of measurement error, then the following is one natural starting point, which is the basis of quantile normalization.

Assumption Q: the marginal distribution of target probes $\{\tilde{U}_i\}$ equals that for the reference $\{\tilde{V}_i\}$.

Alternatively, one can assume that the random variables $\{(\tilde{U}_i, \tilde{V}_i), j = 1, \ldots, n\}$ are independent samples from a joint distribution $\tilde{\Psi}$ whose density centers around the straight line $\tilde{U} = \tilde{V}$, namely,

Assumption R: $E(\tilde{V}|\tilde{U}) = \tilde{U}$, or alternatively, $E(\tilde{U}|\tilde{V}) = \tilde{V}$.

The average deviation from the straight line measures the accuracy of the experiment. If the effective measurement system $\mathbf{h}$ is not an identity one, then the distribution of the output, denoted by Ψ, could be different from $\tilde{\Psi}$. According to the general scheme of BIND, we need to estimate the system function using $\tilde{\Psi}$ and Ψ. An appropriate estimate, $\hat{\mathbf{h}}$, of the transformation should satisfy the following: the distribution $\hat{\mathbf{h}}^{-1}(\Psi)$ matches $\tilde{\Psi}$, which satisfies $E(\tilde{V}|\tilde{U}) = \tilde{U}$. In other words, the right transformation straightens out the distribution of Ψ.

Next, we consider the estimation problem. Roughly speaking, only the component of h_1 relative to h_2 is estimable. Thus, we let $v = h_2(\tilde{v}) = \tilde{v}$. In addition, we assume that

Assumption M: h_1 is a monotone or, specifically, a nondecreasing function.

Denote the inverse of h_1 by g.

Proposition 1. (Quantile normalization). Sort the values of $\{\tilde{u}_i\}$ and $\{\tilde{v}_i\}$ in the ascending order respectively—namely, $\tilde{u}_{(1)} \leq \tilde{u}_{(2)} \leq \cdots \leq \tilde{u}_{(n)}$, and $\tilde{v}_{(1)} \leq \tilde{v}_{(2)} \leq \cdots \leq \tilde{v}_{(n)}$. If we ignore the nonuniqueness of quantiles due to discretization and sampling, then the nondecreasing transformation g satisfying **Assumption Q** is given by $g(\tilde{u}_{(i)}) = \tilde{v}_{(i)}$.

The sorted values define the empirical quantiles for $\tilde{U}$ and $\tilde{V}$. The monotone function that transforms $\tilde{U}$ to $\tilde{V}$ can be determined by their quantiles. This is one way to justify the quantile normalization [3,21,24]. On the other hand, if we consider the first option in **Assumption R**, then we expect the following is valid

$$E[\tilde{V}|\tilde{U}] = \tilde{U}, \text{ or } E[V|g(U)] = g(U). \tag{2.1}$$

Proposition 2. (Regression normalization). Suppose equation (2.1) is valid for some function g. Then it is the minimizer of $\min_l E[V - l(U)]^2$, which equals $E[V|U]$.

According to the well known fact of conditional expectation, $E[V|g(U)] = g(U)$ minimizes $E[V - l_1(g((U))]^2$ with respect to l_1. Next, write $l_1(g(U)) = l(U)$. This fact suggests that we estimate g by minimizing

$$\sum_i [v_i - g(u_i)]^2. \tag{2.2}$$

We can impose restrictions of smoothness or monotonicity on g by appropriate parametric or nonparametric forms. One such example is the use of nonparametric smoother *lowess* [6] in cDNA arrays [27]. However, it is possible that no solution exists for equation (2.1) subject to certain constraints on g. In this case, we define an approximation solution by restricted least squares method equation (2.2).

LITTLE DIFFERENTIATION VERSUS SUBSTANTIAL DIFFERENTIATION

Two typical scenarios of microarray normalization exist. In many situations, the differentiation between the reference and target samples is negligible. Some examples are as follows:

- replicates
- DNA SNP arrays of the same type organism
- samples of little expressions known as *a priori*

In some other cases, the differentiation between the reference and target is significantly large. Some examples are as follows:

- strains of extremely different phenotypes
- expressions of samples at different developmental stages
- samples subject to different environmental conditions
- samples of different but closely related species

In what follows, we focus on the latter case. The degree of differentiation varies from one case to another. It is an important issue to be considered in the design and analysis of microarray experiments.

DIFFERENTIATION FRACTION AND UNDIFFERENTIATED PROBE SET

Next we consider the more complicated situation in which a substantial proportion λ of measured genes are differentially expressed while other genes are not except for random fluctuations. The probes for the undifferentiated genes are referred to as undifferentiated set, and their indices are denoted by I. Thus, the distribution of the input is a mixture of two components. One component consists of those undifferentiated genes and its distribution is similar to $\tilde{\Psi}$. The other component consists of the differentially expressed genes and its distribution is denoted by $\tilde{\Gamma}$.

Although it is difficult to know the specific form of $\tilde{\Gamma}$ *a priori*, its contribution to the input is at most λ. The distribution of the input variables $(\tilde{U}_i, \tilde{V}_i)$ is the mixture $(1-\lambda)\tilde{\Psi} + \lambda\tilde{\Gamma}$. Under the system function $\mathbf{h}$, $\tilde{\Psi}$ and $\tilde{\Gamma}$ are transformed respectively into distributions denoted by Ψ and Γ; that is, $\Psi = \mathbf{h}(\tilde{\Psi})$, $\Gamma = \mathbf{h}(\tilde{\Gamma})$. This implies that the distribution of the output (U_i, V_i) is $(1-\lambda)\Psi + \lambda\Gamma$. If we could separate the two components Ψ and Γ, then the transformation $\mathbf{h}$ of some specific form could be estimated from the knowledge of $\tilde{\Psi}$ and Ψ. From the regression perspective, we modify equation (2.1) and proposition 2 by taking the differentiation into account. One solution to normalization is a function g such that

$$E[V_i \mid g(U_i)] = g(U_i), \text{ for } i \in I. \tag{2.3}$$

We estimate g by minimizing $\Sigma_{i \in I}[u_i - g(v_i)]^2$ subject to necessary constraints. The same principle, equation (2.3), can be used to examine the goodness of normalization. Since the undifferentiated set I is unknown to us, it may be helpful to consider the following form of equation (2.3) for any random subset $\tilde{I} \subset I$:

$$E[V_i \mid g(U_i)] = g(U_i), \text{ for } i \in \tilde{I}. \tag{2.4}$$

In a specific study, we consider one or several subsets to check the goodness of normalization.

SPATIAL PATTERN AND SUBARRAYS

Normalization can be carried out in combination with a stratification strategy. For cDNA arrays, researchers have proposed to group spots according to the layout of array printing so that data within each group share a more similar bias pattern. Then normalization is applied to each group. This is referred to as within-print-tip-group normalization in Yang et al. [27]. On a high-density oligonucleotide array, tens of thousands of probes are laid out on a chip. To take into account any plausible spatial variation in h, we divide each chip into subarrays, or small squares, and carry out normalization for probes within each subarray. To get over any boundary effect, we allow subarrays to overlap. A probe in overlapping regions gets multiple adjusted values from subarrays it belongs to, and we take their average.

LINEAR APPROXIMATION AND ITS APPLICABILITY

We choose to parameterize the function g for each subarray by a simple linear function $\alpha + \beta u$, in which the background α and scale β respectively represent uncontrollable additive and multiplicative effects. This is motivated by several considerations. First, for the most part, normalization aims to remove the difference in the mRNA amount between samples and the difference in the optical background. It has been reported [20] that the relationship between dye concentration and fluorescent intensity is linear only in some range that is determined by experimental and instrumental settings. Suppose the relationships between dye concentration and fluorescent intensity, respectively, in the target and reference array are represented by the following equations:

$$\begin{cases} U = f_1(W) + a_1, \\ V = f_2(\gamma W) + a_2, \end{cases}$$

where f_1 and f_2 are two increasing or nondecreasing functions, W and γW are the amount of mRNA in the target and reference sample, and a_1 and a_2 are the additive optical backgrounds, respectively, for the target and reference. Then, by eliminating W, we have

$$V = f_2(\gamma f_1^{-1}(U - a_1)) + a_2 \approx \alpha + \beta U,$$

where the linear approximation is obtained by the Taylor expansion. As $f_2 \to f_1$ and $\gamma \to 1$ or, in other words, as the experiment conditions and mRNA amount are getting more similar in the target and reference, the linear relation becomes exact $U + a_2 - a_1$, even though f_1 is nonlinear.

Second, each subarray contains only a few hundred or at most a couple of thousand probes, so we prefer a simple function to avoid overfitting. Third, we estimate the parameters by least trimmed squares and thus we achieve an optimal linear approximation in a subset as explained later.

HIGH BREAKDOWN ESTIMATION AND SIMPLE LEAST TRIMMED SQUARES

Our solution consists of two parts: (1) identify the undifferentiated subset of probes; and (2) estimate the parameters in the linear model. We adopt least trimmed squares to solve the problem. Starting with a trimming fraction ρ, set $h = [n(1 - \rho)] + 1$. For any (α,β), define $r(\alpha,\beta) = v_i - (\alpha + \beta u_i)$. Let $H_{(\alpha,\beta)}$ be a size-h index set that satisfies the following property: $|r(\alpha,\beta)_i| \le |r(\alpha,\beta)_j|$, for any $i \in H_{(\alpha,\beta)}$ and $j \notin H_{(\alpha,\beta)}$. Then the LTS estimate minimizes

$$\sum_{i \in H_{(\alpha,\beta)}} r(\alpha, \beta)_i^2 .$$

The solution of LTS can be characterized by either the parameter (α,β) or the size-h index set H. This dual form is ideal for normalization and preserving differentiation. Statistically, LTS is a robust solution for regression problems. On one hand, it can achieve a given breakdown value by setting a proper trimming fraction. On the other hand, it has $\sqrt{n}$-consistency and asymptotic normality under some conditions. In addition, the LTS estimator is regression, scale, and affine equivariant [17]. Despite its good properties, LTS has not been widely used because no practically good algorithm exists to implement computation. Recently, we developed a fast and stable algorithm to compute the exact LTS solution to simple linear problems [15].

On an average desktop PC, it solves LTS for a data set with several thousand points in 2 seconds.

An LTS solution is naturally associated with a size-h index set. By setting a proper trimming fraction ρ, we expect the corresponding size-h set is a subset of the undifferentiated probes explained earlier. Obviously, the trimming fraction ρ should be larger than the differentiation fraction λ.

SYMMETRIC AND TRIMMED SOLUTIONS OF SIMPLE LINEAR REGRESSION

In the treatment of simple linear regression, the roles of response and explanatory variables are asymmetric. In some extreme cases we experience unexpected results using least trimmed squares for simple linear regression [5]. A symmetric treatment of response and explanatory variable is the replacement of the vertical distance of an observation to the regression line by the perpendicular distance. Namely, we minimize the following:

$$\frac{1}{1+\beta^2} \sum_{i \in H_{(\alpha,\beta)}} r(\alpha,\beta)_i^2,$$

with respect to α and β. Two other symmetric and trimmed solutions of simple linear regression and related algorithms can be found in Cheng [5].

MULTIPLE ARRAYS AND REFERENCE

In the case of multiple arrays, the strategy of normalization hinges on the selection of reference. In some experiments, a master reference can be defined. For example, the time zero array can be set as a reference in a time course experiment. In experiments of comparing tumor and normal tissues, the normal sample can serve as a reference. In other cases, the median array or mean array are options for references. Another strategy is first to randomly choose two arrays, one reference and one target, for normalization. Use the normalized target array from the last normalization as the reference for the next normalization; iterate this procedure until all arrays have been normalized once, and repeat this loop for several runs. Hereafter, we adopt the median polishing method in Robust Multichip Average (RMA) [11] to summarize expression levels from multiple arrays.

The direct result of normalization is the calibration of relative expression levels of an array with respect to a reference. Suppose we have an ideal reference array with known concentrations of binding molecules for all probes. Then, in theory, we can measure the absolute expression values of any sample as long as we can normalize its hybridization arrays with the reference.

STATISTICAL BREAKDOWN OF NORMALIZATION

It should be clear at this point that the determination of the undifferentiated probe set is the key in normalization. The larger the differentiation fraction and the differentiation magnitude are, the more challenging normalization is. If the subset selected for normalization is far away from any random subset of the undifferentiated probe set, then the result could break down. The exact definition and determination of breakdown point is neither obvious nor simple due to unknown λ, $\tilde{\Psi}$, and $\tilde{\Gamma}$. Each normalization method has its own breakdown point, whereas for certain data sets some methods are more robust than others. In our opinion, satisfactory normalization may not exist for extreme cases with too large differentiation. Although the meaning of breakdown is specific to certain criterion such as assumption Q or assumption R, this issue should be addressed in all microarray normalization problems.

EXAMPLES AND RESULTS

IMPLEMENTATION AND SUB-SUB NORMALIZATION

We developed a module to implement the normalization method described before, referred to as SUB-SUB normalization. The core code is written in C, and we have an interface in PC Windows. The input of this program is a set of Affymetrix CEL files and output are their CEL files after normalization. Three parameters need to be specified: subarray size, overlapping size, and trimming fraction. The subarray size specifies the size of the sliding window. The overlapping size controls the smoothness of window sliding. Trimming fraction specifies the breakdown value in LTS. An experiment with an expected higher differentiation fraction should be normalized with a higher trimming fraction.

PARAMETER SELECTION

We have tested many combinations of the three parameters on several data sets. In some reasonable range, the interaction between the parameters is negligible. In general the smaller the subarray size is, the more accurately we can capture spatial bias while the fewer number of probes are left for estimation. Thus, we need to trade off bias and variation. From our experiments, we recommend 20 × 20 to 50 × 50 for Affymetrix chips, depending on the availability of computing power. The value of overlapping size is the same for both directions, and we found its effect on normalization is the least among the three parameters. Our recommendation is half of the subarray size. For example, it is 20 if subarray size is 40 × 40. According to our experience, it can even be set to 0 (no overlapping between adjacent subarrays) to speed up computation without obvious change to normalization. The selection of trimming fraction should depend on samples to be compared in the experiments and quality of microarray data. For an experiment with 20% differentiated genes, we should set a trimming fraction larger than 20%. Again, we need a trade-off between robustness and accuracy in the selection of trimming fraction. On one hand, to avoid breakdown of LTS, we prefer large trimming fractions. On the other hand, we want to keep as many probes as possible to achieve accurate estimates of α and β. Without any *a priori* knowledge, we can try different trimming fractions and look for a stable solution. Our recommendation for starting value is 55%.

VARIATION REDUCTION BY SUBARRAY NORMALIZATION

Stratification is a typical statistical technique to reduce variation. Subarray normalization can be regarded as stratification according to spatial locations. From the

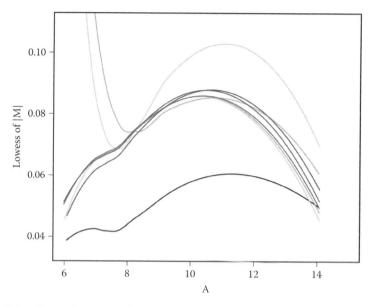

Figure 2.1 **(See color insert following page 80.)** The loess curves of |M| versus A values by various normalization methods. Gray: no normalization; black: SUB-SUB; red: quantiles; green: constant; purple: contrasts; blue: invariant-set; orange: lowess; cyan: qspline.

Affymetrix Webpage, two replicates using yeast array YG-S98 are available: yeast-2-121501 and yeast-2-121502. We use them to study the variation between array replicates. We normalize the yeast array 2-121502 versus 2-121501 by various normalization methods available from Bioconductor. Since the two arrays are replicates, the difference between them is due to experimental variation. In the resulting M-A plots, we fit lowess [6] curves to the absolute values of M, or |M|. These curves measure the variation between the two arrays after normalization (see Figure 2.1). The subarray normalization achieves the minimal variation. As variation is reduced, signal to noise ratio is enhanced and power of significance tests is increased.

Other normalization procedures can be applied at the subarray level as well. For example, if differentiation is not substantial, we can apply quantile, lowess, qspline normalization, etc. We note that different statistical methods, especially those nonparametric ones, may have specific requirements on the sample size. To achieve statistical effectiveness, we need to select the size of subarray for each normalization method.

USAGE OF MISMATCH PROBES

Some studies suggested using perfect match probes only in Affymetrix chips [25]. We checked the contribution of mismatch probes and perfect match probes to the subsets associated with LTS regressions from all subarrays. Figure 2.2 shows the distribution of the percentage of mismatch probes in the subsets identified by LTS. Our result shows that mismatch probes contribute slightly more than perfect match probes in LTS regression, mostly in the range of 46–56%.

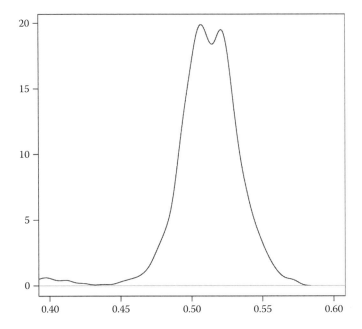

FIGURE 2.2 Histogram of percentages of mismatch probes in the subsets associated with LTS.

AFFYMETRIX SPIKE-IN DATA SET

The Affymetrix spike-in data set includes 14 arrays obtained from Affymetrix HG-U95 chips. Fourteen genes in these arrays are spiked-in at given concentrations in a cyclic fashion known as a Latin square design. The data are available from http://www.affymetrix.com/support/technical/sample_data/datasets.affx. In this chapter, out of the complete data set we chose eight arrays and split them into two groups. The first group contains four arrays: *m99hpp_av06, 1521n99hpp_av06, 1521o99hpp_av06,* and *1521p99hpp_av06*. The second group contains four arrays: *q99hpp_av06, 1521r99hpp_av06, 1521s99hpp_av06,* and *1521t99hpp_av06*. Later we will abbreviate these arrays by M, N, O, P, Q, R, S, and T. As a result, the concentrations of 13 spiked-in genes in the second group are twofold lower. The concentrations of the remaining spike-in gene are respectively 0 and 1024 in the two groups. In addition, two other genes are so controlled that their concentrations are also twofold lower in the second group compared to the first one.

We first investigated the existence of spatial pattern. The HG-U95 chip has 640×640 spots on each array. We divided each array into subarrays of size 20×20. We ran simple LTS regression on the target versus the reference for each subarray. This results in an intercept matrix and a slope matrix of size 32×32, representing the spatial difference between target and reference in background and scale. We first take array M as the common reference. In Figure 2.3, the slope matrices of arrays P and N are respectively shown in the subplots at top left and top right, and their histograms are shown in the subplots at bottom left and bottom right. Two quite

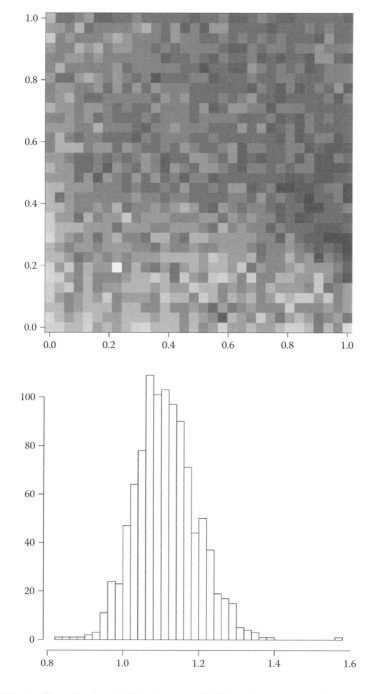

FIGURE 2.3 (See color insert following page 80.) The slope matrices of two arrays show different spatial patterns in scale. The common reference is array M. Top left: array P versus M; top right: array N versus M. Their histograms are shown at the bottom, correspondingly.

Differentiation Detection in Microarray Normalization

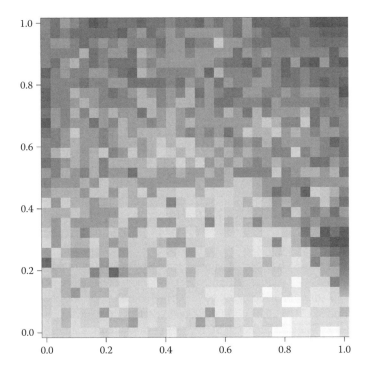

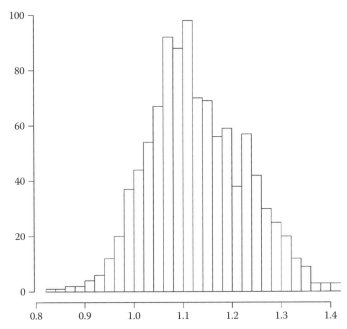

FIGURE 2.3 (continued).

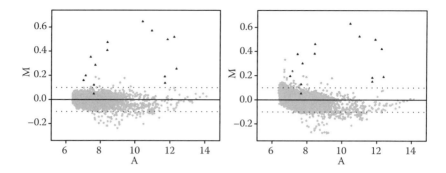

FIGURE 2.4 M-A plots of spike-in data after SUB-SUB normalization: left, trimming fraction is 0.45; right, trimming fraction is 0. Spike-in genes are marked by black triangles.

different patterns are observed. A similar phenomenon exists in patterns of α. The key observation is that spatial patterns are array specific and unpredictable to a great extent. This justifies the need of adaptive normalization.

We carried out SUB-SUB normalization to each of the eight arrays using array M as the reference. We experimented SUB-SUB normalization with different subarray sizes, overlapping sizes, and trimming fractions. Figure 2.4 shows the M-A plots summarized from the eight arrays after normalization—namely, the log ratios of expressions between the two groups versus the average log abundance. The subarray size is 40 × 40, the overlapping size is 20, and the trimming factors are respectively 0.45 and 0 in the left and right subplots. Our result indicates that both the subarray size (data not shown) and trimming fraction matter substantially for normalization. In other words, stratification by spatial neighborhood and selection of breakdown value in LTS do contribute a great deal to normalization. Overlapping size has a little contribution in this data set. Our method identifies 15 of the 16 differentially expressed spike-in gene with only a few false positives; see Figure 2.4, left.

PERTURBED SPIKE-IN DATA SET

SUB-SUB normalization protects substantial differentiation by selecting an appropriate trimming fraction in LTS. To test this, we generate an artificial data set with relatively large fraction of differentiation by perturbing the HG-U95 spike-in data set. Namely, we randomly choose 20% genes and reduce their corresponding probe intensities by 2.0-fold in the four arrays of the second group. We then run SUB-SUB normalization on the perturbed data set with various trimming fractions. The results are shown at the top and in the middle of Figure 2.5 for four trimming fractions: 0, 10, 20, and 40%. The normalization is satisfactory when the trimming fraction is above 20%, or larger than the nominal differentiation fraction.

COMPARATIVE FUNCTIONAL GENOMICS AND PRIMATE BRAIN EXPRESSION DATA SET

Expression profiles offer a way to conduct comparative functional genomics. We consider the data [8] available from http://email.eva.mpg.de/~khaitovi/supplement1.html

Differentiation Detection in Microarray Normalization

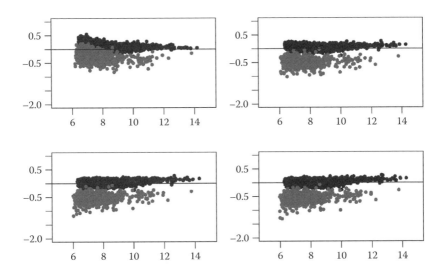

FIGURE 2.5 **(See color insert following page 80.)** M-A plots of perturbed spike-in data set after SUB-SUB normalization. The x-axis is the average of log intensities from two arrays. The y-axis is their difference after normalization. In the top four subplots, 20% randomly selected genes have been artificially down-regulated by 2.0-fold in arrays Q, R, S, and T. The differentiated genes are marked red, and undifferentiated genes are marked black. The trimming fractions in the subplots are: top left, 0%; top right, 10%; bottom left, 20%; bottom right, 40%.

that study the brain activity difference between humans and their closest evolutionary relatives. Two brain samples are extracted from each of three humans, three chimpanzees, and one orangutan. In what follows we only show results on two human individuals, HUMAN 1 and HUMAN 2, one chimpanzee, CHIMP. 1, and the orangutan, ORANG. The mRNA expression levels are measured by hybridizing them with the Affymetrix human chip HG-U95.

Compared to other primate brains such as chimpanzee and orangutan, in human brains a relatively high percentage of genes are differentially expressed, and most of them are up-regulated [4,9]. Moreover, the chimpanzee and orangutan samples are hybridized with human HG-U95 chips, so it is reasonable to assume that, if there were any measurement bias in primate mRNA expressions compared to humans, it would be downward bias. Figure 2.6 shows the density functions of log ratios of gene expressions for four cases: HUMAN 1 versus ORANG.; HUMAN 2 versus ORANG.; HUMAN 1 versus CHIMP. 1; HUMAN 1 versus HUMAN 2. The results from SUB-SUB (trimming fraction is 20%) and quantile normalization are plotted, respectively. When comparing humans with primates, the distribution resulting from the SUB-SUB method is to the right of that resulting from the quantile method. This is more obvious in the cases of humans versus orangutan, which are more genetically distant from each other than in the other cases (see the two subplots at the top). As expected, the distributions skew to the right and the long tails on the right might have a strong influence on the quantile normalization, which aims to match marginal distributions from humans and primates in a global fashion.

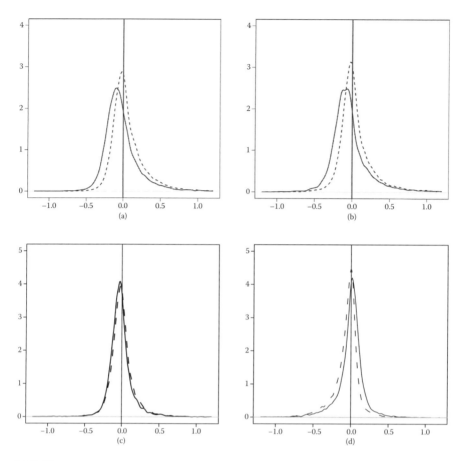

FIGURE 2.6 The densities of expression log ratios between: (top left) HUMAN 1 versus ORANG.; (top right) HUMAN 2 versus ORANG.; (bottom left) HUMAN 1 versus CHIMP. 1; (bottom right) HUMAN 1 versus HUMAN 2. The results from SUB-SUB normalization (trimming fraction is 20%) and quantile normalization are represented by dotted and solid lines, respectively.

However, in the cases of HUMAN 1 versus ORANG. and HUMAN 2 versus ORANG., the modes corresponding to the quantile method are in the negative territory while the modes corresponding to SUB-SUB method are closer to zero. The results from SUB-SUB normalization seems to be more reasonable. Furthermore, the difference in the case of HUMAN 1 versus HUMAN 2 is more distinct than that in the case of HUMAN 1 versus CHIMP. 1 (see the two subplots at the bottom). The analysis in Enard et al. [8] also indicates that HUMAN 2 differs more from other human samples than the latter differ from the chimpanzee samples. We checked the M-A plot of HUMAN2 versus ORANG. after SUB-SUB normalization (figure not shown) and observed that HUMAN 2 has more up-regulated genes than down-regulated genes compared to ORANG.

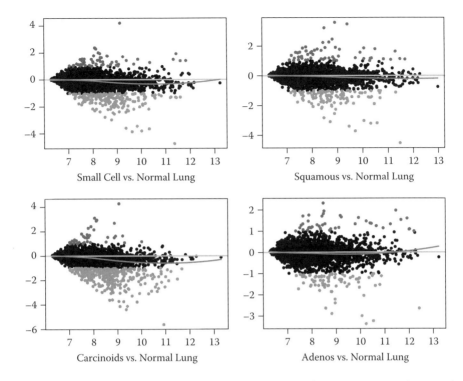

FIGURE 2.7 MA plot of the gene expressions of the four carcinoma types versus the normal lung tissues. The parameters in the SUB-SUB normalization are: 20 by 20 for window size, 10 for shift, and 0.30 for the trimming fraction. The expression levels are summarized by the median polishing method and the average of the two arrays is taken for each sample type.

LUNG CANCER DATA

One important application of the microarray method is the systematical search for molecular markers of cancer classification and outcome prediction in a variety of tumor types. We next consider the expression profiles reported in the lung cancer study [2]. The complete data set contains gene expression profiles of samples from 17 normal lung tissues, six small cell carcinomas, 21 squamous carcinomas, 20 carcinoids, and 127 adenocarcinomas. The mRNA expressions were measured by the Affymetrix human U95A oligonucleotide probe arrays. We randomly choose two arrays from each of the five sample types, and end up with a data set with 10 arrays. We take one array for normal tissue as the reference, and normalize the other nine arrays using the SUB-SUB method. The window size is 20 by 20, the shift size is 10, and the trimming fraction is 0.30. We summarize the expression levels by the median polishing method, and take the average of the two arrays for each sample type. The results are shown in Figure 2.7. The spatial patterns of the scale β for four arrays are shown in Figure 2.8.

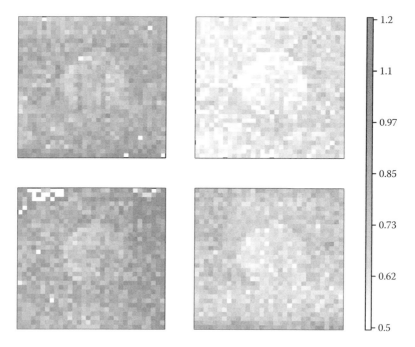

FIGURE 2.8 The spatial patterns of the scale β of four arrays in the lung cancer data.

DISCUSSION

EXTERNAL CONTROLS

In cDNA arrays, some designs use external RNA controls to monitor global messenger RNA changes [23]. In our view, external RNA controls play the role of undifferentiated probe sets. To carry out local normalization, we need quite a number of external controls for each subgrid.

DIFFERENTIATION FRACTION AND BREAKDOWN

In theory, the method can be applied to microarray experiments with differentiation fractions as high as 50%. In addition, our method does not assume an equal percentage of up- and down-regulated genes. In the meantime, LTS keeps the statistical efficiency of least squares. Earlier, we brought the issue of breakdown to readers' attention. We note that the statistical breakdown is different from computational or algorithmic breakdown. Sometimes the biological samples are so different from one another that the normalization results are no longer sensible, although numerically the computation does not show any warning. One option of our normalization only takes external core probes if they are available. The relationship among the statistical breakdown value, the differentiation fraction, and the parameter that controls breakdown in a normalization algorithm such as the trimming fraction ρ in LTS is complicated and needs further investigation.

Nonlinear Array Transformation Versus Linear Subarray Transformation

To eliminate the nonlinear phenomenon seen in M-A plots or Q-Q plots, methods such as lowess, qspline, and quantile normalization use nonlinear transformation at the global level [3,24,26]. In comparison, we apply a spatial strategy in SUB-SUB normalization. One array is split into subarrays and a simple linear transformation is fitted for each subarray. With an appropriate subarray size and trimming fraction, the nonlinear feature observed in M-A plots is effectively removed by linear subarray transformation to a great extent. We hypothesize that the nonlinear phenomenon may be partially caused by spatial variation. One simulation study also supports this hypothesis. Earlier, we argued that even though the dye effect is a nonlinear function of spot intensities, a linear transformation may be a good approximation as long as the mRNA amount and experimental conditions are similar between target and reference. Occasionally, when the amount of mRNA from two arrays is too different, a slight nonlinear pattern is observed even after subarray normalization (see Figure 2.5). To fix the problem, we can apply global lowess after the subarray normalization. Alternatively, to protect substantial differentiation, we can apply a global LTS normalization subject to a differentiation fraction once more.

Transformation

The variance stabilization technique was proposed in relation to normalization [7,10]. We have tested SUB-SUB normalization on the log scale of probe intensities, but the result is not as good at that obtained on the original scale. After normalization, a summarization procedure reports expression levels from probe intensities. We have tried the median polishing method [11] on the log scale. Alternatively, we can do a similar job on the original scale using the method of Model-Based Expression Indexes (MBEI) [13].

Diagnosis

The detection of bad arrays is a practical problem in the routine data analysis of microarrays. In comparison with the obvious physical damages such as bubbles and scratches, subtle abnormalities in hybridization, washing, and optical noise are more difficult to detect. By checking values of α and β in LTS across subarrays, we can detect bad areas in an array and save information from the other areas. Consequently, we can report partial hybridization results instead of throwing away an entire array.

Our blind inversion perspective of normalization and the least trimmed square method apply to other microarrays consisting of oligonucleotide arrays, cDNA arrays, bead arrays, pathway arrays, ChIP-chip experiments, methylation arrays, exon arrays, tissue arrays, and protein arrays. More generally, the comparison of two groups of certain biological measurements, in which data are collected by similar instruments and can be arranged by a vector or an array, requires some kind of normalization. The proposed principle and methods are applicable to such general normalization problems.

ACKNOWLEDGMENTS

This work is supported by the grant R01 GM75308-01 from NIH. Figures 2.1, 2.2, 2.3, and 2.6 were first published in *Nucleic Acids Research*, 2005, 33(17):5565–5573. Discussions with Dr. Yihui Luan and Huanying Ge improved the presentation.

REFERENCES

1. Affymetrix. *Affymetrix microarray suite user guide*, 5th ed. Santa Clara, CA: Affymetrix 2001.
2. A. Bhattacharjee, W. G. Richards, J. Staunton, C. Li, S. Monti, P. Vasa, C. Ladd, et al. Classification of human lung carcinomas by mRNA expression profiling reveals distinct adenocarcinoma subclasses. *PNAS*, 98:13790–13795, 2001.
3. B. M. Bolstad, R. A. Irizarry, M. Astrand, and T. P. Speed. A comparison of normalization methods for high-density oligonucleotide array data based on bias and variance. *Bioinformatics*, 19:185–193, 2003.
4. M. Caceres, J. Lachuer, A. M. Zapala, C. J. Redmond, L. Kudo, H. D. Geschwind, J. D. Lockhart, et al. Elevated gene expression levels distinguish human from nonhuman primate brains. *Proceedings of the National Academy of Science, USA*, 100:13030–13035, 2003.
5. C. Cheng. Symmetric and trimmed solutions of simple linear regression. Master's thesis, University of Southern California, 2006.
6. W. S. Cleveland. Robust locally weighted regression and smoothing scatterplots. *Journal of the American Statistical Association*, 74:829–836, 1979.
7. B. P. Durbin, J. S. Hardin, D. M. Hawkins, and D. M. Rocke. A variance-stabilizing transformation for gene-expression microarray data. *Bioinformatics*, 18:S105–S110, 2002.
8. W. Enard, P. Khaitovich, J. Klose, S. Zöllner, F. Heissig, P. Giavalisco, S. K. Nieselt, et al. Intra- and interspecific variation in primate gene expression patterns. *Science*, 296:340–343, 2002.
9. J. Gu and X. Gu. Induced gene expression in human brain after the split from chimpanzee. *Trends in Genetics*, 19:63–65, 2003.
10. W. Huber, V. A. Heydebreck, H. Sültmann, A. Poustka, and M. Vingron. Variance stabilization applied to microarray data calibration and to the quantification of differential expression. *Bioinformatics*, 18:S96–S104, 2002.
11. R. A. Irizarry, B. Hobbs, F. Collin, Y. D. Beazer-Barclay, K. J. Antonellis, U. Scherf, and T. P. Speed. Exploration, normalization, and summaries of high-density oligonucleotide array probe level data. *Biostatistics*, 4:249–264, 2003.
12. T. Kepler, L. Crosby, and K. Morgan. Normalization and analysis of DNA microarray data by self-consistency and local regression. *Genome Biology*, 3:1–12, 2002.
13. C. Li and W. H. Wong. Model-based analysis of oligonucleotide arrays: Model validation, design issues, and standard error applications. *Genome Biology*, 2:1–11, 2001.
14. L. M. Li. Blind inversion needs distribution (BIND): The general notion and case studies. In D. Goldstein, ed., *Science and statistics: A Festschrift for Terry Speed*, volume 40, pp. 273–293. Institute of Mathematical Statistics, 2003.
15. L. M. Li. An algorithm for computing exact least trimmed squares estimate of simple linear regression with constraints. *Computational Statistics and Data Analysis*, 48:717–734, 2004.
16. D. J. Lockhart, H. Dong, M. C. Byrne, M. T. Follettie, M. V. Gallo, M. S. Chee, M. Mittmann, et al. DNA expression monitoring by hybridization of high density oligonucleotide arrays. *Nature Biotechnology*, 14:1675–1680, 1996.
17. P. J. Rousseeuw and A. M. Leroy. *Robust regression and outlier detection*. New York: Wiley, 1987.

18. E. E. Schadt, C. Li, C. Su, and W. H. Wong. Analyzing high-density oligonucleotide gene expression array data. *Journal of Cellular Biochemistry*, 80:192–202, 2000.
19. E. E. Schadt, C. Li, B. Ellis, and W. H. Wong. Feature extraction and normalization algorithms for high-density oligonucleotide gene expression array data. *Journal of Cellular Biochemistry*, supplement 37:120–125, 2001.
20. L. Shi, W. Tong, Z. Su, T. Han, J. Han, R. K. Puri, H. Fang, et al. Microarray scanner calibration curves: Characteristics and implications. *BMC Bioinformatics*, 6(Suppl 2), 2:S11, 2005.
21. I. A. Sidorov, D. A. Hosack, D. Gee, J. Yang, M. C. Cam, R. A. Lempicki, and D. S. Dimitrov. Oligonucleotide microarray data distribution and normalization. *Information Sciences*, 146:67–73, 2002.
22. G. C. Tseng, M. K. Oh, L. Rohlin, J. C. Liao, and W. H. Wong. Issues in cDNA microarray analysis: Quality filtering, channel normalization, models of variations, and assessment of gene effects. *Nucleic Acids Research*, 29:2549–2557, 2001.
23. J. van de Peppel, P. Kemmeren, H. van Bakel, M. Radonjic, D. van Leenen, and F. C. P. Holstege. Monitoring global messenger RNA changes in externally controlled microarray experiments. *EMBO Reports*, 4:387–393, 2003.
24. C. Workman, J. L. Jensen, H. Jarmer, R. Berka, L. Gautier, B. H. Nielsen, H. Saxild, et al. A new nonlinear normalization method for reducing variability in DNA microarray experiments. *Genome Biology*, 3:1–16, 2002.
25. Z. Wu and R. A. Irizarry. Stochastic models inspired by hybridization theory for short oligonucleotide arrays. RECOMB, 98–106, 2004.
26. Y. H. Yang, S. Dudoit, P. Luu, D. M. Lin, V. Peng, J. Ngai, and T. P. Speed. Normalization for cDNA microarray data: A robust composite method addressing single and multiple slide systematic variation. *Nucleic Acids Research*, 30:e15, 2002.
27. Y. H. Yang, S. Dudoit, P. Luu, and T. P. Speed. Normalization for cDNA microarray data. In M. L. Bittner, Y. Chen, A. N. Dorsel, and E. R. Dougherty, eds., *Microarrays: Optical technologies and informatics*, volume 4266. SPIE, 2001.
28. A. Zien, T. Aigner, R. Zimmer, and T. Lengauer. Centralization: A new method for the normalization of gene expression data. *Bioinformatics*, 17:323–331, 2001.

3 Preprocessing and Normalization for Affymetrix GeneChip Expression Microarrays

Ben Bolstad

CONTENTS

Abstract 41
Introduction 41
Preprocessing Methods 43
    Overview of Preprocessing 43
    Robust Multichip Analysis 44
    GCRMA 48
    MAS5.0 49
Comparing Expression Measures 51
Case Study 53
References 58

ABSTRACT

Affyemtrix GeneChips® are widely used in academia and industry for expression analysis. One of the most intriguing aspects of this high-density array is the presence of multiple probes per array, and the availability of mismatch probes. This rich source of data provides detailed information about hybridization and annealing conditions. Many normalization algorithms use this mismatch information in order to make predictions about ectopic or failed hybridization, and predict the relative value of one probe versus the rest of the probe set. Other algorithms discard mismatch information and instead rely on the high replication of probes to normalize. We present a description of RMA, GC-RMA, and MAS5 normalization techniques with the description of the mathematical differences among the three and the appropriateness of each.

INTRODUCTION

Affymetrix has produced a large number of expression arrays covering a wide range of organisms. This chapter addresses preprocessing techniques relevant to these arrays. Affymetrix GeneChip microarrays consist of millions of specific DNA fragments, chemically synthesized on a coated quartz surface. Each one of these DNA fragments is placed on the array at a specific location and is known as a probe. GeneChips are manufactured via a procedure where all probes are constructed in parallel using photolithography and combinatorial chemistry. Every probe on the array is 25 bases in length. Every probe is designed to target the expression of a particular gene given its sequence. Specifically, the 25-mer sequence for the probe is complementary to the sequence for the targeted gene. Typically, 11 such probes are used to target the expression of a particular gene on the array. These are referred to as perfect match (PM) probes. Each PM probe also has a closely related probe, called the mismatch (MM). An MM probe has the same base sequence as the PM probe except at the 13th base, where a noncomplementary base is used. In theory, the MM probes can be used to quantify and remove nonspecific hybridization. A PM and its related MM are known as a probe pair. The group of probe pairs that targets a specific gene is called a probe set. More details about how probes are selected for expression arrays can be found in Mei et al. [1].

For expression arrays, binding of fluorescently labeled RNA to the array is used to quantify gene expression. The assay process begins with total RNA (or poly-A mRNA) isolated from the source tissue or cell line. The total RNA is then reverse transcribed to produce double-stranded cDNA using a series of reagents. A cleaning procedure is then carried out on the cDNA. Next, biotin labeled cRNA is produced from the cDNA. After another cleaning procedure, the biotin labeled cRNA is fragmented. The fragments are typically 25–200 bases in length.

A number of controls are also used for Affymetrix expression arrays. Control Oligo B2 hybridizes to locations on the edges and corners of the array. BioB, bioC, bioD, and cre are *Escherichia coli* genes that are added at specified concentrations to check how well the hybridization, washing, and staining procedures have performed. An additional five genes from *Bacillus subtilis*—dap, thr, trp, phe, and lys—are also used as controls.

The fragmented biotin labeled cRNA, along with the controls, is mixed to form a hybridization cocktail. The array cartridge is then filled with the mixture and placed in a hybridization oven for 16 hours. During the hybridization procedure the complementary binding properties of DNA ensure that the fragmented cRNA and controls bind to the oligonucleotides on the array. If there is more cRNA for a particular gene in the hybridization cocktail, then after hybridization there should be more material attached to the probes corresponding to that gene.

After hybridization, nonhybridized cRNA is removed from the cartridge and the array is placed in a fluidics station. Then, a series of washing and staining steps are applied to the array. The fluorescent staining agent streptavidin-phycoerythrin (SAPE) binds with the biotin labeling on the cRNA.

After the washing and staining process, the array is removed from the fluidics station and placed in a scanner. Laser light is shone onto the array and excites the fluorescent staining agent. At locations where more cRNA hybridized, a brighter signal should be emitted. The amount of signal emitted is recorded as a value in 16 bits, and by examining the entire chip an image is produced. This image is stored in the DAT file. The control oligo B2 probes are used to superimpose and align a grid upon the image. Once the gridding has taken place, the border pixels are ignored and the internal pixels of each grid square are used to compute a raw probe intensity. In particular, the 75th percentile of the intensities for these pixels gives the probe intensity for each probe cell. These probe intensity values are written into the CEL file. Low-level analysis begins with data read from the CEL file. More details about the assay, hybridization protocol, and scanning can be found in Affymetrix [2].

"Preprocessing" refers to the procedures used to convert raw probe intensities into useful gene expression measures. It occurs before any methods used to directly answer the biological questions of interest, whether this is looking for differential genes, clustering samples and genes, finding biomarkers to build a classifier that predicts tumor status, or something else. It is important to realize that the choice of one preprocessing methodology instead of another can have significant effects on the resultant expression measures and their usefulness in subsequent analysis. The ideal is to have expression measures that are both precise and accurate; that is, they have low bias and low variance. It is important to note that preprocessing is sometimes known as *low-level analysis* or *normalization*. The terminology *high-level analysis* refers to biologically driven analysis and such techniques will not be discussed here. This chapter prefers the term preprocessing rather than normalization, because normalization is only one element of a successful low-level analysis.

In the next section, preprocessing methodology for expression arrays is introduced, and three popular algorithms—robust multichip analysis (RMA), GCRMA (GC[GC content] robust multichip analysis), and MAS5.0 (Microarray Suite 5.0)—are explained in great detail. "Comparing Expression Measures" describes how to compare these algorithms. The final section shows how to carry out a low-level analysis using real data.

PREPROCESSING METHODS

OVERVIEW OF PREPROCESSING

In this chapter, preprocessing for raw probe intensities to expression values is viewed as consisting of three discrete steps: background correction, normalization, and summarization. Each of these will be explained in more detail later in this section. A fourth important step is quality assessment. This is often interlinked with the three other preprocessing steps.

The term *background correction*, also referred to as *signal adjustment*, describes a wide variety of methods. More specifically, a background correction method should perform some or all of the following: correct for background noise and processing effects; adjust for cross-hybridization, which is the binding of nonspecific DNA (i.e., noncomplementary binding) to the array; or adjust expression estimates so that they fall on the proper scale (i.e., linearly related to concentration). It is important to note that this definition is somewhat broader than is often used in the wider microarray community. Many times, only methods dealing with the first problem have been referred to as background correction methods. Unlike other array systems, where pixels surrounding a spot can be used to compute the background adjustment, the probe intensities themselves must be used to determine any adjustment for Affymetrix GeneChips because probe locations are very densely spaced on the array.

Normalization is the process of removing unwanted nonbiological variation that might exist between chips in a microarray experiment. It has long been recognized that variability can exist between arrays—some of biological interest and others of nonbiological interest. These two types of variation are interesting or obscuring, respectively. It is this obscuring variation, sometimes called technical variation, that normalization procedures seek to remove. Sources of obscuring variation could include scanner-setting differences, the quantities of mRNA hybridized, processing order, and many other factors. An important consideration when applying a normalization method to data from a typical comparative microarray experiment is how many genes are expected to change between conditions and how these changes will occur. Specifically, most normalization methods require that either the number of genes changing in expression between conditions be small or that an equivalent number of genes increase and decrease in expression. When neither of these assumptions is true, ideally normalization would only be applied on arrays within each treatment condition group. However, in many cases, experiments are not properly randomized and there is confounding of conditions with sources of nonbiological variation. For this reason, all of the arrays in a particular experiment are typically normalized together as a single group.

Affymetrix GeneChip expression arrays have hundreds of thousands of probes. These probes are grouped together into probe sets. Within a probe set each probe interrogates a different part of the sequence for a particular gene. *Summarization* is the process of combining the multiple probe intensities for each probe set to produce a single expression value.

Another important element of low-level analysis is the ability to distinguish between good- and poor-quality arrays in a data set. This might be done by visualizing the data both before and after various preprocessing steps or based upon quantitative quality control (QC) metrics.

ROBUST MULTICHIP ANALYSIS

Robust multichip analysis (RMA) is a procedure for computing expression measures [3–5]. It consists of three discrete steps: a convolution model-based background correction, a nonlinear probe-level normalization, and a robust multiarray summarization method. This section explains each of these steps in greater detail. Robust multichip analysis is a PM-only algorithm. It completely ignores the MM probes in all steps of the algorithm.

The background correction method use by RMA is based on a convolution model. The model is that the observed probe signal, S, consists of two components, $S = X + Y$, where X is signal, and Y is background. Motivated by the typical smoothed density plot, the convolution model assumes that X is distributed $\exp(\alpha)$ and that Y is distributed $N(\mu, \sigma^2)$. The background corrected value for each probe is given by the expected value of X given the observed signal:

$$E(X|S) = a + b\frac{\phi(a/b)}{\Phi(a/b)}$$

where $a = s - \mu - \sigma^2\alpha$ and $b = \sigma$.

A nonparametric procedure is used to estimate the model parameters. First, a density estimator is fit to the observed PM probe intensities; this is used to estimate the mode of the distribution. Points above the mode are used to estimate the exponential parameter, and a half normal is fit to points below the mode to estimate the normal distribution.

The goal of quantile normalization, as discussed in Bolstad et al. [3], is to give the same empirical distribution of intensities to each array. Suppose that all the probe intensities to be normalized are stored in a matrix, with rows being probes and columns being arrays. The algorithm proceeds as follows: First, the probe intensities on each array are sorted, from lowest to highest. Then, at each quantile, these are averaged across each row in the data set and this average is used as the intensity for that specific quantile in each column of the matrix. Finally, each column is reordered to return each element to its original position. Because the algorithm consists of only sorting and averaging operations, it runs quickly, even with large data sets. Figure 3.1 shows the intensity distributions for 30 HG-U95A arrays before normalization and also the resultant distribution for each array after quantile normalization.

The quantile normalization method is a specific case of the transformation $x'_i = F^{-1}(G(x_i))$, where we estimate G by the empirical distribution of each array and F using the empirical distribution of the averaged sample quantiles.

The RMA approach is to use a robust multiarray summarization method to combine the multiple probes for each probe set on each array to derive expression val-

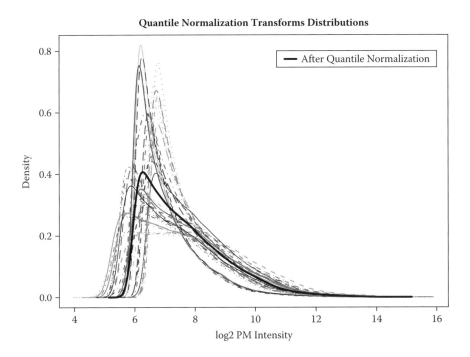

FIGURE 3.1 (See color insert following page 80.) Quantile normalization equalizes the intensity distribution across all arrays.

ues. A little mathematical notation is required to describe these algorithms in more detail. Let β represent an expression value on the $\log_2$ scale and y represent a natural scale probe intensity (which has been previously preprocessed using a background and normalization). Suppose that N_p designates the number of probe sets on the chip, and N_A is the number of arrays in the data set. Furthermore, use I_n to represent the number of probes (or probe pairs) in probe set n. Finally, consider that the subscript n represents the probe set with $n = 1, \ldots, N_p$, the subscript j represents the particular array with $j = 1, \ldots, N_A$, and the subscript i represents the probe in the probe set with $i = 1, \ldots, I_n$.

Summarization is done using a robust multichip linear model fit on a probe set by probe set basis. Specifically, the standard RMA summarization approach is to use median polish to compute expression values. The median polish algorithm [6] is a method for fitting the following model:

$$\log_2\left(y_{nij}\right) = \mu_n + \theta_{nj} + \alpha_{ni} + \varepsilon_{nij}$$

with constraints $\text{median}_j(\theta_{nj}) = \text{median}_i(\alpha_{ni}) = 0$ and $\text{median}_j(\varepsilon_{nij}) = \text{median}_i(\varepsilon_{nij}) = 0$ for any fixed n.

The RMA $\log_2$ expression estimates are given by $\hat{\beta}_{nj} = \hat{\mu}_n + \hat{\theta}_{nj}$. The algorithm proceeds as follows: First, a matrix is formed for each probe set n such that the probes are in rows, and the arrays are in columns. This matrix is augmented with row and column effects, giving a matrix of the form:

$$\begin{bmatrix} \varepsilon_{11} & \cdots & \varepsilon_{1N_A} & a_1 \\ \vdots & \ddots & \vdots & \vdots \\ \varepsilon_{I_n 1} & & \varepsilon_{I_n N_A} & a_{I_n} \\ b_1 & \cdots & b_{N_A} & m \end{bmatrix}$$

where, initially, $\varepsilon_{ij} = \log_2(y_{ij})$ and $a_i = b_j = m = 0$ for $\forall i, j$.

Next, each row is swept by taking the median across columns (ignoring the last column of row effects) and subtracting it from each element in that row and adding it to the final column (the $a_1, \ldots, a_{I_n}, m$). Then the columns are swept in a similar manner by taking medians across rows, subtracting from each element in those rows, and then adding to the bottom row (the $b_1, \ldots, b_{N_A}, m$). The procedure continues, iterating row sweeps followed by column sweeps, until the changes become small or zero. At the conclusion of this procedure, $\hat{\mu}_n = m$, $\hat{\theta}_{nj} = b_j$, and $\hat{\alpha}_i = a_i$. The ε_{ij} elements will be the estimated residuals.

One drawback to the median polish procedure is that it does not naturally provide standard error estimates. Another is that we are restricted to balanced row–column effect models. However, it may be implemented so that it operates very fast and thus large data sets may be processed very quickly.

A second approach to summarization is the PLM (probe level model) approach [7]. This method again fits the model

$$\log_2\left(y_{nij}\right) = \beta_{nj} + \alpha_{mi} + \varepsilon_{nij}$$

where α_{ni} is a probe effect, and ε_{nij} is assumed to be independently and identically distributed errors, with the constraint

$$\sum_{i=1}^{I_n} \alpha_{ni} = 0$$

which makes the model identifiable.

Rather than using ordinary least squares regression, the model is fit by robust regression using M-estimation. Consider the general linear regression model $\mathbf{Y} = \mathbf{Xb} + \varepsilon$, with $\mathbf{X}$ as the design matrix, $\mathbf{Y}$ the vector of dependent observations, $\mathbf{b}$ as the parameter vector, and ε as the vector of errors. The fitting procedure used is iteratively reweighted least squares. Let $\mathbf{W}$ be a diagonal matrix of weights such that

$$w_{ii} = w\left(\frac{y_i - \mathbf{x}_i \mathbf{b}}{\hat{s}}\right)$$

for specified weight functions $w(x)$. $\hat{s}$ is an estimate of scale for which the median absolute deviation of the residuals is used—specifically, $\hat{s} = \text{median}|\varepsilon|/0.6745$. Then, the updated parameter estimates are given by

$$\mathbf{b} = \left(\mathbf{X}^T \mathbf{W} \mathbf{X}\right)^{-1} \mathbf{X}^T \mathbf{W} \mathbf{Y}$$

The process is iterated until convergence. While, in general, there are many possible functions that could be used for the weight function, the PLM summarization method uses a function due to Huber:

$$w(x) = \begin{cases} 1 & |x| < k \\ k/|x| & \text{if } |x| \geq k \end{cases}$$

with $k = 1.345$ by default. In other words, values that are far away from 0 will be given low weights, meaning that this procedure gives some degree of robustness to the resultant expression values. Note that the PLM procedure is somewhat slower than the median polish algorithm.

An advantage of using M-estimation regression for the PLM summarization method is that there are natural standard error estimates. In particular, Huber [8] gives three forms of asymptotic estimators for the variance–covariance matrix of parameter estimates $\hat{\mathbf{b}}$. For PLM we use a form that allows for different SE estimates for each $\hat{\beta}_{nj}$.

In addition, the PLM procedure provides two useful quality assessment metrics. As defined in Brettschneider et al. [9], these are known as NUSE and RLE. The first, NUSE (short for normalized unscaled standard errors), is given by

$$\text{NUSE}\left(\hat{\beta}_{nj}\right) = \frac{\text{SE}\left(\hat{\beta}_{nj}\right)}{\text{median}_j\left(\text{SE}\left(\hat{\beta}_{nj}\right)\right)}$$

The other quality metric is RLE, an abbreviation for relative log expression, and is computed using

$$\text{RLE}\left(\hat{\beta}_{nj}\right) = \hat{\beta}_{nj} - \text{median}_j\left(\hat{\beta}_{nj}\right)$$

Both of these quantities are computed for each probe set on each array and then as aggregate sets used to assess the quality of each array in a data set.

GCRMA

The GCRMA algorithm [10,11] uses both the quantile normalization and median polish summarization methods of the RMA algorithm. Where it differs is in the background procedure. A stochastic model is used to describe the observed PM and MM probe signals for each probe pair on an array. In particular, the model is:

$$PM_{ni} = O_{ni} + N_{1ni} + S_{ni}$$

$$MM_{ni} = O_{ni} + N_{2ni}$$

where O_{ni} represents optical noise, N_1 and N_2 represent nonspecific binding, and S_{nj} is a quantity proportional to the RNA expression in the sample. In addition, the model assumes O follows a normal distribution $N(\mu_o, \sigma_o^2)$ and that $\log_2(N_{1ni})$ and $\log_2(N_{2ni})$ follow a bivariate-normal distribution with equal variances σ_N^2 and correlation 0.7, constant across probe pairs. The means of the distribution for the nonspecific binding terms are dependent on the probe sequence. The optical noise and nonspecific binding terms are assumed to be independent.

The method by which GCRMA includes information about the probe sequence is to compute an *affinity* based on a sum of position-dependent base affinities. In particular, the affinity of a probe is given by:

$$A = \sum_{k=1}^{25} \sum_{b \in \{A,C,G,T\}} \mu_b(k) 1_{B_k=j}$$

where the $\mu_b(k)$ are modeled as spline functions with 5 degrees of freedom.

Figure 3.2 shows the estimated $\hat{\mu}_b(k)$ for a single U133A microarray. In practice, these are either estimated using the observed data for all chips in an experiment or based on some hard-coded estimates from a specific NSB experiment carried out by the creators of GCRMA. The means for the N_1 and N_2 random variables in the GCRMA model are modeled using a smooth function h of the probe affinities.

The optical noise parameters μ_o, σ_o^2 are estimated like this: The variability due to optical noise is much smaller than the variability due to nonspecific binding and thus effectively constant. For simplicity this is set to 0. Then the mean values are estimated using the lowest PM or MM probe intensity on the array, with a correction factor to avoid negatives. Next, all probe intensities are corrected by subtracting this constant $\hat{\mu}_o$. To estimate $h(A_{ni})$ a loess curve is fit to a scatterplot relating the corrected log(MM) intensities to all the MM probe affinities. The negative residuals

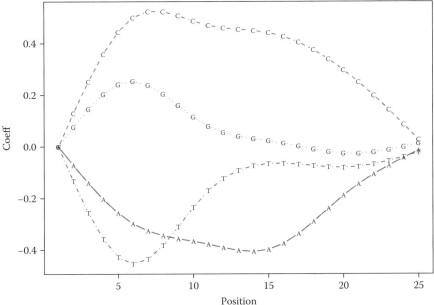

FIGURE 3.2 Effect of each base at given position estimated using data from an HG-U133A chip. These are used by GCRMA to calculate a probe affinity for each probe.

from this loess plot are used to estimate σ_N^2. Finally, the background adjustment procedure for GCRMA is to compute the expected value of S given the observed PM, MM, and model parameters.

Note that although GCRMA uses the median polish summarization of RMA, the PLM summarization approach should not be used in its place if one wants to carry out quality assessment, although the expression estimates generated this way should be otherwise satisfactory.

MAS5.0

The MAS5.0 algorithm [12–14] is designed to work on a single chip basis. Both PM and MM probe intensities are used. It consists of a location-specific background adjustment, correction of the PM probe intensities based on the MM probe intensities, a single chip robust summarization. Normalization consists of a linear adjustment and is carried out after summarization.

The goal of this step is to remove overall background noise. Each array is divided into a set of grid regions. By default this is 16 equally sized regions arranged to form a 4 by 4 grid. For each grid region, a background (b_k) and a noise value (n_k) are estimated using the mean and standard deviations of the lowest 2% of probe intensities in that grid region. Then, every probe intensity is adjusted using the weighted average of each of the background values. The weights are dependent on the distance from the centroid of each of the grids. In particular, the weights are

$$w_k(x,y) = \frac{1}{d_k^2(x,y) + \delta}$$

where $d_k^2(x,y)$ is the Euclidean distance from location (x,y) to the centroid of region k. The default value for δ is 100. To avoid negative values or other numerical problems for low-intensity regions, the noise estimates are used. $B(x,y)$ is the weighted average of the b_k at location (x,y) and $N(x,y)$ is the weighted average of the n_k at location (x,y). Suppose that $P_{x,y}$ is the probe intensity at (x,y) on the array; then, the correct value is given by

$$P_{x,y} = \max\left(P_{x,y} - B(x,y), N_f N(x,y)\right)$$

with N_f a tunable parameter being equal to 0.5 by default.

Originally, the suggested purpose of the MM probes was that they could be used to adjust the PM probes [15] by subtracting the intensity of the MM probe from the intensity of the corresponding PM probe. However, this becomes problematic because, in a typical array, as many as 30% of MM probes have intensities higher than their corresponding PM probe [16]. Thus, when raw MM intensities are subtracted from the PM intensities, it is possible to compute negative expression values. Another drawback is that the negative values preclude the use of logarithms, which have proved useful in many microarray data situations. To remedy the negative impact of using raw MM values, MAS5 introduces the concept of the *ideal mismatch* (IM) [13], which by design is guaranteed to be positive. The goal is to use MM when it is physically possible (i.e., smaller than the corresponding PM intensity) and something smaller than the PM in other cases. First, a quantity referred to as the *specific background* (SB) is calculated for each probe set. This is computed by taking a robust average of the log ratios of PM to MM in each probe pair. If i is the probe and n is the probe set, then for the probe pair indexed by i and n the ideal mismatch IM is given by

$$IM_{ni} = \begin{cases} MM_{ni} & \text{When } MM_{ni} < PM_{ni} \\ \dfrac{PM_{ni}}{2^{SB_n}} & \text{When } MM_{ni} \geq PM_{ni} \text{ and } SB_n > \tau_c \\ \dfrac{PM_{ni}}{2^{\tau_c/(1+(\tau_c - SB_n)/\tau_s)}} & \text{When } MM_{ni} \geq PM_{ni} \text{ and } SB_n \leq \tau_c \end{cases}$$

where τ_c and τ_s are tuning constants. τ_c is referred to as the contrast parameter (default value 0.03), and τ_s is the scaling parameter (default value 10). The adjusted PM intensity is given by subtracting the corresponding IM.

The MAS5.0 algorithm uses a robust single chip summarization method. Specifically, a one-step Tukey biweight, taken on the $\log_2$ scale, provides expression summary values. In particular, the procedure proposed in Bolstad [12] and Hubbell, Liu, and Mei [17] is to compute

$$u_{nij} = \frac{\log_2(y_{nij}) - M}{cS + \varepsilon}$$

where M is the median of the $\log_2(y_{nij})$, and S is the median of the absolute deviation (i.e., MAD) from M. Here, $c = 5$ is a tuning constant, and $\varepsilon = 0.0001$ is chosen to avoid the problem of division by 0. Weights are defined by the bisquare function

$$w(u) = \begin{cases} 0 & \text{when } |u| > 1 \\ (1-u)^2 & \text{when } |u| \leq 1 \end{cases}$$

Finally, the expression measure is given by

$$\hat{\beta}_{nij} = \frac{\sum_{i=1}^{I_n} w(u_{nij}) \log_2(y_{nij})}{\sum_{i=1}^{I_n} w(u_{nij})}$$

A standard error estimate for this expression estimate is also available. Unlike RMA or GCRMA, the expression values returned by MAS5.0 are reported in the original scale (i.e., $\hat{B}_{nj} = 2^{\hat{\beta}_{nj}}$).

In MAS5.0 normalization is carried out after expression values have been computed for each array in the data set. A scaling normalization is used to adjust each array. In particular, for each array a trimmed mean of the expression values is calculated by excluding the highest and lowest 2% of the data. Then a target intensity is chosen, either the mean calculated for one of the arrays in the data set or one previously chosen. Each expression value on array j is adjusted by the following scaling factor:

$$\phi_j = \frac{\text{mean}_n(\hat{B}_{nj})}{T}$$

where T is the target intensity.

COMPARING EXPRESSION MEASURES

Different algorithms produce different expression values, and the chosen algorithm might have significant effect on the ability to identify biologically relevant differentially expressed genes. It can be hard to assess the effect any analysis algorithm will have on an arbitrary data set with potential unknown differential expression changes. For this reason a number of spike-in experiments have been created, and these are useful for assessing the performance of analysis algorithms. A spike-in experiment typically consists of several probe sets that have their concentration deliberately modified on different replicates where the background RNA is otherwise the same. This gives a truth by which to assess algorithmic performance. Affymetrix has supplied two different spike-in studies, one on the U95A microarray platform and another on the U133A platform. These can be downloaded directly from the Affymetrix Website at www.affymetrix.com.

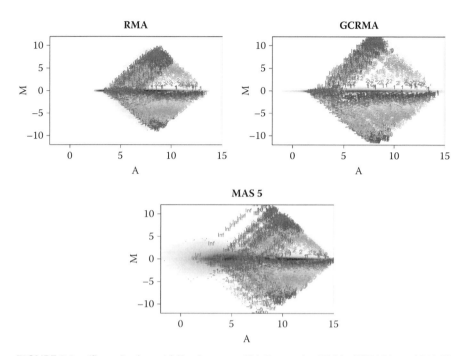

FIGURE 3.3 (See color insert following page 80.) Comparing RMA, GCRMA, and MAS5 expression values using a U133A spike-in data set. MA plots show large level of noise at the low end for MAS5. GCRMA and MAS5 have greater linearity through concentration range.

Assessment is usually on the basis of precision, accuracy, and ability to discriminate the truly differentially expressed probe sets (i.e., the spike-ins) from nondifferential probe sets (i.e., everything else). "Precision" refers to variability and "accuracy" refers to bias. Detailed discussion of how to compare expression measures based on these data sets is given by Cope et al. [18] and Irizarry, Wu, and Jaffee [19] and implemented as part of set of software known as affycomp. Also, the affycomp Website, http://affycomp.biostat.jhsph.edu/, gives comparison results for many additional expression array analysis algorithms and provides functionality for users to upload analysis results from their own algorithms.

This chapter is too brief for a detailed analysis and comparison of all current preprocessing algorithms proposed for Affymetrix expression arrays, or even for a full comparison of RMA, GCRMA, and MAS5.0. However, Figure 3.3 shows a brief comparison of the three algorithms discussed in this chapter using U133A spike-in data. On the left, three composite MA-plots show how well the spike-in probe sets can be differentiated from the nondifferential probe sets in a spike-in data set. The main feature to notice from these plots is that MAS5.0 is much noisier in the low intensities than either RMA or GCRMA. This makes it harder to discriminate true differential genes. On the right is a plot showing the general trends, as estimated using a loess smoother, created by plotting the estimates expression value in $\log_2$ scale versus the true concentration in $\log_2$ scale. This also highlights features of the three different algorithms. MAS5.0 shows the greatest linearity across the

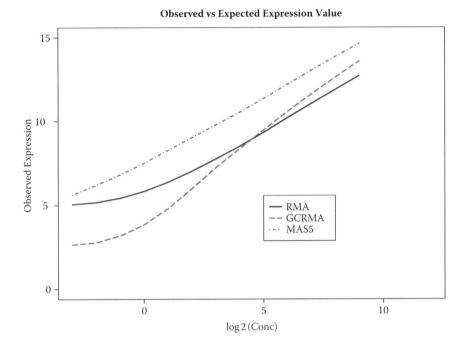

FIGURE 3.3 (continued).

concentration range, whereas RMA has perhaps the least linearity, with the greatest leveling off at the lowest concentration values. The higher slope of GCRMA means that its expression estimates will tend to be less attenuated.

CASE STUDY

This section illustrates the usage of some of the preprocessing techniques outlined in this chapter as applied to real data. It concentrates on a workflow using RMA and related techniques. The data used in this chapter are a small subset of a much larger data set used for studying gene expression in the brain and its relation to bipolar disorder and major depression, among other mental disorders [20]. The raw data may be downloaded from the National Center for Biotechnology Information's Gene Expression Omnibus (GEO, http://www.ncbi.nlm.nih.gov/geo/) [21] and are accessible through GEO series accession number GSE6306. The subset used here consists of 29 HG-U133A microarrays for the anterior cingulate region representing samples from the first two cohorts and hybridized at site I.

The analysis discussed in this section was carried out using RMAExpress (Version 1.0 beta 2). This is a free piece of software available for both Windows and Linux operating systems. It can be downloaded at http://rmaexpress.bmbolstad.com. Another free option for analysis using RMA or PLM would be via the BioConductor software at www.bioconductor.org.

The first step in analysis is to examine the unprocessed probe intensities. Figure 3.4 shows box plots and density curves for all 29 arrays in this data set. These

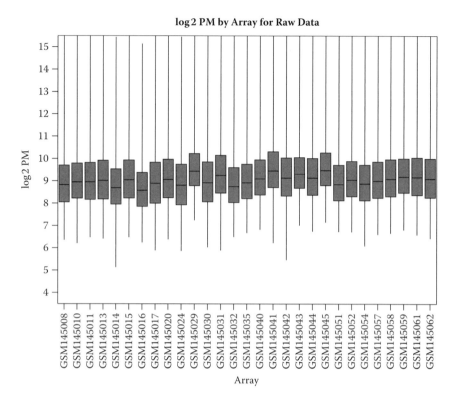

FIGURE 3.4 Examining the raw data using boxplots and density plots. Before normalization with other preprocessing algorithms, the data distributions are not equivalent, as is apparent by boxplot (left) and density plot (right).

plots are useful for identifying potential discordant arrays. In this case, there are some differences in overall intensity level between arrays, including four that are slightly brighter than the others. These differences show the need for normalization to remove technical differences between arrays. Often, arrays that are extremely different from the others in the data set have quality issues. On the density plot this would be indicated by a differently shaped curve and on the box plot by a plot much higher, lower, or more spread than others.

Next, the raw data are preprocessed using the convolution background adjustment, quantile normalization, and the PLM summarization method. RMAExpress also provides the ability to store the estimated residuals from the PLM fitting procedure. A false color image of each chip in the data set can be created using the residuals. These can be used to detect artifacts on the chip. Figure 3.5 shows images of the residuals for two arrays in the data set: GSM14020 and GSM14035. Greater intensity red is used to represent extreme positive residuals—that is, much brighter than expected regions on the array—and a more intense blue is used to represent extreme negative residuals—that is, probes more dim than predicted by the model. White is used for residuals that are close to zero. The software also provides the ability to visualize the positive or negative residuals separately or ignore the magnitude

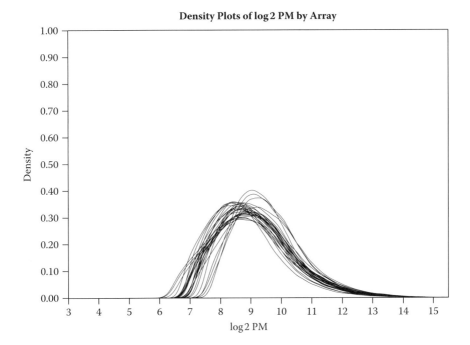

FIGURE 3.4 (continued).

of the residual and just visualize the sign. Both of the two images shown in the figure have artifacts. Note that an artifact does not assure that the expression values from that particular array are of low quality, but these plots are sometimes useful for diagnosing problems with arrays. A gallery of such images from many different data sets is available at http://PLMimagegallery.bmbolstad.com and provides a useful way to get a feel for different potential artifacts.

The RLE and NUSE statistics are used to identify lower quality arrays in a data set. This is done visually by using boxplots, one for each array. Figure 3.6 shows these as provided by RMAExpress. In the case of the RLE statistic, lower quality arrays have median RLE value far from 0 or large IQR. In this figure there are several arrays for which this is true. But because none of the arrays clearly stand out as being extremely different, there is no reason, based on RLE, to choose to exclude any array from subsequent analysis. For the NUSE statistic, lower quality arrays have median NUSE value far above one or larger IQR. In Figure 3.6 the arrays GSM145020, GSM145024, and GSM145030 show lower quality when judging based on the NUSE statistic, but the evidence seems marginal that they are significantly low quality enough to remove from the data set. Note that the array GSM14035, which was previously observed to have an artifact, does not seem to have any significant quality issues as judged by these metrics. Because the probes for any probe set are spread out across the array and robust procedures are used for analysis, smaller artifacts will not tend to cause any significant quality problems. Note that if an array is identified as having a quality issue, there is no need to repeat the low-level analysis with that chip removed. However, it should not be used for any high-level analysis.

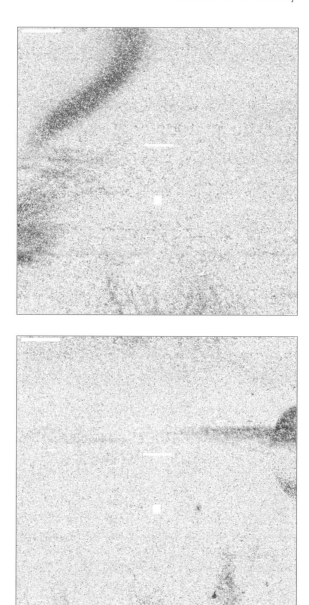

FIGURE 3.5 **(See color insert following page 80.)** Pseudochip images of residuals from probe level models fit using RMAExpress for two arrays: GSM14020 (top) and GSM14035 (bottom).

Affymetrix GeneChip Expression Microarrays

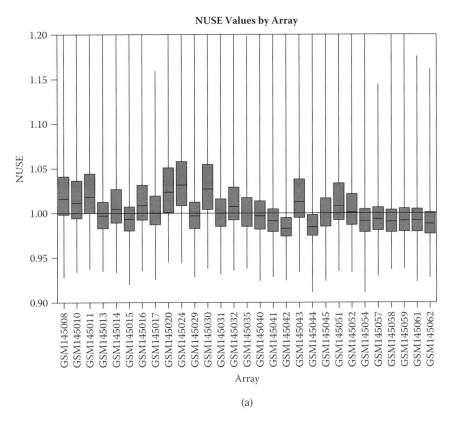

FIGURE 3.6 Normalized unscaled standard error (NUSE) box plots for each array (a). Box plots that have elevated median values or larger IQR are of lower quality. Relative log expression (RLE) box plots for each array (b). Arrays with larger IQR or medians far from zero may be of lower quality.

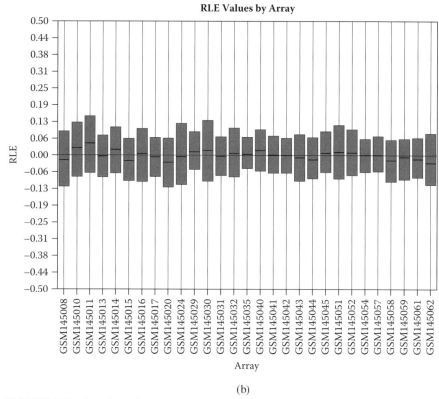

FIGURE 3.6 (continued).

REFERENCES

1. R. Mei, E. Hubbell, S. Bekiranov, M. Mittmann, F. C. Christians, M. M. Shen, G. Lu, et al., Probe selection for high-density oligonucleotide arrays, *Proc Natl Acad Sci USA,* 100, 11237–11242, 2003.
2. Affymetrix, *GeneChip expression analysis technical manual.* Santa Clara, CA: Affymetrix, 2004.
3. B. M. Bolstad, R. A. Irizarry, M. Astrand, and T. P. Speed, A comparison of normalization methods for high-density oligonucleotide array data based on variance and bias, *Bioinformatics,* 19, 185–193, 2003.
4. R. A. Irizarry, B. M. Bolstad, F. Collin, L. M. Cope, B. Hobbs, and T. P. Speed, Summaries of Affymetrix GeneChip probe level data, *Nucleic Acids Res,* 31, e15, 2003.
5. R. A. Irizarry, B. Hobbs, F. Collin, Y. D. Beazer-Barclay, K. J. Antonellis, U. Scherf, and T. P. Speed, Exploration, normalization, and summaries of high-density oligonucleotide array probe level data, *Biostatistics,* 4, 249–264, 2003.
6. J. Tukey, *Exploratory data analysis.* Reading, MA: Addison–Wesley, 1977.
7. B. M. Bolstad, Low-level analysis of high-density oligonucleotide array data: Background, normalization, and summarization, University of California, Berkeley, 2004.
8. P. Huber, *Robust statistics.* New York: John Wiley & Sons, Inc., 1981.
9. J. Brettschneider, F. Collin, B. M. Bolstad, and T. P. Speed, Quality assessment for short oligonucleotide arrays, *Technometrics,* 2007.

10. Z. Wu and R. A. Irizarry, Stochastic models inspired by hybridization theory for short oligonucleotide arrays, *J Comput Biol,* 12, 882–893, 2005.
11. Z. Wu, R. A. Irizarry, R. Gentleman, F. Martinez-Murillo, and F. Spencer, A model-based background adjustment for oligonucleotide expression arrays, *J Am Statistical Assoc,* 99, 909–918, 2004.
12. Affymetrix, Statistical algorithms description document, Santa Clara, CA: Affymetrix, 2002.
13. Affymetrix, *Affymetrix microarray suite users guide,* Version 5.0 ed. Santa Clara, CA: Affymetrix, 2001.
14. Affymetrix, Statistical algorithms reference guide, Santa Clara, CA: Affymetrix, 2001.
15. Affymetrix, *Affymetrix microarray suite users guide,* Version 4.0 ed. Santa Clara, CA: Affymetrix, 1999.
16. F. Naef, D. A. Lim, N. Patil, and M. Magnasco, DNA hybridization to mismatched templates: a chip study, *Phys Rev E Stat Nonlin Soft Matter Phys,* 65, 040902, 2002.
17. E. Hubbell, W. M. Liu, and R. Mei, Robust estimators for expression analysis, *Bioinformatics,* 18, 1585–1592, 2002.
18. L. M. Cope, R. A. Irizarry, H. A. Jaffee, Z. Wu, and T. P. Speed, A benchmark for Affymetrix GeneChip expression measures, *Bioinformatics,* 20, 323–331, 2004.
19. R. A. Irizarry, Z. Wu, and H. A. Jaffee, Comparison of Affymetrix GeneChip expression measures, *Bioinformatics,* 22, 789–794, 2006.
20. M. P. Vawter, H. Tomita, F. Meng, B. Bolstad, J. Li, S. Evans, P. Choudary, et al., Mitochondrial-related gene expression changes are sensitive to agonal-pH state: implications for brain disorders, *Mol Psychiatry,* 11, 615, 663–679, 2006.
21. T. Barrett, D. B. Troup, S. E. Wilhite, P. Ledoux, D. Rudnev, C. Evangelista, I. F. Kim, et al., NCBI GEO: mining tens of millions of expression profiles—database and tools update, *Nucleic Acids Res,* 35, D760–765, 2007.

4 Spatial Detrending and Normalization Methods for Two-Channel DNA and Protein Microarray Data

*Adi Laurentiu Tarca, Sorin Draghici,
Roberto Romero, and Michael Tainsky*

CONTENTS

Abstract 61
Introduction 62
Detecting Spatial Biases 64
    Normalization Methods for Both Spatial and Intensity Biases 67
        Print-Tip Loess Normalization 69
        Composite Normalization 69
        Median Filter 70
        Neural Network-Based Spatial and Intensity Normalization
            (nnNorm) 70
The Benefits of Removing Spatial Biases 72
Implementations Using R and Bioconductor 75
    Installing and Loading Packages 75
    Loading *Apo* Data Set 76
    Detecting Spatial Biases 77
    Spatial Normalization 78
Conclusion 78
Acknowledgments 79
References 79

ABSTRACT

Spatial artifacts may affect cDNA microarray experiments and therefore contribute nonbiological variability to the data. This can affect one's ability to detect the true biological differences between the samples profiled. Such spatial imbalance between the two channel intensities may translate into poor data reproducibility, and it is susceptible to increases in incidence of both false positives and false negatives in microarray experiments. The goals of this chapter are to (1) make the reader aware of the existence of spatial biases with microarray data, (2) introduce a simple method for detection of spatial biases, (3) describe some existing approaches to address spatial biases, (4) illustrate the benefits of spatial normalization, and (5) provide code segments for the R statistical environment that apply these methods on publicly available data sets.

INTRODUCTION

Microarray technology allows simultaneous profiling of the expression of thousands of genes in an mRNA sample [1] and [2]. Gene expression data can be used to make inferences about gene expression on a genomic scale, such as for identifying differentially expressed genes or classifying patients into disease groups [3]. With two-channel cDNA microarrays, the most common sample preparation techniques involve reverse transcription of RNA into cDNA and incorporation of a fluorescent dye that allows quantification of the cDNA amount. Two samples, each labeled with a different dye, undergo a competitive hybridization for the probes spotted on the array surface. The differential labeling permits independent detection of the fluorescent signal associated with each individual sample for each probe on the array. Cy5 (red) and Cy3 (green) dyes are frequently used. When excited by a laser beam, they emit in the red (R) and green (G) ranges of the light spectrum.

The abundance of the corresponding mRNAs in the two samples is assessed by quantifying the images of the array obtained by laser scanning at two different wavelengths that correspond to the two different dyes. The location of each probe on the array images is first identified. A subsequent segmentation step assigns a foreground and background region for each probe. The foreground pixel intensities are considered proportional to the amount of specific hybridization, augmented nonspecifically by an amount equal to the background fluorescence, which is estimated from the pixel intensities in the region surrounding the foreground region. After subtracting the background intensities from both channels, the relative abundance of the two samples is expressed as a ratio (R/G). The ratios are further log transformed to give log-ratios (or log-fold changes), $M = \log_2 R/G$, which are often further preprocessed and eventually analyzed. The log transformation stabilizes variances and converts multiplicative error into additive error [4].

Recently, the same two-channel technology has been deployed for the fabrication of and testing with protein arrays. An illustrative example is described by Chaterjee et al. [5], who used a two-channel protein array to detect ovarian cancer. In this deployment, a T7 phage display library is selectively screened to enrich for epitopes present in cancer and absent in normal sera. Clones with predictive values are printed onto regular microarray slides. The two dyes Cy3 and Cy5 are used to label T7 monoclonal antibodies and antihuman IgG, respectively. Similarly to the DNA microarrays, a competitive antibody reaction takes place when the arrays are hybridized with human sera to be tested for cancer. In this example, the Cy5 channel provides information regarding the abundance of the antibodies to specific cancer epitopes, while the Cy3 provides information regarding the amount of protein present at a given array spot.

As we will see in the following, even though these arrays use an antibody reaction instead of a DNA hybridization reaction, similar spatially correlated biases can occur in both technologies. Such systematic errors are likely to increase the incidence of both false positive and false negative results during data analysis [6]. It is preferred to minimize the introduction of systematic bias during the microarray experiment. However, the effects of systematic bias on the data can also be dealt with on a post hoc basis through data normalization procedures [7,8] or even through appropriate methods of differential gene selection [9,10]. This is possible by making several assumptions. The main assumption here is that most of the genes on the array are not differentially regulated under the condition of study. In other words, the R/G ratios should be close to 1 (and correspondingly $\log_2 R/G \cong 0$) for most of the genes on the array. A second assumption is that the small proportion of genes differentially expressed will be both up- and down-regulated.

The numerous steps of a microarray experiment, which are described in the literature [8,11], often introduce several types of systematic errors and can severely deteriorate the repeatability and reproducibility of microarray experiments [12]. A common type of bias is the *dye bias*, which manifests itself as an overall imbalance in the fluorescent intensities of one sample versus another. When the dye bias appears, most of the spots on the array image will appear either red or green.

Another frequently encountered type of bias is the *intensity bias*, which is manifested by a strong dependence of the log ratios of the spots (M) on their average log intensity,

$$A = \frac{1}{2} \log_2 R \cdot G.$$

This type of bias is illustrated in Figure 4.1. There are also factors that contribute systematic error in a *spatially dependent* fashion [13]. Typical reasons are differences in the print-tips of the robot, which induce systematic differences from one print-tip group to another, or nonhomogeneities of the array surface and/or imperfect washing of the array, etc.

While removal of intensity-dependent bias has been addressed with both linear and nonlinear regression techniques (see Park et al. [14] for a classification of these techniques), there are far fewer published techniques designed to remove both intensity and spatial-dependent bias at the same time. Yang et al. [11] developed a framework called composite normalization that combines spatial local smoothing with intensity-based loess [15] normalization. Wilson et al. [16] proposed a two-step procedure in which intensity-based normalization is applied as a first step and then a median filter is applied to further "polish" the residuals and remove the spatial trends. Tarca, Cooke, and Mackay [17] presented an alternative normalization method, which combines a spatial binning procedure with neural networks fitting, and compared these normalization techniques on real and simulated data sets. Neuvial et al. [18] proposed a spatial normalization for comparative genomic hybridization (CGH) arrays, which first identifies and removes local spatial artifacts and then removes

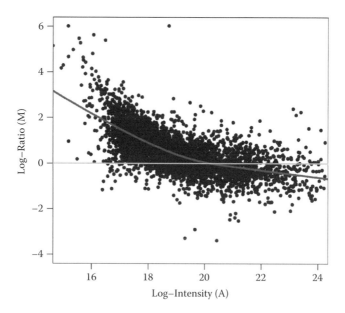

FIGURE 4.1 Intensity bias affecting log ratios from a cDNA array. Log ratios are plotted against average log intensities for each probe on the array. Ideally, the log ratios should be symmetrically scattered around the 0 line for most of the probes on the array. The dark gray line shows the bias as estimated by the loess smoother. It can be seen that log ratios present a strong dependence on the average log intensities, pointing toward a strong intensity bias.

array-level spatial biases. All these approaches are bound to the assumption that most of the genes on the array are not differentially expressed.

A step further in the field of spatial normalization was made by Yuan and Irizarry [19], who suggested a new approach to spatial normalization called conditional residual analysis for microarrays (CRAM). Even though this approach is free from the assumptions that constrain the previously described analytical procedures, CRAM requires a special microarray design that contains technical replicates of the array features, which should be placed in a certain way on the array surface. The array design also requires a certain number of negative controls.

The remainder of this chapter is organized as follows: First, we introduce a simple method to detect spatial bias within the print-tip groups of a two-channel array, followed by introducing several spatial normalization methods that do not require any particular microarray design. The list of normalization methods presented here is not necessarily exhaustive. Then some of the benefits of spatial normalization are given using as case study the *Apolipoprotein* AI data set [20]. Practical illustrations of spatial bias detection and normalization are finally shown using software packages available under the R statistical environment [21].

DETECTING SPATIAL BIASES

There are several reasons why the log ratios in a two-channel cDNA experiment may have a strong spatial dependence. One possible reason may relate to the way in which the probe sequences are spotted on the array surface by the printing robot. We illustrated this kind of problem with an antibody array used for cancer screening [5]. It can be seen in Figure 4.2a that the log ratios obtained on this particular array are correlated to the column index within most of the print-tip groups. This is due to a systematic increase on the red channel as we go from the first to the last column of the print-tips (Figure 4.2b), while the green channel level undergoes a decrease in the same direction (Figure 4.2c). As a result, the log ratio of the two channels increases strongly from the left to the right within the print-tips. In other words, the log ratios are correlated with the column index (row-wise spatial bias).

Other possible reasons for spatial imbalances between the two channels can be related to the physical properties of the array surface, the hybridization, and the washing steps. These kinds of spatial imbalances are expected to appear rather randomly on the different arrays in contrast to the first situation when the cause was the printing of the arrays. This type of artifact is illustrated in Figure 4.3.

An automated procedure to detect the arrays in which significant spatial biases occurred at the print-tip level is currently implemented in the latest version of the *nnNorm* package [22], included in the Bioconductor project (www.bioconductor.org). With this method the correlation coefficient between the log ratios and the column/row index is computed for each row/column. Under the null hypothesis that there is no spatial bias, we expect that most of the rows/columns will have M values weakly correlated with the column/row index both positively and negatively. Observing an important fraction of the rows/columns with a moderate to high correlation will

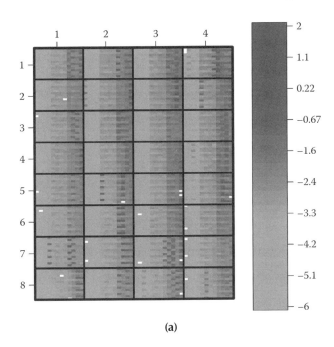

(a)

FIGURE 4.2 (**See color insert following page 80.**) Reconstructed image of an antibody array showing spatial bias most likely due to a print-tip problem: (a) $\log_2$ ratio of the two channels; (b) $\log_2$ intensity on the red channel; and (c) $\log_2$ intensity on the green channel.

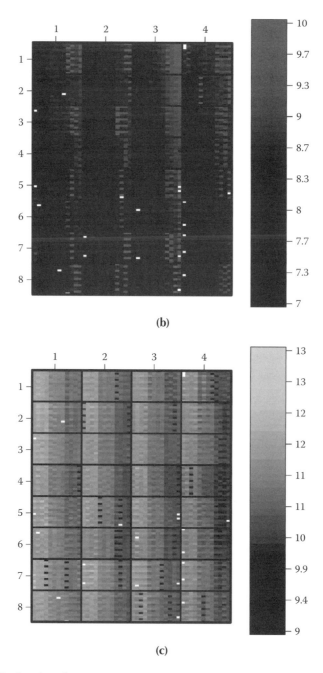

FIGURE 4.2 (continued).

indicate the possible presence of spatial bias. For instance, if a print-tip has 21 rows, it will be unlikely to observe that 7 of its rows (one third of the total) have positive correlation coefficients higher than 0.6.

Spatial Detrending and Normalization Methods

FIGURE 4.3 (See color insert following page 80.) A hybridization/washing-related artifact that artificially augments the red channel intensity in a spatially dependent manner in four print-tips of a protein array.

Table 4.1 shows the result of applying the *detectSpaialBias* method of the *nnNorm* [22] package on the *Apolipoprotein* AI data set [20], referred to here as *Apo* data set and available online at http://www.stat.berkeley.edu/~terry/zarray/Html/apodata.html). This data set contains 16 cDNA arrays.

In Table 4.1, two print-tips of the array c8R showing strong spatial bias are shown in boldface. The corresponding two print-tips are also highlighted in Figure 4.4. Note in Table 4.1 that the array k8R also presents the same kind of column-wise bias.

The method for spatial bias detection presented herein is able to identify primarily linear biases at the print-tip level, and it is mainly designed to ease the data quality inspection process. This method also can be used to assess how different normalization procedures detrend the microarray data and render them independent on the spatial location. This will be shown subsequently in this chapter.

NORMALIZATION METHODS FOR BOTH SPATIAL AND INTENSITY BIASES

This section will introduce four different approaches that can handle both intensity and spatial biases affecting two-channel microarray data. These normalization methods correct the raw log ratio of each probe i on the array M_i by subtracting from it an estimate of the bias $\hat{M}_i$:

$$M_i^* = M_i - \hat{M}_i \tag{4.1}$$

The resulting normalized log ratios M_i^* are to be used for further analysis such as differential expression [23]. What differs from one normalization method to another, among those to be described further, is how the bias $\hat{M}_i$ is estimated.

TABLE 4.1
Percentage of the Number of Columns Having Log Ratio Values Highly Correlated with the Row Index[a]

	c1R	c2R	c3R	c4R	c5R	c6R	c7R	c8R	k1R	k2R	k3R	k4R	k5R	k6R	k7R	k8R
PrintTip1	14	0	0	5	0	0	5	0	38	0	0	19	5	0	0	0
PrintTip2	5	0	10	0	5	5	0	0	5	0	0	0	0	0	0	0
PrintTip3	5	0	24	0	0	0	0	0	0	0	0	5	0	0	0	0
PrintTip4	19	5	5	0	0	5	0	14	5	0	10	5	0	0	0	0
PrintTip5	10	0	5	0	14	10	0	5	10	5	0	5	5	0	0	5
PrintTip6	19	5	5	0	0	0	10	0	10	0	0	10	10	0	5	0
PrintTip7	24	0	14	10	5	24	5	10	29	5	19	24	33	5	10	5
PrintTip8	10	5	5	0	14	19	0	14	24	14	5	5	0	10	10	5
PrintTip9	29	0	5	5	0	0	0	0	33	10	10	43	5	0	0	0
PrintTip10	24	14	5	24	10	5	14	10	19	10	14	19	10	19	5	10
PrintTip11	29	10	14	0	0	10	10	10	14	10	10	33	10	5	0	5
PrintTip12	24	29	10	5	0	33	0	5	14	10	19	29	0	5	0	0
PrintTip13	10	10	19	24	24	10	10	24	0	10	14	0	29	19	24	38
PrintTip14	19	0	24	43	29	29	29	52	33	14	14	0	24	14	33	38
PrintTip15	38	0	19	19	29	10	24	62	19	19	0	14	5	10	24	62
PrintTip16	5	0	0	0	10	0	24	14	0	0	5	0	0	5	5	5

Note: Output obtained with *mnNorm* package on *Apo* data set. The columns correspond to the arrays and the rows correspond to the print-tips.

Spatial Detrending and Normalization Methods

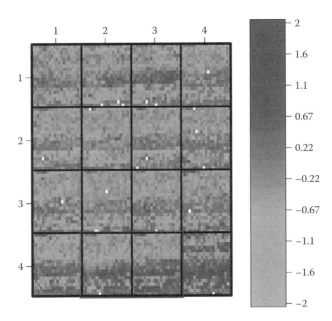

FIGURE 4.4 (See color insert following page 80.) Reconstructed image of the array c8R of the *Apo* data set. Two print-tip regions showing strong spatial bias are highlighted.

Print-Tip Loess Normalization

The earliest normalization method that addressed both intensity and spatial bias is the print-tip loess normalization [10,11]. This method estimates the dependence of the raw log ratios, M, on the average log intensity, A, by fitting a *loess* [15] smoother using the data from probes spotted with the same print-tip on the array surface. The loess estimator is shown in Figure 4.1 using all data from a cDNA array.

The bias estimate computed with print-tip loess normalization is computed as:

$$\hat{M}_i = c_p(A_i) \tag{4.2}$$

where c_p is a function of A that is different from one print-tip group p to another. Not only does print-tip loess allow for physical differences between the actual tips of the printer head, but it can also act as a surrogate for spatial variation across the slide [24].

However, since for all probes within a print-tip group the bias is only dependent on the average intensity, the print-tip loess method can account only for spatial biases at low resolution (i.e., at the level of the entire array).

Composite Normalization

Methods that account for spatial variation within the print-tip groups may easily be constructed using the framework of the composite normalization technique [11]. Here, a finer spatial dependence can be accounted for at the same time as the average log intensity by:

$$\hat{M} = \alpha \cdot c_p(A_i) + (1-\alpha) \cdot c_p(SpotRow_i, SpotCol_i) \quad (4.3)$$

where $c_{pt}(SpotRow_i, SpotCol_i)$ is the loess fit of M using as predictors the spot row and spot column coordinates within the pth print-tip. The weighting coefficient α can be set to 0.5, which gives equal weight to both spatial and intensity biases, or automatically adjusted [11]. This method is a composite between print-tip loess and a pure spatial two-dimensional normalization.

Median Filter

Wilson et al. [16] proposed a modification of the Yang et al. [11] loess normalization method to deal with the intensity and spatial bias. A loess curve $c(A)$ is first computed for the whole array and subtracted from the raw M values. Then, a median filter is applied on the residuals to estimate the spatial trend. The median filter simply subtracts from each residual the median of residuals over its spatial neighborhood (a 3×3 block of spots with the current spot in the center). This method is therefore a combination of a global loess normalization followed by a spatial median filter. Even though this normalization procedure is very effective in removing the spatial bias, it is likely to overnormalize and therefore increases the incidence of false negatives [17].

Neural Network-Based Spatial and Intensity Normalization (nnNorm)

A recent method to handle the intensity- and spatial-dependent biases at the print-tip level was proposed by Tarca et al. [17] and is available in the *nnNorm* package of the Bioconductor project (www.bioconductor.org). This method fits the raw log ratios M within a print-tip group as a function of the average log intensity A, as well as the space coordinates of the spots X and Y.

Unlike the composite normalization described above, the print-tip group (subarray) area is divided into square blocks called bins, usually containing nine spots (3×3) as shown in Figure 4.5. The implementation of the method allows flexibility in choosing the bin size. All the spots within a bin will have the same X and Y coordinates. The motivation for not using the spot row and column indices as spatial variables is to provide some robustness to the bias estimate and avoid approaching too closely the individual spot log ratios. Due to biological variation between the two samples, one expects that some of the spots will have higher or lower log ratios than the surrounding spots. A very common situation is to have two or more replicate spots printed one next to the other. The coordinates of the bin in which the spots are situated are used in the *nnNorm* normalization as surrogates for spatial location of the spots. Therefore, this method accounts for spatial biases at a lower resolution than at the spot level and at a higher resolution than the entire array level.

A second major difference between *nnNorm* and the previously mentioned methods is that the data of a given spot (probe) are not used in computing its bias estimate. This is possible via a 10-fold cross-validation method in which the data within a print-tip group are randomly split into 10 equal subsets. Ninety percent of the data is used to train the model, and the bias estimate of the remaining 10% is predicted using the trained model.

Spatial Detrending and Normalization Methods

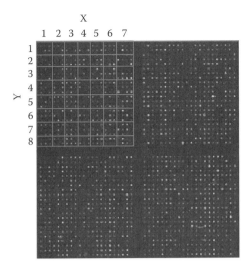

FIGURE 4.5 The binning procedure implemented in *nnNorm* normalization. The area of each print-tip group is divided into small square bins. All spots belonging to the same bin will have the same spatial pseudocoordinates X and Y.

The bias $\hat{M}$ is estimated as

$$\hat{M}_i = c_p\left(A_i, X_i, Y_i\right) \qquad (4.4)$$

where c_p for every print-tip p within an array is a trained neural network function. The type of neural network is a three-layered feed-forward neural network, also known as the multilayer perceptron (MLP) [25]. The parameterized neural network fitting function approximating the bias based on the A, X, and Y would read:

$$c_p\left(\mathbf{x}, \boldsymbol{\omega}\right) = \sigma^{(2)}\left(\sum_{j=1}^{J+1}\left(\alpha_j \cdot \sigma_j^{(1)}\left(\sum_{i=1}^{I+1}\left(x_i \cdot w_{i,j}\right)\right)\right)\right) \qquad (4.5)$$

The symbols of equation 4.5 are defined as follows:

- $\mathbf{x}$: vector having as components the $I = 3$ features X, Y, and A plus a constant value of 1 (accounting for the first layer bias)—that is, $\mathbf{x} = (X, Y, A, 1)$
- $\boldsymbol{\omega}$: fitting parameters called weights: $w_{i,j}$ ($i = 1,\ldots,I + 1$; $j = 1,\ldots,J$) and α_j ($j = 1,\ldots,J + 1$)
- $\sigma_j^{(1)}$ ($j = 1,\ldots,J$) represent the hidden neurons and basically are sigmoid functions

$$\sigma(z) = \frac{1}{1+e^{-z}};$$

- $\sigma^{(2)}$ is the single neuron in the output layer, and again is a sigmoid function
- $\sigma_{J+1}^{(1)} = 1$ accounts for the second layer bias
- J represents the number of neurons in the hidden layer of the network

Once a suitable value for J is selected, training of the neural network resumes, determining the best set of parameters, ω, in such a way that the estimate produced by the neural network, $c_p(\mathbf{x}, \hat{\omega})$, for the N_T spots used for training closely approaches the actual M values.

Training of neural network models with weights associated with the training samples is done using the *nnet* package implemented in R. Details about the *nnet* package may be found in Venables and Ripley [26]. As stated earlier, the number of hidden neurons in the models is set to $J = 3$, giving a total of 16 parameters to estimate using typically a few hundred data points, corresponding to the usual number of spots in a print-tip group. This represents a reasonable ratio between the number of data points available for training and the complexity of the model built, as indicated by the number of free parameters to be fitted.

THE BENEFITS OF REMOVING SPATIAL BIASES

The ideal normalization method produces high true positive and negative rates while minimizing both false negative and false positive findings. Comparing different spatial normalization methods is not an easy task. The ideal data set for this purpose would contain several positive and negative controls and also show important spatial biases. We are not aware of a publicly available data set meeting these requirements. However, a good candidate data set to illustrate the use of spatial normalization methods is the *Apolipoprotein* AI data set [20] available at http://www.stat.berkeley.edu/~terry/zarray/Html/apodata.html. In this data set, 8 out of the 6384 spotted cDNAs represent genes that are known a priori to be down-regulated in mice with a nonfunctional *apo AI* gene (knock-out) compared to "normal" C57B1/6 mice. There are 16 hybridizations comprising this data set, 8 replicates in each condition.

We would expect that a good normalization method would allow distinguishing the eight cDNAs expected to be down-regulated from the rest of nonregulated cDNAs (maximize true positive rate and minimize false positive rate). In this context, this will be possible if the normalization procedure reduces efficiently the within-array and between-array variability without overnormalizing (i.e., removing not only the noise, but also the signal from the data).

A two-sample t-statistic is computed for each element on the array and permutation adjusted p-values [27] are derived to test significance of gene expression changes with this data set. As shown in Figure 4.6, the print-tip loess method brings a significant improvement over the raw data, as it allows identifying all eight differentially regulated cDNAs at 0.05 threshold and seven out of eight at 0.01 threshold. Both the composite and *nnNorm* methods, which take into account spatial variability within the print-tip areas, perform slightly better by identifying all eight cDNAs as significant at 0.01 threshold.

To compare the methods' ability to reduce the within-slides data variability, the density plots of the raw and normalized log ratios were depicted in Figure 4.7. A robust measure of the data variability called the "median of absolute deviations" is computed as well. *nnNorm* reduces the most the within-array variability, followed by print-tip loess and the composite normalization. The reduction of the within-array

Spatial Detrending and Normalization Methods

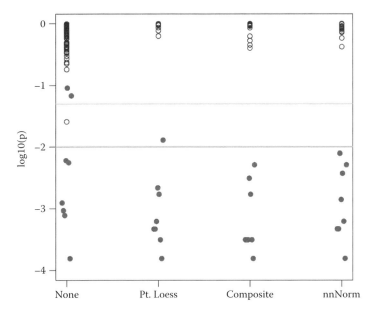

FIGURE 4.6 Permutation adjusted *p*-values obtained with a t-test when comparing gene expression of mutant mice with normal mice in *Apo* data set. The expected down-regulated genes are represented as solid circles, and the remaining ones with empty circles. Low y-axis values correspond to more significant changes. The thin horizontal line represents the 0.05 threshold, and the thick horizontal one the 0.01 threshold.

data variability points toward reduction of the false positive rates, as more probes will have log ratios closer to 0.

To illustrate the impact of the spatial normalizations on the within-group (between-arrays) variability, for each method and each of the two groups (eight arrays each), we have computed the standard deviations for each gene. All three normalization methods reduce significantly the within-group data variability with respect to the non-normalized values (see Figure 4.8). The composite normalization is slightly better than both print-tip loess and *nnNorm* on this data set.

In order to evaluate how effective the different normalization methods are in rendering the data independent on the within-print-tip spatial location (row and column indices), we have computed for each print-tip row/column the correlation coefficient between the log ratios and column/row indices (see Figure 4.9). It can be seen that several columns and rows have moderate to high correlations coefficients, which is evidence for spatially correlated biases. The *nnNorm* normalization is the most effective in reducing the number of row and columns with moderate to high correlations.

Visual inspection of the normalized log ratios can also help to see the differences between the different normalization techniques. Figure 4.10 shows the impact of the different normalization methods on the spatial distribution of the log ratios from four selected print-tip groups of a particular array of the *Apo* data set. The raw data show column-wise spatial bias in all four print-tips; the top rows have in general lower log ratios (green), while the bottom rows have in general higher log ratios (red). In this

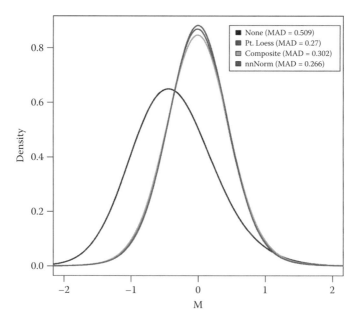

FIGURE 4.7 (See color insert following page 80.) Density plot of log ratios derived from all the genes and arrays of the *Apo* data set before and after normalization. The median absolute deviation gives a robust estimate of the data variability.

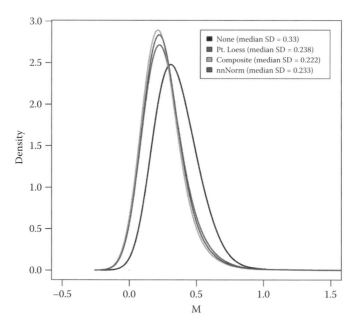

FIGURE 4.8 (See color insert following page 80.) Density plot of within-group (between arrays) standard deviations of log ratios from all the genes in the *Apo* data set. The median value of standard deviations for each method is given in the legend.

Spatial Detrending and Normalization Methods

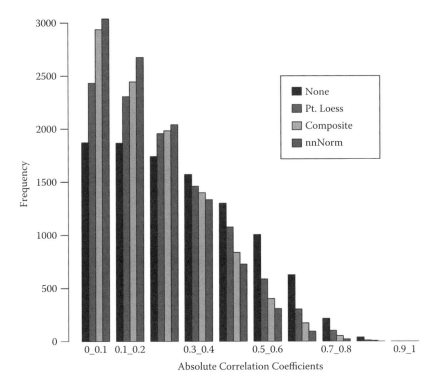

FIGURE 4.9 (See color insert following page 80.) Histogram of absolute values of correlation coefficients between the log ratios and the column or row indices. Ideally, log ratios should not be correlated with the row or column indices. In this figure, the most spatially-biased log ratios are for the raw data while the least spatially-biased are obtained using *nnNorm*.

figure, the median of the log ratio was removed from the raw data to allow using the same color scale for all methods. This corresponds to median normalized data that have the same properties as the raw data for the purpose of this comparison. The *nnNorm* normalization performs best, followed by the composite normalization.

IMPLEMENTATIONS USING R AND BIOCONDUCTOR

In this section, we illustrate detection and correction for spatial and intensity biases with cDNA data using two packages of Bioconductor (www.bioconductor.org): namely, *nnNorm* [22] and *marray* [28]. We will use the publicly available *Apo* data set discussed in the previous sections to illustrate both detection and correction for spatial biases.

INSTALLING AND LOADING PACKAGES

The following R code segments require that R 2.5.0 and Bioconductor 2.0 or later are installed and that an Internet connection is available. First, the required packages are installed by opening an R session and pasting in the following lines:

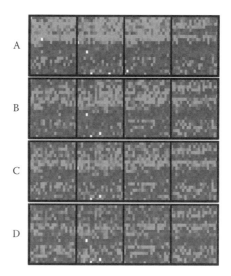

FIGURE 4.10 (See color insert following page 80.) Data from four print-tip groups of the eighth array in *Apo* data set are shown: (A) raw data (median subtracted only); (B) after print-tip loess normalization; (C) after composite spatial and intensity normalization; and (D) after *nnNorm* spatial and intensity normalization. It can be seen that print-tip loess normalization cannot remove much of the spatial dependence on the log ratios. In contrast, *nnNorm* normalization removes most of the spatial trends.

```
source("http://www.bioconductor.org/biocLite.R")
biocLite("nnNorm")
library(marray)
library(nnNorm)
```

The first two lines install the *nnNorm* library. The required packages are loaded with lines 3 and 4.

LOADING *APO* DATA SET

The following will load a publicly available data set and create a data structure that embeds all required information to illustrate the normalization procedures discussed in the previous sections:

```
dweb<-"http://www.stat.berkeley.edu/users/terry/zarray/
Data/ApoA1/rg_a1ko_morph.txt"
mydata<-read.table(dweb,header=TRUE, sep=",", dec=".")
ApoLayout<-new('marrayLayout',maNgr=4,maNgc=4,maNsr=19,
maNsc=21)
spots<-4*4*19*21 # spots per array
```

Spatial Detrending and Normalization Methods

```
Gb<-Rb<-array(c(0),c(spots,16));
Gf<-as.matrix(mydata[,seq(2,33,2)])
Rf<-as.matrix(mydata[,seq(3,33,2)])
labs <- c(paste(c("c"), 1:8, sep=""),paste(c("k"),
1:8, sep=""))
maInfo<-new('marrayInfo',maLabels = labs,maInfo=data.
frame(Names=labs))
Apo<-new('marrayRaw',maRf=Rf,maGf=Gf,maRb=Rb,maGb=Gb,
maLayout=ApoLayout,maTargets=maInfo)
```

Lines 1 and 2 read the data from an Internet location into an R data frame called "mydata." Line 3 creates a layout object that describes the particularities of the cDNA array used. For instance, there are four metarows (grid rows) and four metacolumns (grid columns) for a total of 16 print-tip groups (see Figure 4.4). Each print-tip group has 19 rows and 21 columns. Lines 4 and 5 assign null background values for both channels since mydata contains background-corrected intensities. Lines 6–9 extract the R and G intensities from mydata and create a "marrayInfo" object specifying the names of the different arrays—in this case eight controls and eight mutants. Lines 10 and 11 create a data structure called "Apo" of class "marrayRaw," and assign to it the corresponding layout information. Many data analysis functions that are available in Bioconductor can be directly applied to such a microarray batch.

DETECTING SPATIAL BIASES

One simple way to detect spatial biases is to visually inspect the spatial distribution of the raw log ratios. This can be achieved using the *marray* [28] functionalities:

```
RGcol <- maPalette(low = "green", high = "red", k = 20)
maImage(Apo[,16],x="maM",col=RGcol,zlim=c(-2,2))
```

The first line creates a green-red color scale and reconstructs the composite image of the eighth array of the *Apo* batch as shown in Figure 4.4. The automated spatial bias detection method discussed in "Detecting Spatial Biases" and available in *nnNorm* [28] package will allow one to quickly identify the array and print-tips with the highest spatially correlated biases:

```
tmp<-detectSpatialBias(Apo)
tmp$biasRow
tmp$biasCol
```

The tmp object created by the "detectSpatialBias" function is a list with two components: "biasRow" and "biasCol." "biasRow" is the matrix shown in Table 4.1. See "Detecting Spatial Biases" for the interpretation of these matrices.

SPATIAL NORMALIZATION

We now apply on the *Apo* data set three of the four normalization methods described in the normalization methods section: namely, print-tip loess normalization, composite normalization, and *nnNorm* normalization. The print-tip loess normalization is the default normalization method of the *marray* package, and it can be applied on the *Apo* data set as easily as:

```
pLoApo<-maNorm (Apo)
```

The composite normalization removing both intensity and spatial biases is available in the *marray* package and can be applied as:

```
compApo<-maNormMain(Apo, f.loc = list(maNorm2D(), maNormLoess(x="maA",y="maM",z="maPrintTip")),a.
loc=maCompNormEq())
```

With this method, the intensity and spatial biases receive equal weight via the "maCompNormEq()" argument.

Finally, the *nnNorm* normalization is applied using the function call:

```
nnApo<-maNormNN(Apo, binWidth=3,binHeight=3)
```

This line applies the spatial and intensity normalization method based on neural networks described in the normalization methods section using a 3 × 3 binning procedure.

Optional arguments may be passed to the "maNormNN" function to change its default behavior and/or provide more details. For example,

```
nnApo<-maNormNN(Apo[,8],binWidth=4,binHeight=4,maplots=
TRUE)
```

normalizes the eighth array in the *Apo* batch using a larger binning window and displays the M versus A plots before and after normalization.

More details on each of the functions used here can be obtained within the R session by typing in "?functionName"—for example,

```
?maNormNN
```

CONCLUSION

The goal of this chapter was to discuss methods for detection and correction of spatially correlated biases with two-channel microarray data. Several normalization approaches were discussed focusing on methods able to remove both spatially and intensity-correlated biases while not requiring a special array design.

Using a publicly available data set, we compared, discussed, and illustrated several methods to detect spatial biases and perform spatial normalization. The comparison between the methods was focused on the ability of the methods to

(a) detect true differential expression, (b) reduce within and between arrays data variability, and (c) attenuate the dependence of the data on spatial location with the print-tip areas. It should be clear to the reader that comparing normalization methods in general is a difficult task, because it requires experiments in which the amount of mRNA is controlled for several transcripts by spike-in insertion or by suppression/stimulation of particular genes. Comparing spatial normalization methods is even harder because, in addition to having a controlled experiment, spatial artifacts must be present. More comparisons between these methods on simulated data are available elsewhere [17].

Using two packages of bioconductor—namely, *nnNorm* [22] and *marray* [28] available as R packages within the bioconductor project (www.bioconductor.org), we have illustrated practical use of the available methods to detect and correct for spatial biases.

ACKNOWLEDGMENTS

J. E. K. Cooke and J. Mackay are acknowledged for their input in the original paper presenting the *nnNorm* normalization method.

REFERENCES

1. P. T. Spellman, G. Sherlock, M. Q. Zhang, V. R. Iyer, K. Anders, M. B. Eisen, P. O. Brown, et al., Comprehensive identification of cell cycle-regulated genes of the yeast *Saccharomyces cerevisiae* by microarray hybridization, *Mol. Biol. Cell*, 9(12), 3273–3297, 1998.
2. T. R. Golub, D. K. Slonim, P. Tamayo, C. Huard, M. Gaasenbeek, J. P. Mesirov, H. Coller, et al., Molecular classification of cancer: Class discovery and class prediction by gene expression monitoring, *Science*, 286(5439), 531–537, 1999.
3. R. M. Simon and K. Dobbin, Experimental design of DNA microarray experiments, *Biotechniques*, Suppl, 16–21, Mar. 2003.
4. X. Cui, M. K. Kerr, and G. A. Churchill, Transformations for cDNA microarray data, *Stat. Appl. Genet. Mol. Biol.*, 2, Article 4, 2003.
5. M. Chatterjee, S. Mohapatra, A. Ionan, G. Bawa, R. li-Fehmi, X. Wang, J. Nowak, et al., Diagnostic markers of ovarian cancer by high-throughput antigen cloning and detection on arrays, *Cancer Res.*, 66(2), 1181–1190, 2006.
6. R. Nadon and J. Shoemaker, Statistical issues with microarrays: Processing and analysis, *Trends Genet.*, 18(5), 265–271, 2002.
7. J. Quackenbush, Microarray data normalization and transformation, *Nat. Genet.*, 32 Suppl, 496–501, 2002.
8. S. Draghici, *Data analysis tools for DNA microarrays*, Chapman and Hall/CRC Press, Boca Raton, FL, 2003.
9. S. Draghici, O. Kulaeva, B. Hoff, A. Petrov, S. Shams, and M. A. Tainsky, Noise sampling method: An ANOVA approach allowing robust selection of differentially regulated genes measured by DNA microarrays, *Bioinformatics*, 19(11), 1348–1359, 2003.
10. S. Dudoit, Y. H. Yang, M. J. Callow, and T. P. Speed, Statistical methods for identifying expressed genes in replicated cDNA microarray experiments, *Stat. Sin.*, 12, 111–139, 2002.
11. Y. H. Yang, S. Dudoit, P. Luu, D. M. Lin, V. Peng, J. Ngai, and T. P. Speed, Normalization for cDNA microarray data: a robust composite method addressing single and multiple slide systematic variation, *Nucleic Acids Res.*, 30(4), e15, 2002.

12. S. Draghici, P. Khatri, A. C. Eklund, and Z. Szallasi, Reliability and reproducibility issues in DNA microarray measurements, *Trends Genet.*, 22(2), 101–109, 2006.
13. G. K. Smyth and T. Speed, Normalization of cDNA microarray data, *Methods*, 31, 265–273, 2003.
14. T. Park, S. G. Yi, S. H. Kang, S. Lee, Y. S. Lee, and R. Simon, Evaluation of normalization methods for microarray data, *BMC Bioinformatics*, 4, 33, 2003.
15. W. S. Cleveland, Robust locally weighted regression and smoothing scatterplots, *J. Am. Stat. Assoc.*, 74, 829–836, 1979.
16. D. L. Wilson, M. J. Buckley, C. A. Helliwell, and I. W. Wilson, New normalization methods for cDNA microarray data, *Bioinformatics*, 19(11), 1325–1332, 2003.
17. A. L. Tarca, J. E. Cooke, and J. Mackay, A robust neural networks approach for spatial and intensity-dependent normalization of cDNA microarray data, *Bioinformatics*, 21(11), 2674–2683, 2005.
18. P. Neuvial, P. Hupe, I. Brito, S. Liva, E. Manie, C. Brennetot, F. Radvanyi, et al., Spatial normalization of array-CGH data, *BMC Bioinformatics*, 7, 264, 2006.
19. D. S. Yuan and R. A. Irizarry, High-resolution spatial normalization for microarrays containing embedded technical replicates, *Bioinformatics*, 22(24), 3054–3060, 2006.
20. M. J. Callow, S. Dudoit, E. L. Gong, T. P. Speed, and E. M. Rubin, Microarray expression profiling identifies genes with altered expression in HDL-deficient mice, *Genome Res.*, 10(12), 2022–2029, 2000.
21. R Development Core Team, R: A language and environment for statistical computing, Vienna, Austria: 2006.
22. A. L. Tarca, nnNorm: Spatial and intensity based normalization of cDNA microarray data based on robust neural nets, R package version 2.0, 2007.
23. S. Draghici, Statistical intelligence: Effective analysis of high-density microarray data, *Drug Discov. Today*, 7(11) Suppl, S55–S63, 2002.
24. G. K. Smyth, Y. H. Yang, and T. Speed, Statistical issues in cDNA microarray data analysis, *Methods Mol. Biol.*, 224, 111–136, 2003.
25. D. E. Rumelhart, G. Hinton, and R. Williams, Learning internal representation by error propagation, *Parallel Distributed Processing*, 1, 318–364, 1986.
26. W. N. Venables and B. D. Ripley, *Modern applied statistics with S*, 4th ed. New York: Springer–Verlag, 2002.
27. P. H. Westfall and S. S. Yang, *Resampling-based multiple testing: Examples and methods for p-value adjustment*, Wiley, New York, 1993.
28. Y. H. Yang, A. Paquet, and S. Dudoit, marray: Exploratory analysis for two-color spotted microarray data, version 1.14, software from bioconductor.org, 2007.

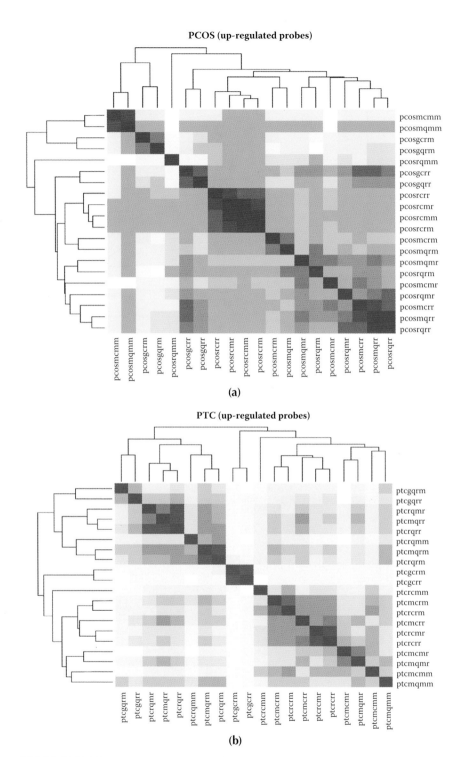

COLOR FIGURE 1.2 SAM clustering of the preprocessing combinations on polycystic ovary syndrome (a) and papillary thyroid cancer (b). Red indicates high similarity and white indicates no similarity.

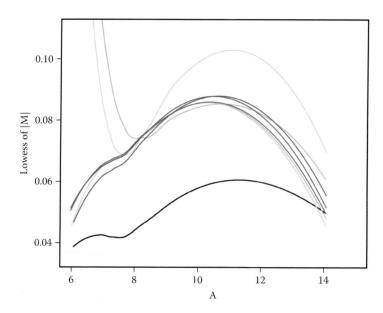

COLOR FIGURE 2.1 The lowess curves of |M| versus A values by various normalization methods. Gray: no normalization; black: SUB-SUB; red: quantiles; green: constant; purple: contrasts; blue: invariant-set; orange: lowess; cyan: qspline.

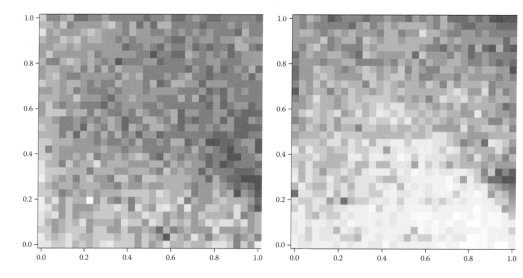

COLOR FIGURE 2.3 The slope matrices of two arrays show different spatial patterns in scale. The common reference is array M. Top left: array P versus M; top right: array N versus M. Their histograms are shown at the bottom, correspondingly.

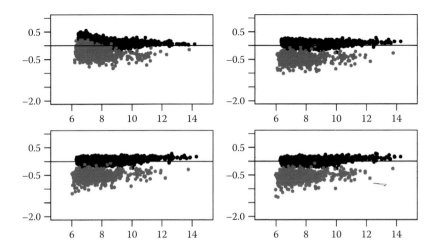

COLOR FIGURE 2.5 M-A plots of perturbed spike-in data set after SUB-SUB normalization. The x-axis is the average of log intensities from two arrays. The y-axis is their difference after normalization. In the top four subplots, 20% randomly selected genes have been artificially down-regulated by 2.0-fold in arrays Q, R, S, and T. The differentiated genes are marked red, and undifferentiated genes are marked black. The trimming fractions in the subplots are: top left, 0%; top right, 10%; bottom left, 20%; bottom right, 40%.

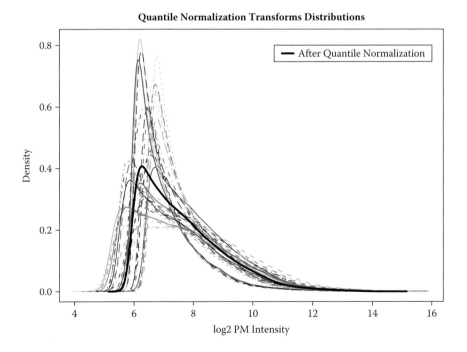

COLOR FIGURE 3.1 Quantile normalization equalizes the intensity distribution across all arrays.

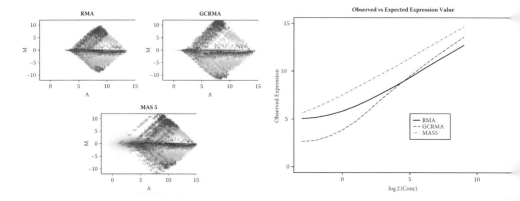

COLOR FIGURE 3.3 Comparing RMA, GCRMA, and MAS5 expression values using a U133A spike-in data set. MA plots show large level of noise at the low end for MAS5. GCRMA and MAS5 have greater linearity through concentration range.

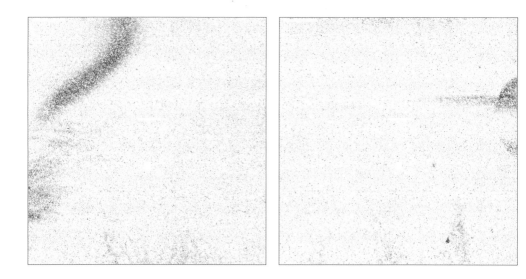

COLOR FIGURE 3.5 Pseudochip images of residuals from probe level models fit using RMAExpress for two arrays: GSM14020 (top) and GSM14035 (bottom).

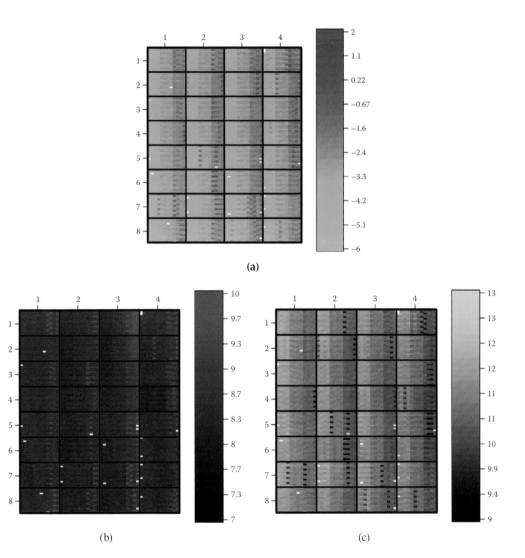

COLOR FIGURE 4.2 Reconstructed image of an antibody array showing spatial bias most likely due to a print-tip problem: (a) $\log_2$ ratio of the two channels; (b) $\log_2$ intensity on the red channel; and (c) $\log_2$ intensity on the green channel.

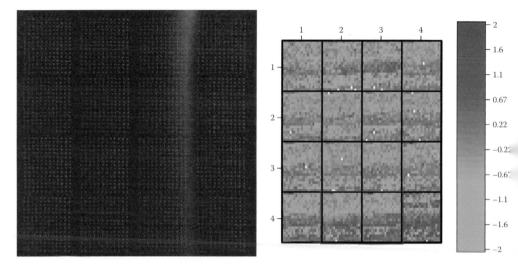

COLOR FIGURE 4.3 A hybridization/washing-related artifact that artificially augments the red channel intensity in a spatially dependent manner in three print-tips of a protein array.

COLOR FIGURE 4.4 Reconstructed image of the array c8R of the *Apo* data set. Two print-tip regions showing strong spatial bias are highlighted.

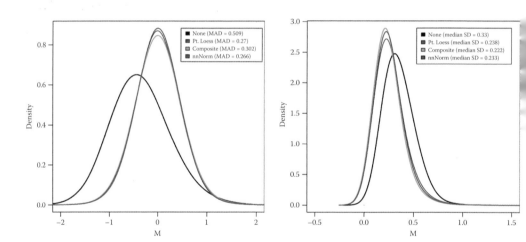

COLOR FIGURE 4.7 Density plot of log ratios derived from all the genes and arrays of the *Apo* data set before and after normalization. The median absolute deviation gives a robust estimate of the data variability.

COLOR FIGURE 4.8 Density plot of within-group (between arrays) standard deviations of log ratios from all the genes in the *Apo* data set. The median value of standard deviations for each method is given in the legend.

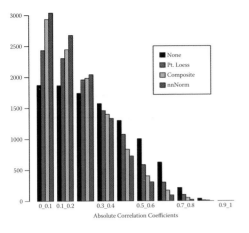

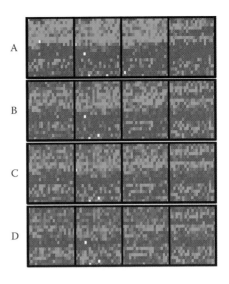

COLOR FIGURE 4.9 Histogram of absolute values of correlation coefficients between the log ratios and the column or row indices. Ideally, log ratios should not be correlated with the row or column indices. In this figure, the most spatially-biased log ratios are for the raw data while the least spatially-biased are obtained using *nnNorm*.

COLOR FIGURE 4.10 Data from four print-tip groups of the eighth array in *Apo* data set are shown.

COLOR FIGURE 5.1 A scanned image of an illustrative cDNA microarray. The configuration (layout of spots) can be described via a previously defined notation encompassing four numbers (ngr, ngc, nsr, nsc).

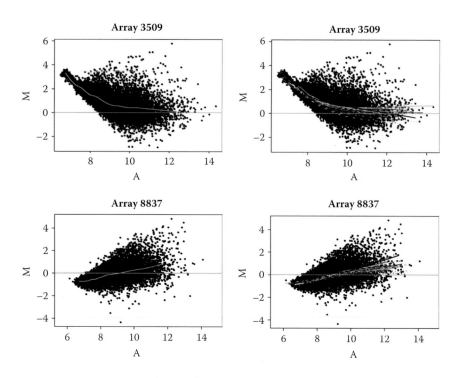

COLOR FIGURE 5.2 MA plots of the non-normalized M values for two microarrays in the LungCancer data set. The plots show intensity-dependent dye effects present in the non-normalized data for microarray 3509 and microarray 8837 in the LungCancer data set. The MA plot is a scatter plot of log ratios $M = \log_2(R/G)$ (abscissa) versus average log intensities $A = \log_2 \sqrt{RG}$ (ordinate). Columns depict the non-normalized (Nonorm) M values for microarrays 3509 and 8837, respectively. Plots in the same row represent same data except that each plot in the left panel shows one lowess curve for all the spots, while that in the right panel shows one lowess curve per PT group. Different colors and line types are used to represent different PT groups from different rows ("ngr," Figure 5.1) and columns ("ngc"), respectively.

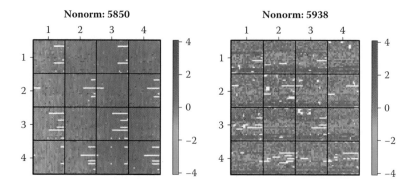

COLOR FIGURE 5.3 The plots show spatial effects present in the non-normalized data for microarrays 5850 and 5938 in the Lymphoma data set. The spatial plot is a spatial representation of spots on the microarray color coded by their M values (marrayPlots/maImage(x="maM"). Spots in white are spots flagged in the original microarray data (missing values).

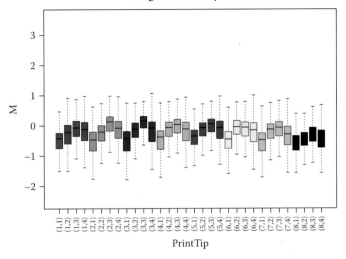
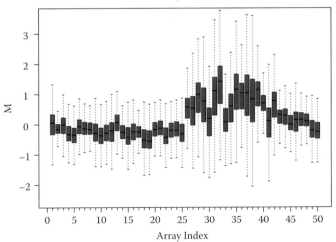

COLOR FIGURE 5.4 Box plots of the distributions of the non-normalized M values for microarrays in the LungCancer data set. The box plots display scale effect for microarray 11581 (top) and other microarrays (bottom) in the LungCancer data set at the PT level (top) and at the global level (bottom), respectively. In each box plot, the box depicts the main body of the data and the whiskers show the extreme values. The variability is indicated by the size of the box and the length of the whiskers (marray/marraymaBoxplot(y="maM")). The top panel shows results for M values at the local level of microarray 11581. The bars are color coded by PT group. The bottom panel shows results for M values at the global level for 50 microarrays chosen at random from the LungCancer data set.

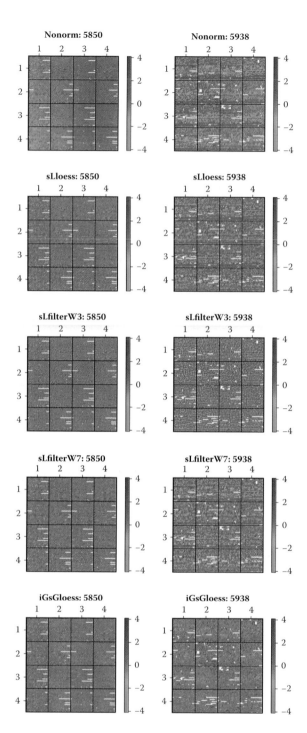

COLOR FIGURE 5.5 Spatial plots of the non-normalized and normalized M values for microarrays 5850 and 5938 in the Lymphoma data set. The plots show the results before and after normalization designed to remove spatial effect. Spots in white are spots flagged in the original microarray data (missing values). Rows depict the non-normalized (Nonorm), and normalized M' values (sLloess, sLfilterW3, sLfilterW7, iGsGloess).

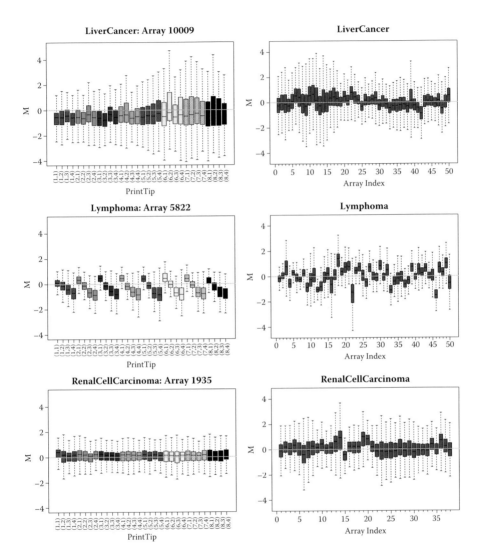

COLOR FIGURE 5.9 Box plots of the distributions of the non-normalized M values for microarrays in the five studies. Each panel in the left-hand column shows results for M values at the local level of a microarray chosen at random from a given data set. The bars are color coded by PT group. Each panel in the right-hand column shows results for M values at the global level for 50 microarrays chosen at random from a given data set (the total number of microarrays in RenalCellCarcinoma is 38). Each row corresponds to a particular study.

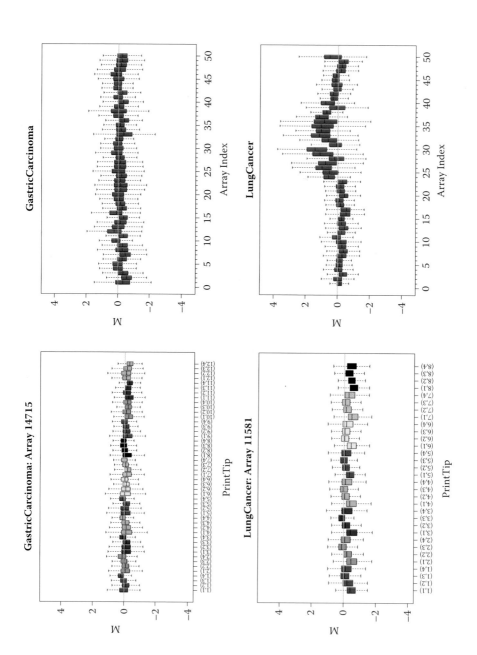

COLOR FIGURE 5.9 (continued)

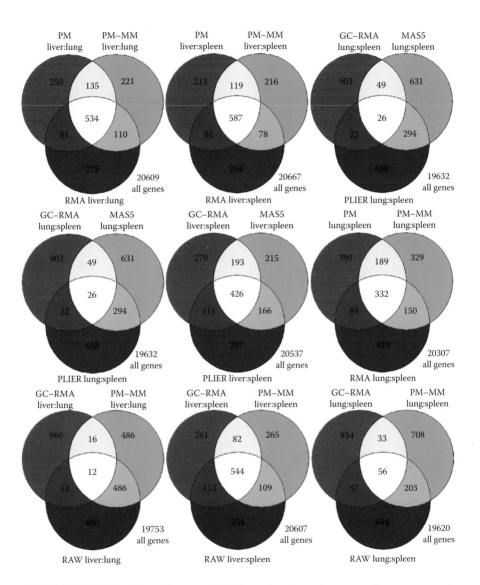

COLOR FIGURE 7.2 Venn diagrams of 1000 overlapping probes across the seven normalizations. Three-way common genes appear in the white section.

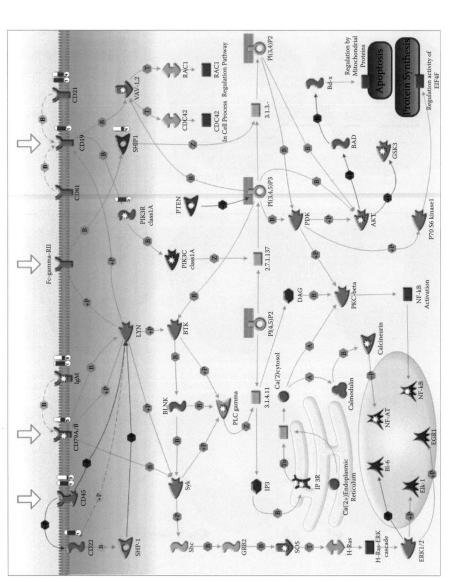

COLOR FIGURE 8.4 BCR pathway.

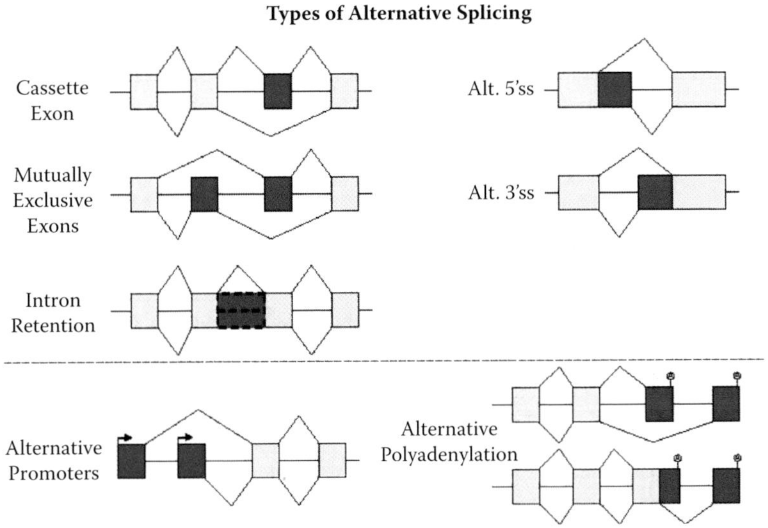

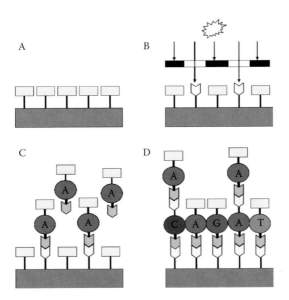

COLOR FIGURE 9.1 Types of alternative splicing events. Potential exons are represented by boxes. Red boxes are exons that might be included or skipped in any particular transcript. The angled lines above and below the exons represent possible alternatives for joining the exon borders into a mature transcript.

COLOR FIGURE 9.2 Affymetrix's photolithographic method to synthesize oligonucleotide probes directly on the chip substrate.

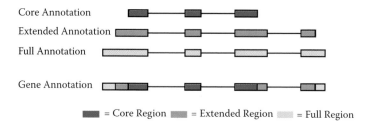

COLOR FIGURE 9.4 Assembly of exon annotations into a gene-level annotation. Exonic sequences from all three confidence levels (core, extended, or full) are reassembled into transcript level sequences according to a defined set of rules. Each region is annotated according to the highest confidence level that supports it, and a gene may contain sequences from all three sets. Genes in the core metaprobe set will contain at least one core exon; conversely, a gene from the full metaprobe set might only contain highly speculative exons from the full set.

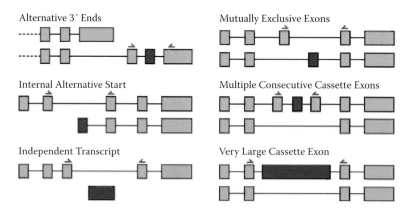

COLOR FIGURE 9.13 Examples of challenging cases for RT-PCR validation. In these cases, the standard approach to designing primers (half arrows) will not produce PCR products that represent both alternatives of the splice variation. Therefore, an alternative validation strategy must be employed.

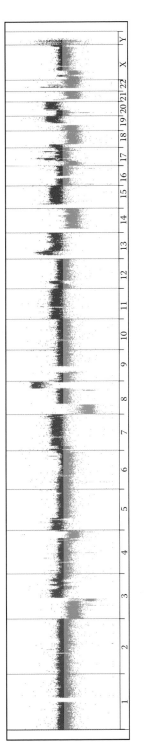

COLOR FIGURE 10.1 Log$_2$ ratios of an HT29 cell line versus normal female DNA, after primary normalization and centralization to the dominant (triploid) peak.

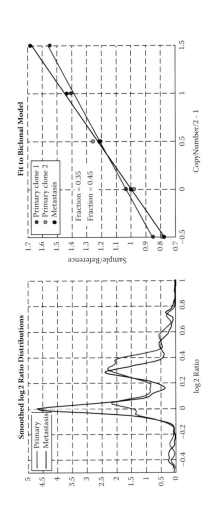

COLOR FIGURE 10.4 Centralization curves and copy number plots of a primary (blue) and metastatic (black) tumor sample from the same individual versus a matched normal reference. The best fit to the metastatic tissue is consistent with a predominantly diploid genome, with successive peaks assigned copy numbers of 1, 2, 3, 4, and 5, but the copy number slope is only 0.45. We interpret this to mean that the malignant cells are about 50% diluted with normal cells. The primary tumor (blue) is more interesting, as each of the major peaks is doubled. This indicates the presence of more than one aberrant clone in the sample. One peak of each pair can be assigned to the dominant clone, so as to produce a linear copy number curve, which suggests that about 35% of the sample consists of the dominant aberrant clone. These interpretations are consistent with cytological data.

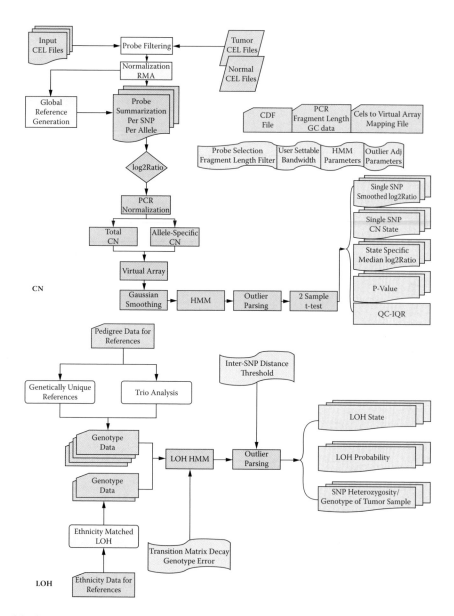

COLOR FIGURE 11.1 Workflow diagrams of CN and LOH.

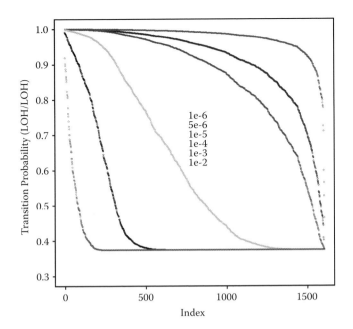

COLOR FIGURE 11.4 Transition decay (β) as a function of inter-SNP distance.

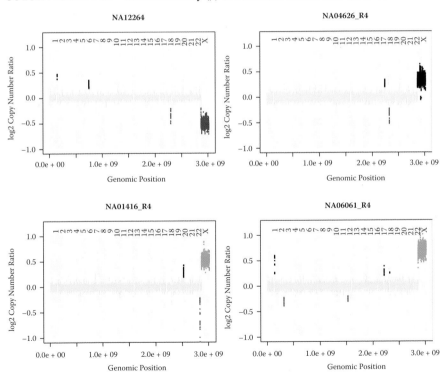

COLOR FIGURE 11.6 CNAT4 CN estimates (smoothed over a 100K window), color coded by the HMM call for four samples with different X chromosome CNs: a normal HapMap male (NA12264; 1X), 3X (NA04626), 4X (NA01416), and 5X (NA06061) samples. Color coding: red: CN = 0, blue: CN = 1; yellow: CN = 2; black: CN = 3; green: CN = 4+.

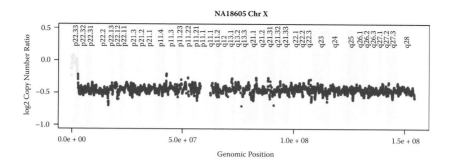

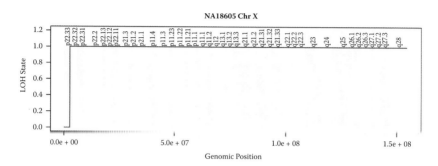

COLOR FIGURE 11.7 CNAT4 CN estimates (top; with smoothing at 100K) and LOH calls (bottom) for the X chromosome of a HapMap male sample. The LOH state is represented by the gray line, which jumps from 0 to 1 at the end of the pseudo-autosomal region near the p-terminus.

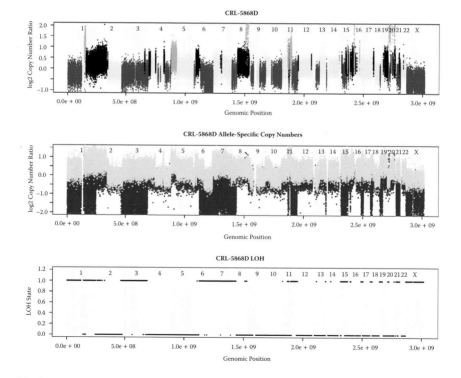

COLOR FIGURE 11.8 Unsmoothed CNAT4 output produced for a tumor sample.

5 A Survey of cDNA Microarray Normalization and a Comparison by *k*-NN Classification

Wei Wu and Eric P. Xing

CONTENTS

Introduction 82
Introduction to cDNA Microarrays 83
Unwanted Variations in cDNA Microarray Data 84
    Intensity-Dependent Dye Effect 85
    Spatial-Dependent Dye Effect 86
    Scale Effect 87
Strategies for Minimizing Unwanted Variations in cDNA Microarray Data 89
    Data Transformation 89
        Logarithmic Transformation 89
        Ratios or Intensity Values 89
    Normalization Methods for cDNA Microarray Data 90
        Global Normalization Techniques 90
        ANOVA Methods 91
        Normalization Strategies That Minimize Intensity-Dependent
            Dye Effect 92
        Normalization Strategies Using External Controls 95
        Other Normalization Methods 96
        Normalization Strategies That Minimize Spatial-Dependent
            Dye Effect 97
        Normalization Strategies That Minimize Intensity- and Spatial-
            Dependent Dye Effects 98
        Normalization Strategies That Minimize Scale Effect 99
Criteria for Comparison of Normalization Methods 99
    Ability to Reduce Unwanted Variations 100
    Ability to Retain Biological Variations 100
Evaluation of Normalization Methods for cDNA Microarray Data 101
Comparison of Normalization Methods for cDNA Microarray Data by *k*-NN
    Classification 101

Comparison of Intensity- and Spatial-Dependent Normalization Strategies by
  k-NN Classification 104
    Assessing the Ability to Remove Dye Effects Using Diagnostic Plots 104
    Functional Assessment Using the $LOOCV$ Classification Error
      of k-NNs 106
      Single-Bias-Removal Methods 106
      Double-Bias-Removal Methods 109
      Qspline-Related Approaches 110
Comparison of Scale Normalization Strategies by k-NN Classification 111
  Assessing the Ability to Remove Dye Biases Using Diagnostic Plots 112
  Functional Assessment Using the k-NN $LOOCV$ Error 112
  Summary of k-NN Evaluation of Normalization Methods 112
Future Directions 116
Conclusions 116
References 117

INTRODUCTION

With the advent of microarray technology about a decade ago [1,2] and the recent completion of genome sequences of many different species, microarrays have become powerful and commonly used tools for measuring the abundance of tens of thousands of genes or gene products in a single experiment. They can be used to measure steady-state mRNA expression levels of genes (e.g., gene expression microarrays), to detect changes of copy numbers and polymorphisms of genomic regions (e.g., array-based comparative genomic hybridization (CGH), and single nucleotide polymorphism (SNP) microarrays), to identify promoters that bind transcription factors (e.g., ChIP-chip (chromatin immunoprecipitation on chip) microarrays), to detect epigenetic modification of genes such as methylation and acetylation (e.g., methylation and acetylation arrays), and to measure protein levels of genes (e.g., antibody microarrays).

Among these different types of microarrays, gene expression arrays are the most commonly used ones. They can be classified in different ways based on different criteria. For example, depending on the fabrication format, gene expression arrays can be classified into spotted arrays and in situ synthesized oligonucleotide arrays. Depending on the probe types printed on the array, they can be classified into cDNA and oligonucleotide microarrays. Additionally, depending on the number of fluorescent dyes used to label the samples hybridized to a single array, they can be classified into single-channel, two-channel, and multichannel arrays.

In this chapter, we will focus our discussion on the issues related to two-channel spotted cDNA microarrays. First, we will discuss systematic dye effects in the data generated from these arrays, and normalization methods that have been designed to minimize these dye effects. Then we will discuss criteria and relevant work for evaluation of normalization methods for cDNA microarrays. Finally, we will review our recent work on comparison of normalization strategies using the k-NN leave-one-out cross-validation ($LOOCV$) classification error. In this work, we re-examined

TABLE 5.1
Multiclass Microarray Data Sets Obtained from SMD and Reanalyzed in Wu et al. [46]

Data set name	Description
LiverCancer [53]	Microarrays: 181; K = 2 http://genome-www5.stanford.edu/cgi-bin/publication/viewPublication.pl?pub_no=107
Lymphoma [54]	Microarrays: 81; K = 3 http://genome-www5.stanford.edu/cgi-bin/publication/viewPublication.pl?pub_no=79
RenalCellCarcinoma [55]	Microarrays: 38; K = 4 http://genome-www5.stanford.edu/cgi-bin/publication/viewPublication.pl?pub_no=210
GastricCarcinoma [56]	Microarrays: 130; K = 2 http://genome-www5.stanford.edu/cgi-bin/publication/viewPublication.pl?pub_no=232
LungCancer [57]	Microarrays: 60; K = 5 http://genome-www5.stanford.edu/cgi-bin/publication/viewPublication.pl?pub_no=100

Notes: For each of the five published studies, the fluorescent intensities, microarray images, and associated information were downloaded from the URLs indicated in the table. The abbreviations are as follows: microarrays: number of cDNA microarrays; K: total number of categories to which a sample could be assigned. Other information on the data sets, such as sample types and configuration of a microarray, can be found in detail in Wu et al. [46].

five cDNA microarray data sets (relevant information of which is summarized in Table 5.1) from the Stanford Microarray Database (SMD) [3]. We will use the microarrays in these data sets as illustrative running examples in this chapter.

A variety of computational tools are available for manipulating, analyzing, and visualizing microarray data. These include open source implementations based on the R language for statistical computing (http://www.r-project.org) [4] such as the Bioconductor (http://www.bioconductor.org) packages [5]. As a reference to the readers, we will refer to some useful Bioconductor R packages and functions for cDNA microarray analysis in this chapter.

INTRODUCTION TO CDNA MICROARRAYS

The cDNA microarrays are manufactured by immobilizing spotted polymerase chain reaction (PCR) products amplified from cDNA clones on the arrays. These PCR products are approximately 300–1500 base pairs in length, and are often referred to as "probes," "spots," or "genes" in the microarray literature, depending on the context. In a typical two-channel cDNA microarray experiment, RNAs are extracted from two biological samples (which can be samples from different tissues, development stages, disease types or those subject to different drug treatments). The RNAs of the studied samples are reverse transcribed into cDNAs and labeled with distinct fluorescent dyes (usually red (Cy5) and green (Cy3) fluorescent dyes) respectively; then the labeled cDNAs are competitively hybridized to a cDNA microarray using an appropriate design strategy (see reviews on cDNA microarrays in Quackenbush [6] and Aittokallio [7]).

The design strategy for microarray experiments should facilitate data analysis and interpretation to make them simple yet powerful. The most common design strategies for two-channel cDNA microarrays include the direct design and the indirect one known as the reference design (see reviews in Yang and Speed [8] and Churchill [9]). In the direct design, the samples (e.g., a test sample and a control sample) in which the expressions of genes to be compared are hybridized directly on the same array. In the reference design, the test sample and the control sample are hybridized on two arrays, and each of them is hybridized with a common reference sample on the same array; the expressions of genes in the test and control samples are thus compared indirectly through those of the reference sample.

Under both design strategies, after hybridization of the microarray with the labeled cDNAs, the microarray is scanned using a scanner; separate images are acquired for Cy5 and Cy3, and then processed to yield an intensity measurement for each dye at every spot. A typical microarray image is shown in Figure 5.1.

For each array spot, the foreground intensities of red (R_f) and green (G_f) quantitated fluorescent signals and the local background values (B_g) are acquired using image analysis software (e.g., GenePix and SPOT). By convention, the local background B_g should be subtracted from the foreground intensities R_f and G_f to obtain an estimate of the expression levels of genes in each sample hybridized to an array [10]. However, due to the concern that inappropriate local background adjustments may add additional noise to the microarray data [11], the local background values are often not subtracted from their corresponding foreground values R_f and G_f. Therefore, if we let R and G represent the spot-specific, quantitated, fluorescent intensities of the expression signals of the two samples hybridized to the array respectively, then $R = R_f$ and $G = G_f$.

UNWANTED VARIATIONS IN CDNA MICROARRAY DATA

Many processes in microarray experiments can contribute to unwanted variations (or noise) in cDNA microarray data. For example, noise can arise from sources such as sample preparation, RNA extraction, hybridization, printing or scanning of microarrays, or other technical factors, and it can mask signals arising from genuine biological variations in gene expression. The unwanted variations include systematic variations and unidentified sources of variations; identifying them and the sources that generate them is important in that (1) it enables appropriate design and selection of normalization methods that can minimize systematic variations, and (2) it can provide guidelines for biologists to avoid improper experimental procedures that can give rise to the unwanted variations.

A Survey of cDNA Microarray Normalization

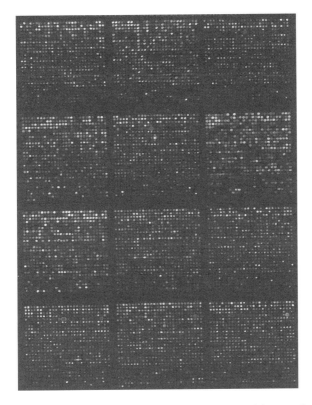

FIGURE 5.1 (See color insert following page 80.) A scanned image of an illustrative cDNA microarray. The configuration (layout of spots) can be described via a previously defined notation encompassing four numbers (ngr, ngc, nsr, nsc) [38]. A print-tip (PT) group is a set of spots arranged in a grid with "nsr" rows and "nsc" columns. A microarray is a set of PT groups arranged in a pattern of "ngr" rows and "ngc" columns. The configuration of the microarray shown is (ngr = 5, ngc = 3, nsr = 26, nsc = 28)—that is, 4 × 3 PT groups each composed of 26 × 28 spots. The terms "local" and "global" level refer to the spots in a PT group and the entire microarray, respectively.

The major sources of systematic variations in cDNA microarray data include intensity-dependent dye effect, spatial-dependent dye effect, and scale effect. In the next section, we will illustrate and discuss these systematic variations in detail.

INTENSITY-DEPENDENT DYE EFFECT

Suppose that an array is hybridized with two samples—for example, sample a and sample b, which are labeled with Cy5 and Cy3 fluorescent dyes, respectively. Let R represent the Cy5-intensity values and G represent the Cy3-intensity values for genes on an array. Then, ideally, the relative expressions of genes on the array, measured as log intensity ratios M of genes ($M = \log_2 R - \log_2 G$), should be independent of their average log-intensity values A ($A = (\log_2 R + \log_2 G)/2$). However, in reality, the dependence of M on A is often observed in microarray data, and this is the so-called *intensity-dependent dye effect*. Intensity effect can be commonly seen in data

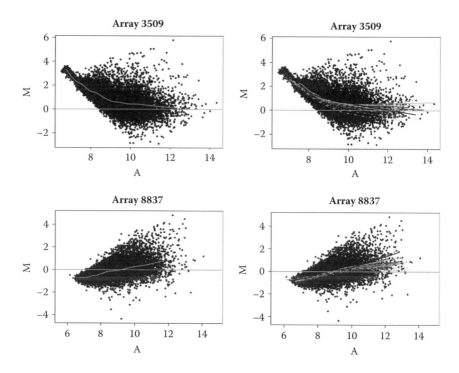

FIGURE 5.2 (**See color insert following page 80.**) MA plots of the non-normalized M values for two microarrays in the LungCancer data set. The plots show intensity-dependent dye effects present in the non-normalized data for microarray 3509 and microarray 8837 in the LungCancer data set. The MA plot is a scatter plot of log ratios $M = \log_2(R/G)$ (abscissa) versus average log intensities $A = \log_2 \sqrt{RG}$ (ordinate). Columns depict the non-normalized (Nonorm) M values for microarrays 3509 and 8837, respectively. Plots in the same row represent same data except that each plot in the left panel shows one lowess curve for all the spots, while that in the right panel shows one lowess curve per PT group. Different colors and line types are used to represent different PT groups from different rows ("ngr," Figure 5.1) and columns ("ngc"), respectively.

generated from many microarray platforms, including cDNA microarrays, Affymatrix GeneChips, and CodeLink Bioarrays [12–14].

Intensity effect can be best illustrated and identified using scatter plots of M versus A, which are also called the *MA plots* [15]. Figure 5.2 shows two typical types of intensity-dependent dye effects. The curvature in the MA plot indicates the presence of nonlinear intensity effect in the non-normalized M values (Nonorm) at the levels of array (left) and point-tip (PT) group (right) for microarrays 3509 and 8837. It can be seen that for microarray 3509, the values of M decrease as those of A increase, whereas the opposite trend is apparent for microarray 8837.

SPATIAL-DEPENDENT DYE EFFECT

The log ratios M of genes can also be dependent on the physical locations of the spots where the genes reside on the array, and this type of dye effect is called the

A Survey of cDNA Microarray Normalization

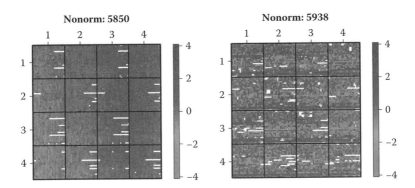

FIGURE 5.3 (See color insert following page 80.) The plots show spatial effects present in the non-normalized data for microarrays 5850 and 5938 in the Lymphoma data set. The spatial plot is a spatial representation of spots on the microarray color coded by their M values (marrayPlots/maImage(x="maM")). Spots in white are spots flagged in the original microarray data (missing values).

spatial-dependent dye effect. Spatial effect is especially common in cDNA microarray data. Although they are sometimes present in the data generated from Affymatrix GeneChips, the spatial effects in these data are not as severe as those in cDNA microarray data.

Spatial effects can be identified using *spatial plots*, which are a color-coded representation of each spot on a microarray that depicts the log ratios M or the intensity values R or G. Figure 5.3 shows the spatial plots of the M values for two specific Lymphoma microarrays, illustrating spatial effects in the data. The non-normalized M values (Nonorm) for microarray 5850 display a global spatial effect (left-to-right, green-to-red pattern), whereas those for microarray 5938 exhibit a local spatial effect (top-to-bottom, green-to-red pattern in each PT group).

SCALE EFFECT

While intensity- and spatial-dependent dye effects in cDNA microarray data occur within a microarray, scale effect is the variation that can occur both within and between microarrays. Since each array in a data set is hybridized with equal amounts of mRNAs from biological samples, if we assume that only a few genes are differentially expressed among different samples, variabilities (e.g., variances or scale) in data from different arrays should presumably be similar; however, in reality such variabilities across different arrays are often very different. This is the so-called *scale effect*.

Scale effect can be illustrated and identified using box plots. Figure 5.4 depicts the distribution of the non-normalized M values for different microarrays in the LungCancer data set (Table 5.1). In each box plot, the box depicts the main body of the data and the whiskers show the extreme values. The variability is reflected by the size of the box and the length of the whiskers. The variations in the sizes of the boxes and in the lengths of the whiskers in these box plots illustrate an obvious scale effect across different microarrays (*between-microarray scale effect*) in the LungCancer data set.

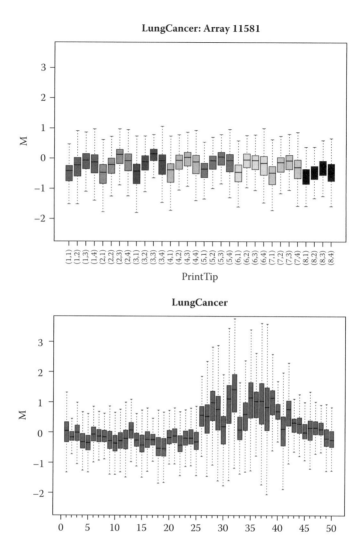

FIGURE 5.4 **(See color insert following page 80.)** Box plots of the distributions of the non-normalized M values for microarrays in the LungCancer data set. The box plots display scale effect for microarray 11581 (top) and other microarrays (bottom) in the LungCancer data set at the PT level (top) and at the global level (bottom), respectively. In each box plot, the box depicts the main body of the data and the whiskers show the extreme values. The variability is indicated by the size of the box and the length of the whiskers (marray/marraymaBoxplot(y="maM")). The top panel shows results for M values at the local level of microarray 11581. The bars are color coded by PT group. The bottom panel shows results for M values at the global level for 50 microarrays chosen at random from the LungCancer data set.

Scale effect can also occur within a microarray. The non-normalized M values for microarray 11581 in the LungCancer data set are shown in Figure 5.4 (top). The data are stratified by the PT group; the bars are color coded by rows, each of which contains four PT groups. Slight variabilities, again depicted by different sizes of the boxes and lengths of the whiskers in the box plots, indicate that a scale effect is present at the PT level. This effect is called the *within-microarray scale effect*.

STRATEGIES FOR MINIMIZING UNWANTED VARIATIONS IN CDNA MICROARRAY DATA

Once the fluorescent intensities R and G of expression signals are obtained for many samples, downstream analyses such as classification, feature selection, clustering, and network estimation can be carried out. A variety of algorithms seeking to address these problems have been developed and applied to interpret expression patterns of genes and to generate biological predictions. However, the accuracy of these predictions depends critically on how effectively the unwanted variations can be minimized from microarray data. The key preprocessing steps for minimizing unwanted variations in measurements include data transformation and normalization. In the following sections, we will review commonly used methods and strategies for data transformation and normalization for cDNA microarray data.

DATA TRANSFORMATION

Data transformation is a common practice in microarray data analysis. Appropriate transformation of microarray data can simplify the subsequent computational and statistical analysis, and even reduce some unwanted variations in the data. In the next section, we will discuss two common types of transformation for cDNA microarray data.

Logarithmic Transformation

Raw cDNA microarray data (i.e., Cy5-intensity values, R, and Cy3-intensity values, G) are usually log transformed (usually based 2) before downstream data analyses are carried out. This transformation can facilitate data analysis in the following ways: (1) the calculation of the fold changes is simplified, since the ratios of expressions of genes between different samples are converted into differences after log transformation; (2) the estimation of error models of dye effects is simplified, since multiplicative models of dye effects (which we will describe later) can be turned into additive models; and (3) variances of genes with high-intensity values can become more homoscedastic (i.e., stabilized) after log transformation.

Ratios or Intensity Values

For two-channel cDNA microarrays, if we use an indirect reference design to measure differential expression of genes in different biological samples (say, sample t and sample c), we can competitively hybridize mRNAs of each of these samples labeled with one dye (e.g., Cy5) and those of a common reference sample r labeled with another dye (e.g., Cy3) on the same array [8]. The abundance of genes in samples t and c can be estimated using intensity values of genes from a single channel—in our example the Cy5-intensity values R, or using ratios R/G. A question can arise regarding which measurement for the expressions of genes, intensity values or ratios, is better for downstream analyses.

Chen, Dougherty, and Bittner first used intensity ratios R/G to compare the relative expression levels of genes in different samples [16]. A major advantage of using intensity ratios is that the spot effect associated with individual spots can be cancelled out. Due to reasons such as manufacturing, printing, and hybridization, dye effect can be associated with individual spots, and this effect is called *spot effect*. Spot effect affects intensity values obtained from both channels (i.e., intensity R and G values) and can be cancelled out when intensity ratios are used to represent the abundance of genes in the samples. The effects of ratios and intensity values were also compared in the context of classification [17]. Using data generated from model-based simulations, it was shown that spot effect introduced by immobilization of probes on the array gave high leave-one-out cross-validation errors in intensity-based k-NN classification, while they gave little or no error in ratio-based k-NN classification. These studies suggest that the ratio measurements can minimize the impact of spot effects and thus are a better choice than intensity values for ensuring the accuracy of downstream analysis. However, one should be aware that when using ratios, the same reference sample needs to be used to ensure comparability of data obtained from different studies [17].

NORMALIZATION METHODS FOR cDNA MICROARRAY DATA

While the data transformation scheme described previously can be useful for stabilizing variances and minimizing spot effect in cDNA microarray data, it cannot reduce systematic dye effects in the data. In order to minimize these effects, normalization methods need to be used. Based on different biological assumptions and design principles, many normalization approaches have been proposed. In the next section, we will discuss some commonly used normalization strategies for cDNA microarray data.

Global Normalization Techniques

Global normalization techniques are among the first proposed normalization methods [16,18–20]. Global normalization adjusts the center (e.g., mean or median) of the distribution of the log ratios M on each microarray to a constant—that is,

$$M' = M - C,$$

where M' is the normalized log ratios, and C is a normalization constant.

Various strategies have been proposed to identify the normalization constant C. Among them, the globalization method assumes that biological samples hybridized on different arrays all have the same amount of mRNAs, and therefore the sum (or mean) of intensity values of all genes in the pair of samples hybridized to the same array should be the same. As a result, the mean of the log ratios M of all genes in the two samples should be centered around zero, and thus the normalized log ratio M' values can be obtained as $M' = M - mean(M)$.

Other global normalization techniques include an iterative procedure proposed by Chen et al. to determine the normalization constant C [16]. This procedure assumes that the number of mRNAs of each gene is stochastically fluctuated. Later, a centralization method was proposed by Zien et al. [19]; it further relaxes the assumption of the globalization normalization method and assumes that, for the pair of samples hybridized to each array, only a few genes in these samples are differentially expressed or the average expression levels of the up- or down-regulated genes are normally distributed (i.e., roughly the same). This method determines the normalization constant C as the central tendency (e.g., mean, median) of log ratios M on each array; when $C = median(M)$, the normalized log ratios M' become:

$$M' = M - median(M).$$

This is a special case of the centralization method, and is also called the *median* (Gmedian) normalization method. Because the Gmedian method is easy to implement and the biological assumptions it requires are reasonable, it is recommended by many software packages as a default approach for normalization of cDNA microarray data (and also for data obtained from other microarray platforms—for example, CodeLink microarrays).

These global normalization approaches, however, are limited in that they do not correct any intensity- or spatial-dependent dye effects. Therefore, after global normalization, intensity and spatial effects are still present in cDNA microarray data and can still compromise the quality of the data.

ANOVA Methods

Statistical approaches based on the analysis of variance (ANOVA) have been proposed to account explicitly for variations arising from multiple sources [21,22]. Following Kerr et al. [21], if we let y_{gijk} denote intensity values for gene g labeled by dye j in variety (or sample) k on array i, an ANOVA model can be written as:

$$\log(y_{gijk}) = \mu + A_i + D_j + V_k + G_g + (AG)_{gi} + (VG)_{gk} + (DG)_{gj} + \varepsilon_{gijk}.$$

In this model, variations arising from both biological differences and unwanted variations are taken into consideration. In particular, biological differences between different sample types are accounted for by gene-variety interaction effect (VG_{gk}); unwanted variations are accounted for by array effect (A_i), dye effect (D_j), variety effect (V_k), and gene effect (G_g), as well as gene–array interaction effect (AG_{gi}) and gene-dye effect (DG_{gj}), respectively. To be more specific:

- The array effect (A_i) accounts for variations between arrays, which may arise from differential hybridization efficiencies across arrays.
- The dye effect (D_j) accounts for variations between fluorescent dyes, which may arise from distinct properties of different dyes that cause genes labeled with one dye to have differential intensities with respect to the same genes labeled with another dye.

The variety effect (V_k) accounts for the overall differences between different sample types (e.g., there are more mRNAs in sample *a* than those in sample *b*), which can be caused by intrinsic biological differences between different samples (e.g., the transcription machinery in sample *a* is more active than that in sample *b*) or by pipetting errors in the microarray experiment.

The gene effect (G_g) accounts for variations of individual genes spotted on the arrays.

The gene–array interaction effect (AG_{gi}) accounts for the effect of individual spots on a particular array, which is the spot effect we have discussed previously.

The gene–dye interaction effect (DG_{gi}) accounts for variations caused by gene–dye interaction (e.g., some genes may incorporate one dye more efficiently than another dye).

As shown, the linear ANOVA model allows a joint estimation of biological differences between different sample types and unwanted variations arising from many sources simultaneously in a single model. The ANOVA model was later extended to account for additional effects [22,23]. However, like global normalization techniques, the linear ANOVA models proposed by Kerr et al. and the extended, mixed-linear ANOVA models proposed by Wolfinger et al. are equivalent to correcting dye effect using the normalization constant *C*. Therefore, these ANOVA models are also limited in that they do not correct intensity- and spatial-dependent dye biases in cDNA microarray data.

Normalization Strategies That Minimize Intensity-Dependent Dye Effect

Total Gene Approaches

The global lowess (iGloess) method: Intensity-dependent dye effect is a major source of variations in cDNA microarray data that can seriously affect the accuracies of the measurements of intensities of genes on the array. Intensity effect is often nonlinear and can vary from slide to slide [24]; as a result, linear methods can be inadequate for removing intensity effect completely from the data, and a normalization treatment is needed for each individual slide to remove slide-dependent dye effect.

A variety of techniques based on local regression have been proposed to remove nonlinear intensity effect [12,25], among which global lowess (or iGloess) is one of the first and most commonly used methods. This method first employs a robust locally weighted regression (*lowess*) [26] to estimate the dependences of log ratios *M* on log intensities *A* for all genes on each array (i.e., intensity-dependent dye biases). More specifically,

$$M = f_{lowess}(A),$$

iGloess attempts to estimate the lowess function f_{lowess} locally; that is, the log ratio M_g corresponding to a specific intensity value A_g (for the gene *g*) is estimated using the spots in the neighborhood of A_g (the size of the neighborhood is also called the *span* of the lowess function). Then intensity-dependent dye biases B_i for all genes on each array can be estimated as:

$$Bi = \hat{M} = \hat{f}_{lowess}(A),$$

and their normalized log ratios M' can be obtained as:

$$M' = M - Bi = M - \hat{f}_{lowess}(A).$$

Because the lowess smooth curve is fitted locally, in order to estimate it accurately, there should be sufficient number of spots in the whole range of A. The span parameter in the lowess function provides an option for choosing the number of spots for estimating the curve. For convenience, it is often set heuristically as 40% of total spots on the array (*span* = 0.4); it can also be chosen optimally using the leave-one-out cross-validation [27,28].

The iGloess approach requires that only a few genes in the pair of samples hybridized to each array are differentially expressed, or that the average expression levels of the up- and down-regulated genes in these samples are similar. Thus, the basic assumptions of iGloess are the same as those of the centralization method. Since iGloess relies on total genes on the array for normalization, it is also considered as a total-gene strategy.

The local lowess (iLloess) method: As shown in Figure 5.2, the intensity-dependent dye effect can be present not only at the global level (left), but also at the PT level (right). Although the iGloess method can remove intensity effects effectively at the global level, it often fails to do so at the local (i.e., PT) level (see later discussion). A normalization method, *local lowess* or iLloess, was designed to adjust intensity effect at the PT level [12]. In this approach, intensity-dependent dye bias for each gene is estimated using genes in the same PT group. Let Bi_b denote the intensity-dependent dye biases for all genes in the PT group b; it can be estimated as follows:

$$M_b = f_{lowess}(A_b),\ Bi_b = \hat{M}_b = \hat{f}_{lowess}(A_b),\ b = 1,\ \ldots,\ B,$$

where A_b represents the average log intensities of genes in the PT group b. The normalized log ratios M'_b for genes in the PT group b can thus be obtained as:

$$M'_b = M_b - Bi_b = M_b - \hat{f}_{lowess}(A_b).$$

The iLloess strategy appears to remove the intensity effect effectively at the PT level. However, the usefulness of this method can be limited due to the following reasons. In iLloess, the same regression assumptions adopted by iGloess need to be satisfied for genes at the PT level. Since the number of genes in each PT group is far fewer than that of genes on each array (e.g., in many cases, ~3–5%), it is difficult to ensure that the symmetric property of the up- and down-regulated genes in each PT group is satisfied; it is also difficult to guarantee an accurate estimate of intensity-dependent dye bias Bi_b for each gene, since there may not be a sufficient number of genes in the entire range of A_b.

Both the global normalization techniques and the intensity-dependent normalization approaches described previously use all genes for normalization. These genes include both differentially expressed and nondifferentially expressed genes on the array or in the PT group of each array. However, a blind inclusion of all genes can seriously limit the applicability of these methods, since they all assume that only a few genes are differentially expressed or that the expression levels of the up- and down-regulated genes are similar, in many biological situations, however, the up- and down-regulated genes can exhibit unbalanced expression patterns. To reduce the dye effect in the unbalanced situations, a suite of normalization strategies that involve the utilization of "housekeeping" genes, "spike-in" controls, and replicate spots on the array has been proposed [24,25,29]. As we discuss next, these approaches offer useful alternatives for normalizing cDNA microarray data in more diverse biological settings.

Housekeeping-gene-related approaches: These approaches are related to the housekeeping gene approach, which is often used in traditional biological experiments, such as northern blotting. The housekeeping gene approach assumes that some genes, such as glyceraldehyde-3-phosphate dehydrogenase (GAPDH) and β-actin, are stably expressed at constant levels in different biological conditions; therefore, they can be used to estimate the relative expression levels of other genes. However, due to the concern that housekeeping genes may not actually exist in higher organisms [30], this approach is rarely used in practice for normalizing microarray data.

To overcome the limitations of the housekeeping gene approach, several housekeeping-gene-related methods were proposed [24,25,29]. Instead of using a pre-specified set of housekeeping genes, these methods first identify genes that are not differentially expressed across different biological samples in a data set, and then use these genes for normalization. Among these methods, a rank invariant approach was employed to identify nondifferentially expressed genes [24]. Following Tseng et al., the rank invariant method first ranks genes by their Cy5- and Cy3-intensity values, R and G, respectively. Let $Rank(R_g)$ and $Rank(G_g)$ denote the ranks of R and G values respectively for gene g. Then, if

$$Rank(R_g) - Rank(G_g) < d, \text{ and } 1 \leq g \leq G,$$

where d is a predetermined threshold value, and $Rank(R_g)$ and $Rank(G_g)$ are not among the highest or lowest l ranks of total genes on the array, then gene g is determined as a nondifferentially expressed gene. This procedure can be iterated until a final stable set of the nondifferentially expressed genes is identified; the resulting set of the genes is then used to estimate intensity-dependent dye effect using lowess or spline smoothing techniques.

The rank invariant method does not require that the expression levels of the up- and down-regulated genes be similar in the pair of biological samples hybridized to each array. However, since the number of the rank invariant genes used for normalization is usually small (e.g., ranging from a few hundred to a thousand genes in some data sets that we have examined) and hence they can hardly cover the entire range of intensity values of all genes on the array, it is therefore difficult to estimate intensity effect accurately in the array. This may explain why the performance of the rank invariant method was not stable in some studies [14,31].

Normalization Strategies Using External Controls

In order to overcome the limitations of the iLloess method (described previously), a composite method was proposed to remove intensity effects at the local (PT) level using both external control samples and genes in each PT group of a microarray [32]. In this method, external control samples were first serially diluted and printed in each PT group; the titrated control samples were made to have intensity values cover the entire intensity range of all genes spotted on the array. Then, for normalization, a composite normalization curve for estimating the intensity effect at the local level is constructed using the weighted average of the lowess normalization curves estimated using external control samples and the rest of the genes in each PT group, respectively. Following Yang et al., let Bi_b denote intensity-dependent dye biases for all genes in the PT group b. Then,

$$Bi_b = w_A \hat{f}_{lowess}(A_{bEC}) + (1-w_A)\hat{f}_{lowess}(A_{bG}), \ b=1,\ldots,B,$$

where $\hat{f}_{lowess}(A_{bEC})$ represents intensity-dependent dye biases estimated using external control samples in the PT group b, $\hat{f}_{lowess}(A_{bG})$ represents intensity-dependent dye biases estimated using the rest of genes in the same PT group, and w_A is defined as the proportional of genes less than a given intensity value A. The normalized log ratios M'_b for genes in the PT group b can be obtained as $M'_b = M_b - Bi_b$.

External control samples can be useful for constructing reliable normalization curves; however, this method has the following limitations. First, since genes in each PT group need to be used to construct the composite normalization curve at the local level, they still need to satisfy the assumptions of the iLloess method, and hence the limitations of the iLloess method still apply here. Second, a significant number of external control samples need to be printed in each PT group of the array, which can be expensive in practice.

Semilinear models (SLMs): While many normalization strategies seek to employ housekeeping genes or external (spike-in) control samples to estimate dye effects in the data, a suite of promising normalization approaches has recently been proposed to tackle the problem from different angles. In these methods, semilinear models were elegantly designed to jointly estimate dye effects in the array and biological differences between different samples simultaneously in a single model [33–36]. The SLM methods do not require either that few genes are differentially expressed or that the expression levels of the up- and down-regulated genes are similar. Instead, they estimate dye effects using replicate spots and/or replicate arrays. These methods thus represent an advance in normalization methodology that can be suitable for minimizing dye effects in microarray data obtained from many biological settings.

Next, we will describe a simple SLM model to illustrate how these methods work. Following the notations in Fan, Peng, and Huang [35] and Huang, Wang, and Zhang [36] with minor modifications, an SLM model can be represented as:

$$M_{gi} = f(A_{gi}) + \beta_{gi} + \varepsilon_{gi}, \; g = 1, \ldots, n, \; i = 1, \ldots, I,$$

where M_{gi} is the log ratio for gene g in array i, A_{gi} is the average log-intensity value for the same gene, and β_{gi} is the treatment effect (i.e., biological differences) for gene g. The normalized log ratios M'_g can be obtained as:

$$M'_g = M_g - \hat{f}(A_g)$$

for all genes. This model can also be extended to allow an estimate of intensity effects at the PT level using replicate spots and/or replicate arrays.

The SLM models can be considered as extensions to the ANOVA models proposed previously (described earlier) [21,22]. The differences of these models are as follow: The SLM models are semiparametric, they jointly estimate biological differences and intensity-dependent dye effect in a single model, and they minimize nonlinear intensity effects. The ANOVA models are either linear [21] or mixed linear [22] and are thus equivalent to global normalization techniques that normalize data using a constant C; they do not correct nonlinear intensity effects.

Other Normalization Methods

A Baseline-Array Approach (Qspline)

The Qspline method [37] assumes that the distributions of Cy5-intensity values (R_T) and Cy3-intensity values (G_T) on each array (also called the target array) are similar to intensity values I_B of genes on a common baseline array B; it thus adjusts R_T and G_T values on each array separately using the baseline array B for all arrays. The baseline array B can be chosen as the geometric mean of the R (or G) channels of all arrays. Then, for normalizing R_T (and G_T), Qspline first ranks all the genes on the target array by R_T and ranks those on the baseline array by I_B, respectively. Then it employs the quantiles of the ranked genes to construct a normalization curve using spline smoothing techniques, and obtains an estimate of intensity-dependent dye biases in R_T (and G_T) with respect to I_B. Finally, the normalized intensity R'_T (and G'_T) values are calculated by subtracting the estimated intensity-dependent dye biases from the non-normalized R_T (and G_T), and the normalized log ratios M'_T for the target array can be obtained as:

$$M'_T = \log_2(R'_T/G'_T).$$

Dye-Swap Normalization

Dye-swap experiments are carried out in the following way. If an array is hybridized with a pair of biological samples labeled with two distinct fluorescent dyes (e.g., sample a labeled with Cy5, and sample b labeled with Cy3), then in a dye-swap experiment, the second array is hybridized with the same two samples labeled with swapped dyes (i.e., sample a is labeled with Cy3 and sample b with Cy5).

The motivation for performing dye-swap experiments is that intensity effect in microarray data can be minimized using a simple dye-swap normalization method. The method works as follows. If we let M'_1 denote the normalized log ratios for the first array, then

A Survey of cDNA Microarray Normalization

$$M'_1 = M_1 - \hat{f}_1(A),$$

where $\hat{f}_1(A)$ is the estimated intensity-dependent dye biases in the data for the first array.

Similarly, if we let M'_2 denote the normalized log ratios for the dye-swapped array, then

$$M'_2 = M_2 - \hat{f}_2(A),$$

where $\hat{f}_2(A)$ represents the estimated intensity-dependent dye biases in the data for the dye-swapped array. Since the same samples are hybridized to the dye-swapped arrays, the normalized log ratios M'_1 and M'_2 should have the same absolute values but with opposite signs, and thus the normalized log ratios M' can be derived as:

$$M' = (M'_1 - M'_2)/2$$
$$= (M_1 - M_2)/2 - (\hat{f}_1(A) - \hat{f}_2(A))/2.$$

If the intensity effects in the two arrays are similar, then $\hat{f}_1(A) \approx \hat{f}_2(A)$; the normalized log ratios M' can be simply obtained as:

$$M' = (M_1 - M_2)/2.$$

From this equation, one can see that the dye-swap normalization method "corrects" intensity effects without estimating and removing them explicitly. In reality, however, intensity effects are often slide dependent and can vary considerably from slide to slide (as shown previously), and hence $\hat{f}_1(A) \neq \hat{f}_2(A)$. Therefore, dye-swap normalization alone is usually not sufficient for removing intensity effects in the microarray data.

Normalization Strategies That Minimize Spatial-Dependent Dye Effect

In addition to intensity-dependent dye effect, spatial-dependent dye effect (illustrated in Figure 5.3) is another major source of unwanted variations in cDNA microarray data. In order to minimize spatial effect, a spatial-dependent normalization method, *spatial lowess* (or sGloess), employs lowess to estimate spatial-dependent dye biases, which are the dependence of M values on physical positions of spots [38]. That is,

$$M = f_{lowess}(rloc, cloc),$$

where $f_{lowess}(rloc, cloc)$ represents the lowess curve fitted as a function of the row location (*rloc*) and column location (*cloc*) of spots on the array. The spatial-dependent dye biases *Bs* for all genes on the array can be estimated as:

$$Bs = \hat{M} = \hat{f}_{lowess}(rloc, cloc).$$

The normalized log ratios M' for genes on the array can thus be obtained as:

$$M' = M - Bs = M - \hat{f}_{lowess}(rloc, cloc).$$

Like intensity-dependent normalization strategies, spatial-dependent normalization can also be performed at the PT level. In the sLloess method, spatial-dependent dye biases are estimated using genes in the same PT group, and thus the normalized log ratios M'_b for genes in PT group b can be obtained as:

$$M'_b = M_b - Bs_b = M_b - \hat{f}_{lowess}(rloc_b, cloc_b), \; b = 1, \ldots, B,$$

where M_b represents the log ratios of genes in PT group b before normalization, and $\hat{f}_{lowess}(rloc_b, cloc_b)$ represents the lowess curve fitted as a function of the row location ($rloc_b$) and column location ($cloc_b$) of spots in PT group b.

Since the sLloess method estimates spatial-dependent dye biases by fitting log ratios M to the two-dimensional (2D) location parameters ($rloc_b$, $cloc_b$), theoretically it fits and removes spatial effects effectively if such effects appear in a continuous pattern. This proves to be true in practice. However, when the spatial effect exhibits a noncontinuous pattern, the method may not work well (see later discussion).

To remove spatial effects more completely, the *local median filter* method (or sLfilter) estimates spatial-dependent dye bias for each gene using spots within a small neighborhood of the gene (e.g., using a window size of 3 × 3 or 7 × 7 spots) [29]. If we let Bs_{gw} denote spatial-dependent dye bias for gene g in a window w of $N_w \times N_w$ spots, sLfilter estimates Bs_{gw} as:

$$Bs_{gw} = median(M_w),$$

where M_w is the log ratios of the genes in window w before normalization. The normalized log ratio M'_{gw} for gene g in window w can thus be obtained as:

$$M'_{gw} = M_{gw} - Bs_{gw} = M_{gw} - median(M_w).$$

When the window w is set sufficiently small, the sLfilter method can remove continuous and noncontinuous spatial effects effectively (see later discussion). However, appying this method blindly can be risky, because it estimates spatial-dependent dye bias Bs_{gw} using the median of M_w, and thus it would require that fewer than 50% of differentially expressed genes be co-localized in the same window w; this criterion can be difficult to meet when the size of window is small.

Normalization Strategies That Minimize Intensity- and Spatial-Dependent Dye Effects

The intensity- and spatial-dependent normalization strategies described earlier were designed to remove either intensity or spatial effect, and can also be called the *single-dye-removal* strategies. These strategies can be applied in a sequential order (first an intensity-dependent method, followed by a spatial-dependent method) to remove both intensity and spatial effects in two steps.

Considering that intensity and spatial effects may be mutually dependent, a method called *joint-loess* (or iGsGloess) was proposed to remove global intensity and spatial effects in a single step [39]. Basically, iGsGloess models the dependence of M values on the average log intensities A and the physical locations of spots (*rloc*, *cloc*) simultaneously—that is,

$$M = f_{lowess}(A, rloc, cloc).$$

The normalized log ratios M' for all genes on the array can thus be obtained as:

$$M' = M - Bis = M - \hat{f}_{lowess}(A, rloc, cloc),$$

where *Bis* represents both intensity- and spatial-dependent dye biases for genes on the array.

Normalization Strategies That Minimize Scale Effect

Scale normalization methods adjust differences in the scale of M values within and/or between microarrays. These methods assume that the majority of genes are not differentially expressed between different samples in a data set; therefore, the scale of their M values should be constant. For a given microarray, if we denote Ms' as the scale-normalized log ratios, then $Ms' = M/s$, where M is the log ratios of genes before normalization, and s is the scale factor. The scale factor s can be calculated using median absolute deviation (*MAD*), which is a robust estimate of s [32]. Following Yang et al. [12], let MAD_i denote the median absolute deviation for array i. Then

$$MAD_i = median_i \left\{ \left| M_i - median(M_i) \right| \right\},$$

where M_i represents the log ratios of genes on array i before normalization, and the scale factor s_i for array i can be estimated as follows:

$$s_i = MAD_i \Big/ \sqrt[I]{\prod_{i=1}^{I} MAD_i}.$$

Since this scale normalization method aims to remove scale differences in the data from different microarrays, it is also called *between-slide normalization* (or Bscale).

Scale normalization can also be applied to data within a microarray, and this method is called within-slide normalization, Wscale. In Wscale, scale normalization is performed at the PT level, and thus can remove scale effect at the local level.

CRITERIA FOR COMPARISON OF NORMALIZATION METHODS

As reviewed previously, there exists a big repertoire of normalization schemes for preprocessing microarray data; however, a challenge for evaluation of these normalization methods is the lack of a gold standard (i.e., a "perfect" validation data set, about which true intensity values of genes are known and that can reflect the complexity of real data). Ideally, normalization approaches seek to ensure that dye effects are removed while biological variations are retained. Therefore, two criteria need to be employed to compare normalization strategies: One is the ability to reduce unwanted variations in the data, and the other is the ability to retain biological variations in the same data. In the next section, we will discuss how these criteria have been used to compare normalization methods.

ABILITY TO REDUCE UNWANTED VARIATIONS

Diagnostic plots provide a qualitative measure to visualize and identify dye effects in microarray data. As described earlier, MA plots, spatial plots, and box plots have been used to identify intensity-dependent, spatial-dependent, and scale dye effects, respectively in real and simulated cDNA microarray data [12,39].

Various quantitative measures have also been employed for validating the effectiveness of normalization methods in removing variability. For example, Kroll et al. used rank variations of intensities of spots to compare variability in the non-normalized and normalized data [20]. Rank variations are defined by two measures: absolute rank deviation (ARD), which is the standard deviation of intensity values of a given rank, and relative rank deviation (RRD), which is ARD divided by the mean of intensity values of that given rank. In addition, variances [37,40] and coefficient of variation [14,41] of log ratios of genes in replicate arrays have been used to compare variability in real microarray data. Simulation data have also been used to compare normalization approaches quantitatively. Since the true log ratios of genes on the array were known in simulation data, the mean square errors, variances, and biases were calculated and compared for data normalized with different normalized methods [40].

ABILITY TO RETAIN BIOLOGICAL VARIATIONS

Evaluation of the effectiveness of normalization methods in signal retention is relatively more ambiguous, because there is no ground truth for total biological signals in real data. To ensure that biological variations are retained after normalization, several functional measures have been employed. The prevailing approaches determine the ability to predict a fixed number of differentially expressed genes in real or simulation data using quantitative measures based on t-statistics [12,15,37,39], adjusted p-values [15], or false-discovery rates [42]. Spike-in data were also used to assess normalization approaches for Affymetrix GeneChip data [13,43,44]. However, all these functional measures are limited for the following reasons: (1) Since the true number of differentially expressed genes is unknown in the real data, there is uncertainty associated with these measures; and (2) although the fixed number of differentially expressed genes is known in simulation or spike-in data, these data need to be generated using certain statistical models and biological/statistical assumptions that may not reflect the complexity of real data.

To overcome the limitations of the functional measures described before, we and others estimated biological variations by calculating the overabundance or the number of differentially expressed genes or informative genes in real data sets [14,45]. We assumed that the more differentially expressed genes can be revealed by a normalization method, the more biological variations should be retained in the data. Recently, we have also employed the k-NN leave-one-out cross-validation ($LOOCV$) classification error as the quantitative functional measure for comparing normalization methods in real data [46].

EVALUATION OF NORMALIZATION METHODS FOR CDNA MICROARRAY DATA

Although there are many normalization methods designed to remove systematic dye effects in cDNA microarray data, few studies have evaluated them systematically. In a comparative study, Park et al. assessed several normalization methods for cDNA microarray data using both replicate data and simulation data [40]. First, the variances of real replicate data that are either normalized or non-normalized by different methods were calculated and compared using a two-way ANOVA model. Then synthetic data that simulate microarray data with four different patterns of dye effects were generated, and normalized with different methods. Since the true log ratios of the non-normalized data were known in the simulation data, the mean square errors (MSE), variances and biases ($MSE = variance + bias$) were calculated respectively and compared with those of the normalized data. It was found that (1) the global intensity-dependent normalization methods (both linear and nonlinear iGloess methods) could reduce more biases and variability from non-normalized data than the *global median* (Gmedian) normalization method; and (2) compared to iGloess, the local intensity-dependent normalization method (iLloess) and scale normalization do not have beneficial effects in reducing variability in the data.

The study described above compared the normalization methods based on their ability to remove unwanted variations in cDNA microarray data; however, it did not assess how effectively these methods could retain biological variations in the data. As we have discussed in the previous section, this can limit the generalizability of the conclusions of this study.

COMPARISON OF NORMALIZATION METHODS FOR CDNA MICROARRAY DATA BY K-NN CLASSIFICATION

As described before, many studies evaluated normalization strategies based on their ability to reveal differentially expressed genes; however, using this specific task as a comparative criterion can also lead to oversight of uncertainty. Moreover, normalization methods performing well in one downstream task may not perform equally well on other tasks, such as classification and clustering.

In order to investigate how normalization strategies affect other downstream analyses, such as classification, we have evaluated a broad spectrum of normalization methods for cDNA microarray data using the k-NN $LOOCV$ classification error of biological samples characterized by the gene expression profiles. The $LOOCV$ classification error of k-NNs provides an alternative quantitative functional measure that is relatively unambiguous, objective, and readily computable. We used k-NN classifiers because (1) their sensitivity to noise and variance enables one to discriminate between and hence evaluate normalization techniques, (2) they perform

TABLE 5.2
Single-Bias-Removal Normalization Techniques Used in Wu et al. [46]

Name[a]	Description: effect/level	Bioconductor R package/function(parameters)
Nonorm	No normalization $M' = M$	marray/maNorm(norm="none")
Gmedian	Global $M' = M - median(M)$	marray/maNorm(norm="median")
sLloess	Spatial/local lowess $M' = M - f_{lowess}(rloc_b, cloc_b)$	marray/maNormMain (f.loc= list(maNorm2D(g="maPrintTip"; span=0.4))
sLfilterW3	Spatial/local median filter $M' = M - median(M_w)$, $W = 3 \times 3$	tRMA/SpatiallyNormalize (M, width = 3, height = 3)
sLfilterW7	Spatial/local median filter $M' = M - median(M_w)$, $W = 7 \times 7$	tRMA/SpatiallyNormalize (M, width = 7, height = 7)
iGloess	Intensity/global lowess $M' = M - f_{lowess}(A)$	marray/maNorm(norm="loess"; span=0.4)
iLloess	Intensity/local lowess $M' = M - f_{lowess}(A_b)$	marray/maNorm (norm="printTipLoess"; span=0.4)
iSTspline	Intensity/global spline $M' = M - spline(A_{iset})$	affy/normalize.invariantset (prd.td=c(0.003, 0.007))
QsplineG	Intensity/global Qspline $R'_T = R_T - qspline(G_b)$, $G'_T = G_T - qspline(G_b)$ $M' = \log_2(R'_T/G'_T)$	affy/ $R'_T \leftarrow$ normalize.qspline(RT, rowMeans(Gb)) $G'_T \leftarrow$ normalize.qspline(GT, rowMeans(Gb))
QsplineR	Intensity/global qspline $R'_T = R_T - qspline(R_b)$, $G'_T = G_T - qspline(R_b)$ $M' = \log_2(R'_T/G'_T)$	affy/ $R'_T \leftarrow$ normalize.qspline(RT, rowMeans(Rb)) $G'_T \leftarrow$ normalize.qspline(GT, rowMeans(Rb))

Notes: These strategies remove spatial or intensity effects in a single step. Most abbreviations in the table can be found in detail in the text. The remaining abbreviations are as follows: for a given microarray, $spline(A_{iset})$: spline curve fitted to an MA plot of spots in the invariant set, *iset*; $qspline(G_b)$: Qspline smoothing using the geometric mean of the G channels of all arrays as the baseline array; $qspline(R_b)$: Qspline smoothing using the geometric mean of the R channels of all arrays as the baseline array.

[a] We adopted the terminology given in the table to avoid confusion in our discussion in this chapter. Readers should refer to Wu et al. [46] for the names of these methods that are known elsewhere.

well in practice, and (3) their nonparametric nature ensures that the distribution of data would have relatively less influence on the performance of classification, compared to other parametric or model-based classifiers.

In this study, we investigated 10 normalization methods that adjust intensity effects and/or spatial effects (Table 5.2), and three scale methods that adjust scale differences (Table 5.4). These methods, individually and in combination (41 strategies in all, Tables 5.2–5.4), were applied to five diverse, published cDNA microarray

TABLE 5.3
Double-Bias-Removal Normalization Techniques Used in Wu et al. [46]

Name	Description: method/effect/level
iGsGloess[a]	Joint intensity/global and spatial/global $M' = M - f_{lowess}(A, rloc, cloc)$
iGloess–sLloess	Step 1: iGloess/intensity/global lowess Step 2: sLloess/spatial/local lowess
iLloess–sLloess	Step 1: iLloess/intensity/local lowess Step 2: sLloess/spatial/local lowess
iGloess–sLfilterW3	Step 1: iGloess/intensity/global lowess Step 2: sLfilterW3/spatial/local median filter
iGloess–sLfilterW7	Step 1: iGloess/intensity/global lowess Step 2: sLfilterW7/spatial/local median filter
iSTspline–sLloess	Step 1: iSTspline/intensity/global spline Step 2: sLloess/spatial/local lowess
iSTspline–sLfilterW3	Step 1: iSTspline/intensity/ global spline Step 2: sLfilterW3/spatial/local median filter
iSTspline–sLfilterW7	Step 1: iSTspline/intensity/global spline Step 2: sLfilterW7/spatial/local median filter
QsplineG–sLloess	Step 1: QsplineG/intensity/global Qspline Step 2: sLloess/spatial/local lowess
QsplineG–sLfilterW3	Step 1: QsplineG/intensity/global Qspline Step 2: sLfilterW3/spatial/local median filter
QsplineG–sLfilterW7	Step 1: QsplineG/intensity/global Qspline Step 2: sLfilterW7/spatial/local median filter
QsplineR–sLloess	Step 1: QsplineR/intensity/global Qspline Step 2: sLloess/spatial/local lowess
QsplineR–sLfilterW3	Step 1: QsplineR/intensity/global Qspline Step 2: sLfilterW3/spatial/local median filter
QsplineR–sLfilterW7	Step 1: QsplineR/intensity/global Qspline Step 2: sLfilterW7/spatial/local median filter

Notes: These strategies remove both intensity and spatial effects either in a single step (iGsGloess) or in two steps (the remaining 13 approaches) by combining methods listed in Table 5.2.

[a] iGsGloess was implemented in the following Bioconductor R package/function: MAANOVA/smooth (method = "rlowess"; $f = 0.4$; degree = 2). Elsewhere, iGsGloess is known as joint loess [39].

data sets related to cancer studies (Table 5.1), and the data sets with the spatial effect, intensity effect, and scale differences removed to varying degrees were generated. The *LOOCV* classification error of k-NNs estimated on these data sets allows one to investigate which and how much of the dye effect is removed by the 41 strategies.

COMPARISON OF INTENSITY- AND SPATIAL-DEPENDENT NORMALIZATION STRATEGIES BY K-NN CLASSIFICATION

Assessing the Ability to Remove Dye Effects Using Diagnostic Plots

First, we used diagnostic plots to examine the ability of spatial-dependent normalization methods (sLloess, sLfilterW7, sLfilterW3, iGsGloess) to remove spatial effects (Tables 5.2 and 5.3). Figure 5.5 shows that the non-normalized M values (Nonorm) for microarray 5850 display a global spatial effect (left-to-right, green-to-red pattern) whereas those for microarray 5938 exhibit a local spatial effect (top-to-bottom, green-to-red pattern in each PT group). Removal of spatial effect should result in a "random" red and green pattern of M values.

Since the sLloess method estimates spatial-dependent dye biases by fitting log ratios to the two-dimensional location parameters ($rloc_b$, $cloc_b$) (discussed previously), theoretically it should remove spatial effect in a continuous pattern more effectively than that in a noncontinuous pattern. Indeed, the method does so in practice; after sLloess normalization, the global spatial pattern in array 5850 disappears (Figure 5.5, sLloess: 5850), while the local spatial pattern in array 5938 can still be seen at the PT level (i.e., red spots lining at the bottoms of each PT group) (Figure 5.5, sLloess: 5938), albeit sLloess operates locally.

Both sLfilterW3 (sLfilter with $w = 3 \times 3$) and sLfilterW7 (sLfilter with $w = 7 \times 7$) minimize the continuous spatial pattern in array 5850 effectively (Figure 5.5). However, only sLfilterW3 removes the noncontinuous spatial pattern in array 5938 effectively (sLfilterW3: 5938); this is not surprising because it uses a median filter of a small window size (3×3 spots) for normalization. Spatial effect can still be seen in the sLfilterW7-normalized data (sLfilterW7: 5938). These results suggest that sometimes a small window size may need to be used to remove spatial effects completely with the sLfilter method. But doing so can also be risky, because the sLfilter method requires that fewer than 50% of differentially expressed genes are co-localized within the window w (see previous discussion). This criterion can be difficult to meet when w is small. This concern proved to be valid when sLfilterW3 and sLfilterW7 were later compared using the functional measure, the $LOOCV$ classification error of k-NNs (see later discussion).

The iGsGloess method (designed to remove spatial and intensity effects simultaneously) removes most, but not all, global and local spatial effects (a strip of red spots on the right side of 5850 and on the bottom of the PT groups in the first row of 5938 remains).

FIGURE 5.5 *(facing page)* (See color insert following page 80.) Spatial plots of the non-normalized and normalized M values for microarrays 5850 and 5938 in the Lymphoma data set. The plots show the results before and after normalization designed to remove spatial effect. Spots in white are spots flagged in the original microarray data (missing values). Rows depict the non-normalized (Nonorm), and normalized M' values (sLloess, sLfilterW3, sLfilterW7, iGsGloess).

A Survey of cDNA Microarray Normalization

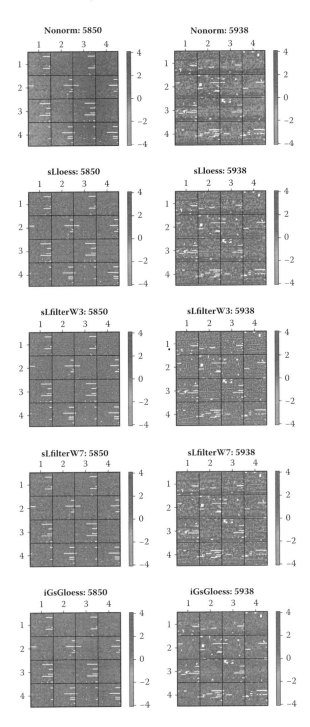

FIGURE 5.5

Then we examined the ability of intensity-dependent normalization methods (iGloess, iLloess, iSTspline, QsplineG, QsplineR, iGsGloess) to remove intensity effect (Tables 5.2 and 5.3).

The MA plots for one specific Lymphoma microarray overlaid with one lowess curve (Figure 5.6, left) or one lowess curve per PT group (Figure 5.6, right) reveal that intensity effect is present in the non-normalized M values (Nonorm) at both the array (left) and PT (right) levels. All six examined normalization methods designed to correct intensity effect remove global intensity effect completely (flat lowess curves, left), but only iLloess and iGsGloess remove local intensity effect thoroughly (right).

Visual inspection of the diagnostic plots in Figures 5.5 and 5.6 suggest that sLfilterW3 is an effective method for removing both global and local spatial effects, and that iLloess is good at removing intensity effect.

FUNCTIONAL ASSESSMENT USING THE *LOOCV* CLASSIFICATION ERROR OF *k*-NNS

For a functional, quantitative evaluation of normalization methods that remove intensity and spatial effects, we first computed the k-NN *LOOCV* classification error for data normalized by the examined methods individually and/or in combination. Then, for each data set, we ranked the normalization methods based on their *LOOCV* error. The smaller the *LOOCV* error was, the lower was the rank of the normalization strategy (see our results in Wu et al. [46] for details).

In the following section, for the convenience of discussion, we group the examined normalization strategies that adjust spatial and intensity effects into three categories: single-bias-removal methods, double-bias-removal methods, and Qspline-related approaches.

Single-Bias-Removal Methods

These normalization strategies include those minimizing either spatial- or intensity-dependent dye effect (Table 5.2). Although the three spatial-dependent normalization methods (sLloess, sLfilterW3, sLfilterW7) remove spatial effect to different extents (Figure 5.5), they reduce k-NN *LOOCV* errors to a similar level (Figure 5.7) and have similar ranks (Figure 5.8). On the other hand, the three intensity-dependent methods (iGloess, iLloess, iSTspline) appear to reduce k-NN *LOOCV* errors to slightly different degrees. The k-NN *LOOCV* error and the rank of iGloess are similar to those of

---

FIGURE 5.6 *(facing page)* MA plots of microarray 5812 in the Lymphoma data set. The plots show the results before and after normalization designed to remove intensity effect. Columns depict the non-normalized (Nonorm) and normalized M' values (iGloess, iLloess, iSTspline, QsplineG, QsplineR, iGsGloess). Plots in the same row represent same data except that each plot in the left panel shows one lowess curve for all the spots (marrayPlots/maPlot(data)), while that in the right panel shows one lowess curve per PT group (marrayPlots/maPlot(x="maA," y="maM," and z="maPrintTip")). Different colors and line types are used to represent different groups from different rows ("ngr," Figure 5.1) and columns ("ngc"), respectively.

A Survey of cDNA Microarray Normalization

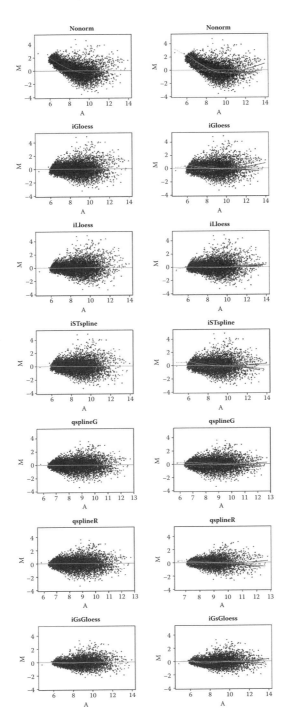

FIGURE 5.6

108 Methods in Microarray Normalization

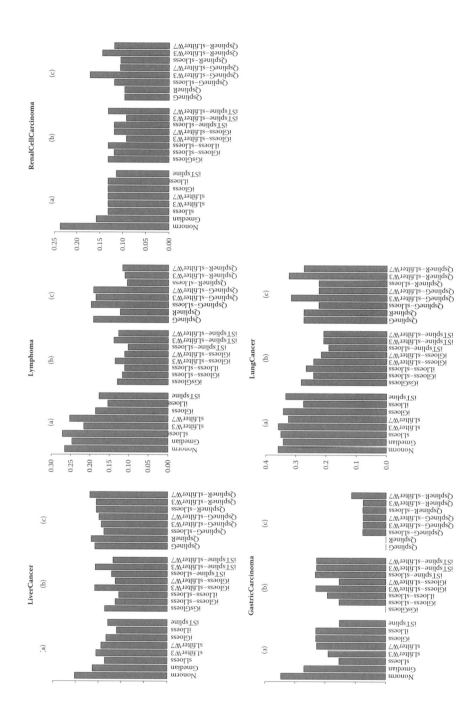

the three spatial-dependent methods (sLloess, sLfilterW3, sLfilterW7) (Figures 5.7 and 5.8), whereas iLloess (which removes intensity effect at the PT level and thus more completely than iGloess) and iSTspline (which uses a rank invariant set for normalization) seem to have slightly smaller k-NN $LOOCV$ errors than iGloess in some data sets (Figures 5.7 and 5.8). Overall, the single-bias-removal normalization methods that remove either intensity or spatial effect consistently yield smaller $LOOCV$ errors than methods that do not remove either dye effect, Nonorm and Gmedian.

Double-Bias-Removal Methods

iGsGloess removes both spatial and intensity effect in one step, whereas the remaining seven approaches are two-step strategies that consist of single-bias-removal methods applied sequentially (first a method to remove intensity effect, followed by a method to remove spatial effect).

In general, double-bias-removal methods (Table 5.3) have smaller k-NN $LOOCV$ errors and thus better classification performance than single-bias-removal methods, and all perform better than Nonorm and Gmedian (Figures 5.7 and 5.8). Overall, three two-step strategies that remove intensity effect globally and then spatial effect locally (iGloess–sLfilterW7, iSTspline–sLloess, iGloess–sLloess) appear to be the best methods; they not only have the lowest ranks among all normalization methods and across all data sets (Figure 5.8), but also show the most consistent and significant improvement in classification performance over both Nonorm and Gmedian across all five data sets (Figure 5.7). The benefits of using them over no normalization (Nonorm) range from approximately 30 to 60% improvement in classification performance [46]. Compared to the two-step approaches, the rank of the one-step method iGsGloess is higher, indicating that the one-step iGsGloess has no apparent advantage over the two-step double-bias-removal strategies.

We also found that the ranks of the sLfilterW3-related two-step approaches (first intensity-dependent normalization followed by sLfilterW3 normalization) are higher than their sLfilterW7 counterparts (Figure 5.8), suggesting that a window size of 7×7 is more preferable than that of 3×3. A smaller window size may over-normalize the data, and thus conceal real biological variations.

FIGURE 5.7 *(facing page)* Bar plots of leave-one-out cross-validation error rates for k-NNs. The classifiers were estimated from five data sets (Table 5.1) either without normalization (Nonorm) or normalized using 23 normalization techniques that remove spatial and/or intensity effects to varying degrees (Tables 5.2 and 5.3). In each plot, the normalization methods are arranged in the following order: (A) Methods that remove no spatial or intensity dye bias (Gmedian), or remove a single dye bias (sLloess, sLfilterW3, sLfilterW7, iGloess, iLloess, iSTspline). (B) Methods that remove two dye biases (iGsGloess, iGloess–sLloess, iLloess–sLloess, iGloess–sLfilterW3, iGloess–sLfilterW7, iSTspline–sLloess, iSTspline–sLfilterW3, iSTspline–sLfilterW7). (C) Qspline-related methods (QsplineG, QsplineR, QsplineG–sLloess, QsplineG–sLfilterW3, QsplineG–sLfilterW7, QsplineR–sLloess, QsplineR–sLfilterW3, QsplineR–sLfilterW7).

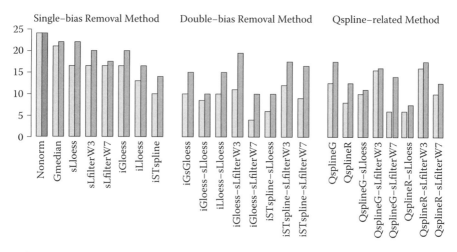

FIGURE 5.8 Rank summary for intensity- and spatial-dependent normalization methods. The bar plots present the median and upper quantile ranks of each method, which are defined as the median and upper quantile values of the ranks of each method across all five data sets [46]. Median ranks are shown in light gray; upper quantile ranks are shown in dark gray. (From W. Wu, E. P. Xing, C. Myers, I. S. Mian, and M. J. Bissell, *BMC Bioinformatics*, 6, 191, 2005.)

Overall, the classification performances of data normalized with the double-bias-removal methods are better than that of Nonorm, and the benefits of doing so range from approximately 21 to 100% improvement in classification performance [46].

Qspline-Related Approaches

Unlike the intensity- and spatial-dependent normalization methods discussed earlier, the Qspline-related approaches require a baseline array for normalization. We compared two Qspline methods, QsplineG and QsplineR, which differ only in that the former uses G, while the latter uses R, to construct the baseline array. Both QsplineG and QsplineR are single-bias-removal techniques. The reduction in k-NN *LOOCV* errors for these methods is comparable to, and sometimes significantly better than, the other single-bias-removal methods (Figure 5.7).

However, the Qspline-related approaches seem to be "incompatible" with other spatial-dependent normalization methods—that is, when QsplineG or QsplineR is combined with one of the examined spatial-dependent methods, the performances of the resulting double-bias-removal techniques are often worsened (Figures 5.7 and 5.8). These results may be explained by the following reason: After the log ratios M are normalized using the baseline array, the spatial effect in M is "disturbed," and thus subsequent spatial normalization cannot further improve, but rather worsen, the quality of data. Overall, none of the Qspline-related methods outperforms the two-step strategies that remove intensity effect globally and spatial effect locally (Figure 5.8).

COMPARISON OF SCALE NORMALIZATION STRATEGIES BY K-NN CLASSIFICATION

We have evaluated three scale normalization methods (see Table 5.4 for details): Wscale (which adjusts the scale of M values at the PT level), Bscale (which adjusts the scale of M values globally across all microarrays in a data set), and WBscale (which adjusts the scale locally followed by globally, in two steps). These scale

TABLE 5.4
Extant Scale Normalization Techniques Used in Wu et al. [46]

Name[a]	Description	Bioconductor R package/function (parameters)
Wscale	Within-microarray scale normalization $$s_b = MAD_b \Big/ \sqrt[B]{\prod_{b=1}^{B} MAD_b}$$ $$Ms'_b = M/s_b$$	marray/maNormScale(norm="printTipMAD"; geo=T)
Bscale	Between-microarray scale normalization $$s_i = MAD_i \Big/ \sqrt[I]{\prod_{i=1}^{I} MAD_i}$$ $$Ms'_i = M/s_i$$	marray/maNormScale(norm="globalMAD"; geo=T)
WBscale	Step 1: Within-microarray scale normalization $$s_b = MAD_b \Big/ \sqrt[B]{\prod_{b=1}^{B} MAD_b}$$ $$Ms'_b = M/s_b$$ Step 2: Between-microarray scale normalization $$s_i = MAD_i \Big/ \sqrt[I]{\prod_{i=1}^{I} MAD_i}$$ $$Ms''_i = Ms'_b/s_i$$	marray/maNormScale(norm="printTipMAD"; geo=T)marray/ maNormScale(norm="globalMAD"; geo=T)

Notes: Most abbreviations in the table can be found in detail in the text. The remaining abbreviations are as follows: for a given microarray, MAD_b: median absolute deviation of M values of spots in PT group b of the array; s_b: the scale factor for PT group b; Ms'_b: the Wscale-normalized log ratios for genes in PT group b; Ms'_i: the Bscale-normalized log ratios for genes on array i; Ms''_b: the WBscale-normalized log ratios for genes in PT group b on array i.

[a] We adopted the terminology given in the table to avoid confusion in our discussion in this chapter. Readers should refer to Wu et al. [46] for the names of these methods that are known elsewhere.

normalization methods have been applied to the non-normalized data (Nonorm) and to data that had been normalized using the five intensity and/or spatial normalization methods (sLloess, iGloess, iLloess, iLloess–sLloess, iGloess–sLfilterW7); these methods were selected to represent normalization strategies that remove intensity and/or spatial effects at different levels. Then we compared the three scale normalization methods combined with the five intensity- and/or spatial-dependent normalization methods (18 strategies in all) using diagnostic plots and the $LOOCV$ classification error of k-NNs.

Assessing the Ability to Remove Dye Biases Using Diagnostic Plots

The distributions of the non-normalized M values for microarrays in the five studies reveal scale differences present within (Figure 5.9, left) and between (Figure 5.9, right) microarrays. The Lymphoma and LungCancer data sets show considerable variations in box size and whisker length between microarrays, indicating that severe scale effect exists in these data. Removal of scale effect should result in box size and whisker length with little variations. As expected, Wscale removes local scale effect but does not remove global scale effect (Figure 5.10, iLloess–sLloess + Wscale), whereas the opposite is true for Bscale normalization, which removes global scale effects effectively (Figure 5.10, iLloess–sLloess + Bscale). WBscale removes both local and global scale effect completely (Figure 5.10, iLloess–sLloess + WBscale). Figure 5.10 illustrates the effectiveness of the three scale normalization methods in removing scale effect using data normalized with iLloess–sLloess (which removes both local intensity and spatial effects in two steps); the same results were observed using the non-normalized data or data normalized with the other examined intensity and/or spatial normalization methods.

Functional Assessment Using the k-NN $LOOCV$ Error

Next, we compared the three scale normalization methods combined with the five other intensity- and/or spatial-dependent normalization methods (18 strategies in all) using the k-NN $LOOCV$ classification error. For data normalized first with spatial- and/or intensity-dependent methods, little or no reduction in $LOOCV$ errors was observed when within-microarray scale normalization (Wscale) was applied later (see Wu et al. [46] for details). Furthermore, when between-microarray scale normalization (Bscale) was used alone, or when both scale normalization techniques were used sequentially (WBscale), there was an increase in both the median and the upper quantile ranks (poorer performance, Figure 5.11), suggesting that Bscale should not be applied on the studied data sets.

Summary of k-NN Evaluation of Normalization Methods

Using the $LOOCV$ classification error of k-NNs as the evaluation criterion, we assessed a variety of normalization methods that remove spatial effect, intensity effect, and scale differences from cDNA microarray data. Overall, the single-bias-removal normalization approaches (which remove either spatial or intensity effect from the data) perform better than Gmedian and Nonorm, while the double-bias-removal

A Survey of cDNA Microarray Normalization 113

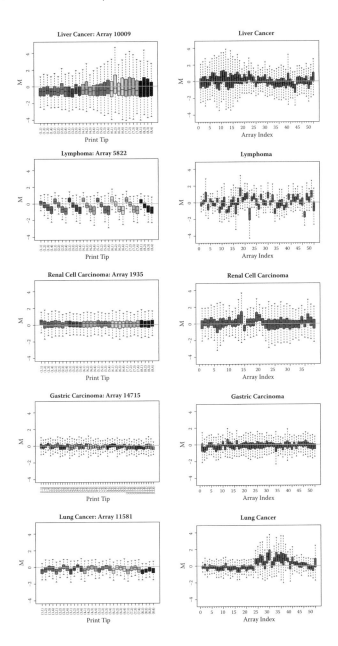

FIGURE 5.9 (See color insert following page 80.) Box plots of the distributions of the non-normalized M values for microarrays in the five studies. Each panel in the left-hand column shows results for M values at the local level of a microarray chosen at random from a given data set. The bars are color coded by PT group. Each panel in the right-hand column shows results for M values at the global level for 50 microarrays chosen at random from a given data set (the total number of microarrays in RenalCellCarcinoma is 38). Each row corresponds to a particular study.

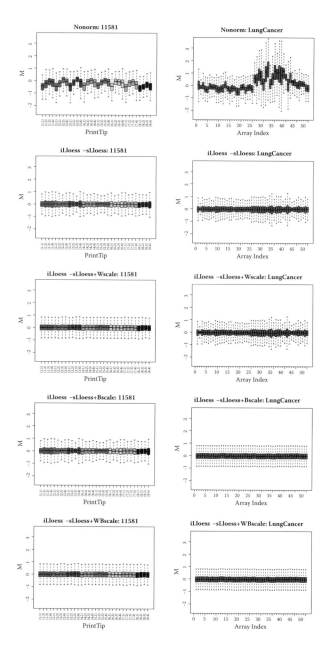

FIGURE 5.10 Box plots of microarrays in the LungCancer data set. The box plots display results for microarray 11581 (left) and 50 microarrays (right) in the LungCancer data set before and after scale normalization designed to remove scale effect. Columns depict the non-normalized M values (Nonorm, also shown in Figure 5.4), normalized M' values (iLloess–sLloess), and normalized Ms' values (iLloess–sLloess +Wscale, +Bscale, +WBscale). Each panel in the left-hand column shows results for M values at the local level of microarray 11581. The bars are color coded by PT group. Each panel in the right-hand column shows results for M values at the global level for the 50 microarrays chosen at random from the LungCancer data set.

A Survey of cDNA Microarray Normalization

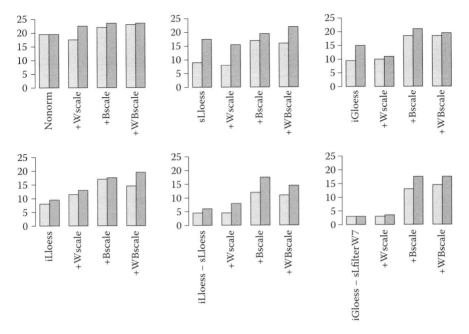

FIGURE 5.11 Rank summary for scale normalization methods. The median ranks and upper quantile ranks are defined as described in Figure 5.8. Mean ranks are shown in light gray; median ranks are shown in dark gray. In each plot, normalization strategies are arranged in the following order: an intensity and/or spatial normalization method, an intensity and/or spatial normalization method followed by Wscale (+Wscale), an intensity and/or spatial normalization method followed by Bscale (+Bscale), an intensity and/or spatial normalization method followed by WBscale (+WBscale).

normalization strategies (which remove both spatial and intensity effects) perform better than the single-bias-removal normalization approaches. Of the 41 different strategies examined, three two-step approaches, all of which remove both intensity effect at the global level and spatial effect at the local level, appear to be the most effective at reducing the *LOOCV* classification error.

The investigated scale normalization methods do not have beneficial effect on classification performance; Bscale and WBscale sometimes even result in an overall increase in *LOOCV* error and ranking (poorer performance, Figure 5.11). These results are not unexpected. It is known that differences in the scales between microarrays can arise from both unwanted technical factors (differences in experimental reagents, equipment, personnel, and so on), as well as from genuine biological variations. The scale normalization techniques applied here aim to remove unwanted technical factors, and assume the existence of little biological variations between samples. The results showing that scale normalization worsens classification performance suggest that in the examined cancer-related data sets, considerable genuine biological variations are present (which is plausible because genomic aberrations found in cancer cells [47,48] can alter the number and nature of expressed genes compared to normal cells), and that these variations are masked by the applied scale normalization.

In summary, our results indicate that intensity and spatial effects have profound impact on downstream data analyses, such as classification, and that removing these effects can improve the quality of such analyses.

FUTURE DIRECTIONS

There are several potential directions that can be pursued regarding the utilization of classification performance as the functional measure for comparing normalization methods. First, error estimators that have low variance, such as bootstrapping and k-fold cross-validation [49,50], are worth further investigation. The *LOOCV* error provides an unbiased estimate of the generalizability of a classifier, especially when the number of the available training samples is severely limited, and is thus highly desirable for model selection or other relevant algorithm evaluation [51,52]. However, the *LOOCV* error is also known to have high variance in some situations [49], which could in turn affect the accuracy of the rankings of the normalization methods. Error estimators that have low variance could be good alternatives for assessing classification performance.

Second, future studies should explore auxiliary techniques, such as feature selection, when normalization methods are evaluated using classification. Many classification algorithms suffer from the curse of dimensionality, or overfitting, in classification tasks using high-dimensional data (e.g., microarray data). To avoid overfitting, dimensionality reduction techniques, such as feature selection, can be used to select a small number of highly informative features for classification. More investigations are thus desirable to understand the interplay between normalization approaches, feature selection techniques, and classification algorithms to ascertain normalization strategies that can lead to an optimal classifier.

CONCLUSIONS

In this chapter, we have discussed major systematic dye effects in cDNA microarray data, normalization strategies that have been commonly used to remove the systematic dye effects from these data, criteria for assessment of normalization methods, and our studies on evaluation of normalization strategies using the *LOOCV* classification error of k-NNs. Our work has shown that intensity- and spatial-dependent dye effects have significant impact on k-NN classification of biological samples; minimizing these dye biases from microarray data can reduce the *LOOCV* error of k-NNs significantly, and thus improve classification performance of k-NNs greatly.

We have shown that different normalization approaches require different biological and statistical assumptions, and the performance of these methods can vary considerably in different biological scenarios. Moreover, different normalization methods may have different impacts on downstream analyses, such as identification of differentially expressed genes, and classification and clustering of data. Therefore, to ensure the accuracy of downstream analyses, it is crucial to choose the most effective and appropriate methods for normalizing data obtained from a specific biological and experimental context for a specific task.

REFERENCES

1. E. M. Southern, DNA chips: Analyzing sequence by hybridization to oligonucleotides on a large scale, *Trends Genet*, 12, 110–115, 1996.
2. G. Ramsay, DNA chips: State of the art, *Nat Biotechnol*, 16, 40–44, 1998.
3. J. Demeter, C. Beauheim, J. Gollub, T. Hernandez-Boussard, H. Jin, D. Maier, J. C. Matese, et al., The Stanford Microarray Database: Implementation of new analysis tools and open source release of software, *Nucleic Acids Res*, 35, D766–770, 2007.
4. R Development Core Team, R: A language and environment for statistical computing, R Foundation for Statistical Computing, 2007.
5. R. C. Gentleman, V. J. Carey, D. M. Bates, B. Bolstad, M. Dettling, S. Dudoit, B. Ellis, et al., Bioconductor: Open software development for computational biology and bioinformatics, *Genome Biol*, 5, R80, 2004.
6. J. Quackenbush, Microarray data normalization and transformation, *Nat Genet*, 32 Suppl 2, 496–501, 2002.
7. T. Aittokallio, M. Kurki, O. Nevalainen, T. Nikula, A. West, and R. Lahesmaa, Computational strategies for analyzing data in gene expression microarray experiments, *J Bioinform Comput Biol*, 1, 541–586, 2003.
8. Y. H. Yang and T. Speed, Design issues for cDNA microarray experiments, *Nat Rev Genet*, 3, 579–588, 2002.
9. G. A. Churchill, Fundamentals of experimental design for cDNA microarrays, *Nat Genet*, 32 Suppl, 490–495, 2002.
10. Y. H. Yang, M. J. Buckley, and T. P. Speed, Analysis of cDNA microarray images, *Brief Bioinform*, 2, 341–349, 2001.
11. L. X. Qin and K. F. Kerr, Empirical evaluation of data transformations and ranking statistics for microarray analysis, *Nucleic Acids Res*, 32, 5471–5479, 2004.
12. Y. H. Yang, S. Dudoit, P. Luu, and T. P. Speed, Normalization for cDNA microarray data, presented at Microarrays: Optical Technologies and Informatics, San Jose, CA, 2001.
13. B. M. Bolstad, R. A. Irizarry, M. Astrand, and T. P. Speed, A comparison of normalization methods for high density oligonucleotide array data based on variance and bias, *Bioinformatics*, 19, 185–193, 2003.
14. W. Wu, N. Dave, G. Tseng, T. Richards, E. P. Xing, and N. Kaminski, Comparison of normalization methods for CodeLink Bioarray data, *BMC Bioinformatics*, 6, 309, 2005.
15. S. Dudoit, Y. H. Yang, M. J. Callow, and T. P. Speed, Statistical methods for identifying differentially expressed genes in replicated cDNA microarray experiments, presented at Tech Report, 578, 2000.
16. Y. Chen, E. R. Dougherty, and M. L. Bittner, Ratio-based decisions and the quantitative analysis of cDNA microarray images, *J Biomed Optics*, 2, 364–374, 1997.
17. S. Attoor, E. R. Dougherty, Y. Chen, M. L. Bittner, and J. M. Trent, Which is better for cDNA-microarray-based classification: Ratios or direct intensities? *Bioinformatics*, 20, 2513–2520, 2004.
18. M. B. Eisen, P. T. Spellman, P. O. Brown, and D. Botstein, Cluster analysis and display of genome-wide expression patterns, *Proc Natl Acad Sci USA*, 95, 14863–14868, 1998.
19. A. Zien, T. Aigner, R. Zimmer, and T. Lengauer, Centralization: A new method for the normalization of gene expression data, *Bioinformatics*, 17, 323S–331S, 2001.
20. T. C. Kroll and S. Wolfl, Ranking: a closer look on globalization methods for normalization of gene expression arrays, *Nucleic Acids Res*, 30, e50, 2002.
21. M. K. Kerr, M. Martin, and G. A. Churchill, Analysis of variance for gene expression microarray data, *J Comput Biol*, 7, 819–837, 2000.
22. R. D. Wolfinger, G. Gibson, E. D. Wolfinger, L. Bennett, H. Hamadeh, P. Bushel, C. Afshari, and R. S. Paules, Assessing gene significance from cDNA microarray expression data via mixed models, *J Comput Biol*, 8, 625–637, 2001.

23. K. F. Kerr, C. A. Afshari, L. Bennett, P. Bushel, J. Martinez, N. J. Walker, and G. A. Churchill, Statistical analysis of a gene expression microarray experiment with replication, *Statistica Sinica*, 12, 203–217, 2002.
24. G. C. Tseng, M. K. Oh, L. Rohlin, J. C. Liao, and W. H. Wong, Issues in cDNA microarray analysis: quality filtering, channel normalization, models of variations and assessment of gene effects, *Nucleic Acids Res*, 29, 2549–2557, 2001.
25. T. B. Kepler, L. Crosby, and K. T. Morgan, Normalization and analysis of DNA microarray data by self-consistency and local regression, *Genome Biol*, 3, RESEARCH0037, 2002.
26. W. S. Cleveland and S. J. Devlin, Locally weighted regression: An approach to regression analysis by local fitting, *J Am Stat Assoc*, 83, 596–610, 1988.
27. W. S. Cleveland and C. Loader, Smoothing by local regression: principles and methods, in *Statistical theory and computational aspects of smoothing*. New York: Springer, 1996, pp. 10–49.
28. M. Futschik and T. Crompton, Model selection and efficiency testing for normalization of cDNA microarray data, *Genome Biol*, 5, R60, 2004.
29. D. L. Wilson, M. J. Buckley, C. A. Helliwell, and I. W. Wilson, New normalization methods for cDNA microarray data, *Bioinformatics*, 19, 1325–1332, 2003.
30. M. J. Bissell, The differentiated state of normal and malignant cells or how to define a "normal" cell in culture, *Int Rev Cytol*, 70, 27 100, 1981.
31. S. W. Chua, P. Vijayakumar, P. M. Nissom, C. Y. Yam, V. V. Wong, and H. Yang, A novel normalization method for effective removal of systematic variation in microarray data, *Nucleic Acids Res*, 34, e38, 2006.
32. Y. H. Yang, S. Dudoit, P. Luu, D. M. Lin, V. Peng, J. Ngai, and T. P. Speed, Normalization for cDNA microarray data: a robust composite method addressing single and multiple slide systematic variation, *Nucleic Acids Res*, 30, e15, 2002.
33. J. Huang, H. C. Kuo, I. Koroleva, C. H. Zhang, and M. B. Soares, A semilinear model for normalization and analysis of cDNA microarray data, statistics and actuarial science, The University of Iowa, Iowa City 321, October 2003.
34. J. Fan, P. Tam, G. V. Woude, and Y. Ren, Normalization and analysis of cDNA microarrays using within-array replications applied to neuroblastoma cell response to a cytokine, *Proc Natl Acad Sci USA*, 101, 1135–1140, 2004.
35. J. Fan, H. Peng, and T. Huang, Semilinear high-dimensional model for normalization of microarray data: A theoretical analysis and partial consistency, *J Am Stat Assoc*, 100, 781–796, 2005.
36. J. Huang, D. Wang, and C. H. Zhang, A two-way semilinear model for normalization and analysis of cDNA Microarray Data, *J Am Stat Assoc*, 100, 814–829, 2005.
37. C. Workman, L. J. Jensen, H. Jarmer, R. Berka, L. Gautier, H. B. Nielser, H. H. Saxild, et al., A new nonlinear normalization method for reducing variability in DNA microarray experiments, *Genome Biol*, 3, research0048, 2002.
38. S. Dudoit and Y. H. Yang, Bioconductor R packages for exploratory analysis and normalization of cDNA microarray data, in *The analysis of gene expression data: Methods and software*, G. Parmigiani, E. S. Garrett, R. A. Irizarry, and S. L. Zeger, eds. New York: Springer, 2003, pp. 73-101.
39. X. Cui, M. K. Kerr, and G. A. Churchill, Transformations for cDNA microarray data, *Stat Appl Genet Molec Biol*, 2, Article 4, 2003.
40. T. Park, S. G. Yi, S. H. Kang, S. Lee, Y. S. Lee, and R. Simon, Evaluation of normalization methods for microarray data, *BMC Bioinformatics*, 4, 33, 2003.
41. R. Shippy, T. J. Sendera, R. Lockner, C. Palaniappan, T. Kaysser-Kranich, G. Watts, and J. Alsobrook, Performance evaluation of commercial short-oligonucleotide microarrays and the impact of noise in making cross-platform correlations, *BMC Genomics*, 5, 61, 2004.

42. X. Wang, M. J. Hessner, Y. Wu, N. Pati, and S. Ghosh, Quantitative quality control in microarray experiments and the application in data filtering, normalization, and false positive rate prediction, *Bioinformatics*, 19, 1341–1347, 2003.
43. L. M. Cope, R. A. Irizarry, H. A. Jaffee, Z. Wu, and T. P. Speed, A benchmark for Affymetrix GeneChip expression measures, *Bioinformatics*, 20, 323–331, 2004.
44. R. Irizarry and L. Cope, Bioconductor expression assessment tool for Affymetrix oligonucleotide arrays (affycomp), 2004.
45. Y. Barash, E. Dehan, M. Krupsky, W. Franklin, M. Geraci, N. Friedman, and N. Kaminski, Comparative analysis of algorithms for signal quantitation from oligonucleotide microarrays, *Bioinformatics*, 20, 839–846, 2004.
46. W. Wu, E. P. Xing, C. Myers, I. S. Mian, and M. J. Bissell, Evaluation of normalization methods for cDNA microarray data by k-NN classification, *BMC Bioinformatics*, 6, 191, 2005.
47. J. W. Gray and C. Collins, Genome changes and gene expression in human solid tumors, *Carcinogenesis*, 21, 443–452, 2000.
48. D. G. Albertson, C. Collins, F. McCormick, and J. W. Gray, Chromosome aberrations in solid tumors, *Nat Genet*, 34, 369–376, 2003.
49. U. Braga-Neto and E. R. Dougherty, Bolstered error estimation, *Pattern Recognition*, 37, 1267–1281, 2004.
50. G. J. McLachlan, K. Do, and C. Ambroise, *Analyzing microarray gene expression data,* Wiley-Interscience, New York, 2004.
51. V. Vapnik and O. Chapelle, Bounds on error expectation for SVM, in *Advances in large margin classifiers (neural information processing series)*, A. J. Smola, P. J. Bartlett, B. Scholkopf, and D. Schuurmans, eds., MIT Press, Cambridge, MA, 2000, pp. 261–280.
52. O. Chapelle, V. Vapnik, O. Bousquet, and S. Mukherjee, Choosing multiple parameters for support vector machines, *Mach Learning*, 46, 131–159, 2002.
53. X. Chen, S. T. Cheung, S. So, S. T. Fan, C. Barry, J. Higgins, K. M. Lai, et al., Gene expression patterns in human liver cancers, *Molec Biol Cell*, 13, 1929–1939, 2002.
54. A. A. Alizadeh, M. B. Eisen, R. E. Davis, C. Ma, I. S. Lossos, A. Rosenwald, J. C. Boldrick, et al., Distinct types of diffuse large B-cell lymphoma identified by gene expression profiling, *Nature*, 403, 503–511, 2000.
55. J. P. Higgins, R. Shinghal, H. Gill, J. H. Reese, M. Terris, R. J. Cohen, M. Fero, et al., Gene expression patterns in renal cell carcinoma assessed by complementary DNA microarray, *Am J Pathol*, 162, 925–932, 2003.
56. X. Chen, S. Y. Leung, S. T. Yuen, K. M. Chu, J. Ji, R. Li, A. S. Chan, et al., Variation in gene expression patterns in human gastric cancers, *Mol Biol Cell*, 14, 3208–3215, 2003.
57. M. E. Garber, O. G. Troyanskaya, K. Schluens, S. Petersen, Z. Thaesler, M. Pacyna-Gengelbach, M. van de Rijn, et al., Diversity of gene expression in adenocarcinoma of the lung, *Proc Natl Acad Sci USA*, 98, 13784–13789, 2001.

6 Technical Variation in Modeling the Joint Expression of Several Genes

Walter Liggett

CONTENTS

Abstract 121
Introduction 122
Measurement Error Model 124
Results for CodeLink Platform 126
    Normalization 126
    Linearity 127
    Measurement Error 128
    Interlaboratory Variation 132
Results for Affymetrix Platform 136
Discussion 141
Statistical Analysis 142
    Normalization 142
    Linearity 144
    Measurement Error 146
    Interlaboratory Variation 147
Disclaimer 149
References 149

ABSTRACT

Biological studies based on expression microarrays often depend crucially on comparison of expression measurements that are subject to technical variation associated with changes in time, laboratory, or other measurement conditions. If a biological study involves different measurement conditions, then an experiment should be performed to gauge the associated technical variation and to compare it with the technical variation evident in replicate measurements made under the same measurement conditions. Such an experiment might be similar in design to the interlaboratory study performed in the Microarray Quality Control (MAQC) project.

This chapter shows how one can infer, from MAQC-like data, the technical variation to be expected for materials of immediate biological interest. Using single-platform data from the MAQC project, we build a model of technical variation that includes assay-to-assay normalization, microarray response linearity, variance stabilization, variances of individual genes and covariances between pairs of genes, and site-to-site variation in microarray response.

INTRODUCTION

As their contribution to biological research, mRNA microarrays provide simultaneous measurements of the expression levels of thousands of genes, which in turn provide the basis for building biological models of the joint expression of groups of genes. Generally, expression changes with biological condition are of interest, and assays corresponding to different biological conditions are performed (Rocke 2004). Choice of an approach to the statistical analysis of the assay results begins with the study design—that is, with the assignment of assays to biological conditions and to biological replicates. Among other considerations in the choice of approach are assumptions about the technical variation. The purpose of this chapter is to show how one can develop the information on the technical variation that is necessary for study design and choice of statistical analysis.

An important example of the need to model the technical variation arises in evaluation of the expression of each gene individually. Consider the use of analysis of variance (ANOVA) (Wolfinger et al. 2001; Cook et al. 2007). The ANOVA is usually based on a simple model of the technical variation—namely, that the measurement errors in the log transformed intensity levels are independent and have the same variance. This assumption may lead to incorrect inferences when, for example, the assays are performed in two different laboratories. Sources of technical variation associated with differences in the two laboratories cause the independence assumption to be violated. Obviously, the worst case would be assays for one biological condition performed in one laboratory and assays for the other condition performed in another laboratory. Irizarry et al. (2005) discuss problems caused by interlaboratory variation. This chapter provides another perspective.

Gene expression studies often involve changed measurement conditions such as different laboratory, different month of measurement, different microarray manufacturing batch, and different reagent lot. Because a microarray assay is a high dimensional measurement, the effect of changed measurement conditions has various manifestations. The actual mechanisms through which these differences affect assay results are of interest but may not be easily discovered. Such changed conditions are part of the definition of reproducibility, but not of repeatability (ISO 1993). In the case of a univariate measurement, the effect of changed measurement conditions is usually portrayed in terms of variances. Comparison of repeatability and reproducibility by means of variances is called gauge R&R (Burdick, Borror, and Montgomery 2005). Greater complexity is associated with the high dimensional case. For example, gene expression studies often involve the generation of gene lists, which might vary in complex ways under reproducibility conditions (Qiu et al. 2006). The high dimensional case also provides more possibilities for understanding the technical variation linked to changed measurement conditions.

In a gene expression study, comparison of assays performed under different measurement conditions may be unavoidable. For example, measurement of an individual and remeasurement of the same individual 6 months later may be essential to the goals of the study. Obviously, if one could perform all the assays under repeatability conditions, one would. The question addressed in this chapter is, "How does one assess the particular type of reproducibility needed for one's study?" Assessing reproducibility is a commonly recommended first step in industrial experiments (Hahn et al. 1999). As an answer in the gene expression case, this chapter suggests performing a reproducibility study like the MAQC project (MAQC Consortium 2006), but with the particular change in measurement conditions relevant to the study substituted for the MAQC change of laboratories. This chapter provides a guide to the analysis of the results of such a reproducibility study. The guide is illustrated with data from the MAQC project.

Our approach is to develop a model of technical variation in expression levels—that is, a measurement error model. The purpose of this model is to provide answers to three questions. The first question is whether the particular type of reproducibility needed for the biological study is sufficiently good. This question is related to the question of whether improvements in the assay protocols are needed before the biological study can proceed. The second question is whether there are unsuspected attributes of the technical variation that must be considered in design and analysis of the biological study. An example is the perhaps surprising amount of interlaboratory variation evident in the MAQC data analyzed in this chapter. Ignoring this variation could lead to incorrect study conclusions.

The third question involves quantification of the effect of technical variation on complicated functions of the assay results such as derived gene lists. For univariate measurements, the process of inferring uncertainties from replicate measurements is familiar, and the range of measured values that might be observed is usually expressed as a plus-or-minus quantity. In the case of expression assays, both the biological context and the measurement error model are more complex. For this reason, only simulation based on the measurement error model may be feasible. Consider, for example, an approach to forming a gene list. Starting with hypothesized gene expression levels and the measurement error model, one might simulate the process of measuring a material with the hypothesized levels and then compile a collection of gene lists that shows how the technical variation causes observed gene lists to vary. Simulation is a demanding application of the measurement error model.

This chapter proposes a measurement error model for expression levels produced by one-color microarrays and argues for the validity of this model in the case of two of the microarray platforms included in the MAQC project. For the uses of our measurement error model, the most important aspect of validity is valid extrapolation from the four MAQC materials to the materials of interest in the biological study. Measurement error studies usually face this challenge of extrapolating from reference materials to materials of actual interest. The arguments for validity in this chapter involve two statistical methods. The first is a method for comparison of a group of univariate test statistics, one for each gene, with the probability distribution these statistics would have were the model to be valid (Venables and Ripley 2002). The second is a method based on discriminant analysis (Gnanadesikan 1997). This method focuses attention on the linear combinations of expression levels that best distinguish variation under reproducibility conditions from variation under repeatability conditions.

In the next section, we specify our measurement error model with enough mathematical detail to allow judging the adequacy of the model. There are two sections on results. In the first, we investigate the applicability of our measurement error model to the MAQC measurements obtained with the CodeLink platform. In the second, we perform the same investigation using the MAQC measurements obtained with the Affymetrix platform. Next, we discuss the use of the characterizations in the results sections. Finally, we show how to perform an analysis such as those in the results sections.

MEASUREMENT ERROR MODEL

Consider the MAQC assays performed with a single platform. With this platform, four materials were each measured five times at each of three sites for a total of 60 assays. The mRNA levels in the four materials are related in that two original materials were mixed in known proportions to form the other two materials (Shippy et al. 2006). The original materials are labeled A and B, and the mixtures are labeled C and D. In this section, we specify a model for measurements from such a design.

As defined in the International Vocabulary of Metrology (ISO 1993), the measurement error is the difference between the measured quantity value and the true quantity value. The true value quantities, which we call model intensities, have only two properties that are essential to the modeling. First, for assays of the same material, the model intensities are the same. Second, the mathematical relations among mRNA concentrations that are the consequence of mixing materials A and B to obtain materials C and D apply to the model intensities. Although the data analysis establishes an overall intensity scale, which scale is established is not important because calibration curves for individual mRNA species are not available. If they were, analytical chemists would prefer identification of a true value quantity with a true concentration and identification of a measured quantity value with a concentration derived by means of a calibration curve.

We equate measured quantity values with observed intensities that have been normalized. In keeping with general experience in gene expression data analysis, normalization is part of our measurement error model. However, not every normalization approach is appropriate in connection with the two properties we expect the model intensity to have. As detailed in the statistical analysis section, our normalization approach, which we call parametric normalization, preserves the mixing relations. Parametric normalization was proposed by Liggett (2006).

Our approach consists of a linear transformation of all the intensities of an assay. Thus, after normalization, the intensity for probe g obtained in assay i performed at site l is given by

$$\frac{y_{ig}(l) - \eta_{0i}(l)}{\eta_i(l)},$$

where $y_{ig}(l)$ is the observed intensity, and $\eta_{0i}(l)$ and $\eta_i(l)$ are the normalization coefficients for assay i performed at site l. Compared to some normalization methods that have been proposed, this normalization allows only a comparatively restricted transformation of the intensities observed in an assay (W. Wu et al. 2005; Irizarry, Wu, and Jaffee 2006).

The 20 assays performed at a single site, which consist of five hybridizations of material A, five of material B, five of material C, and five of material D, respectively, are indexed by i. Because materials C and D are mixtures of materials A and B, we can write the model intensities as

$$\mu_{ig} = x_{Ai}\theta_{Ag} + x_{Bi}\theta_{Bg}, \tag{6.1}$$

where for materials A, B, C, and D, x_{Ai} equals 1, 0, γ_C, or $(1 - \gamma_D)$, respectively, and x_{Bi} equals 0, 1, $(1 - \gamma_C)$, or γ_D, respectively. This equation represents the model intensities in terms of θ_{Ag}, the model intensity for material A, and θ_{Bg}, the model intensity for material B. The amounts of RNA material mixed to form materials C and D were in 3 to 1 and 1 to 3 ratios. Were the proportion of mRNA in materials A and B to be the same, we would have $\gamma_C = \gamma_D = 0.75$. However, these proportions are not exactly equal. Based on our analysis of the MAQC data and a complementary analysis (Shippy et al. 2006), we have adopted $\gamma_C = 0.80$ and $\gamma_D = 0.69$ as values for the mixture fractions.

The measurement error model is given by

$$\frac{y_{ig}(l) - \eta_{0i}(l)}{\eta_i(l)} = \mu_{ig} + \alpha_{ig}(l) + \sqrt{\mu_{ig}^2 + \delta^2}\left(\lambda_g(l)f_i(l) + \sigma_g(l)e_{ig}(l)\right). \tag{6.2}$$

The second and third terms on the right-hand side of this equation represent the measurement error. The second term contains no randomness, but the third term does. The term $\alpha_{ig}(l)$ is a gene-specific and site-specific deviation in the model intensity parameters θ_{Ag} and θ_{Bg}. We assume that $\alpha_{ig}(l)$ satisfies

$$\alpha_{ig}(l) = x_{Ai}\phi_{Ag}(l) + x_{Bi}\phi_{Bg}(l) \tag{6.3}$$

for some $\phi_{Ag}(l)$ and $\phi_{Bg}(l)$. Note that as i varies, the term $\alpha_{ig}(l)$ varies with material but not with replicate for the same matrial. Essentially, the term $\alpha_{ig}(l)$ portrays differences between sites in terms of differences in the linear calibration curves that connect normalized intensities with species concentrations. The third term contains a multiplier $\sqrt{\mu_{ig}^2 + \delta^2}$ that depicts increase in the measurement error variance with the intensity. The quantity $\lambda_g(l)f_i(l)$ portrays a source of technical variation that affects the intensities for many genes. We take the $\lambda_g(l)$ to be parameters and the $f_i(l)$ to be independent random variables with mean 0 and variance 1. The source of variation produces a value of $f_i(l)$, and this value affects the intensity for gene g according to the value of $\lambda_g(l)$. The quantity $\sigma_g(l)e_{ig}(l)$ represents the independent error as composed of independent error with mean 0 and variance 1, denoted $e_{ig}(l)$, and a coefficient that portrays dependence of the variance of the independent measurement error on the gene targeted. For a particular site, the quantities $\lambda_g(l)f_i(l)$ and $\sigma_g(l)e_{ig}(l)$ depict measurement error for an assay as having a covariance matrix that is neither diagonal nor constant along the main diagonal.

The main question addressed in the remainder of this chapter is whether equation 6.2 is a suitable model for two of the MAQC platforms. There are three facets to this question. First is whether this model contains unnecessary terms such as provision for a source of variation that affects the expression intensities for many genes. Second is whether this model is oversimplified. Perhaps the interlaboratory differences require a more complicated representation. Third is whether this model leads to simulation that is useful in the design of gene expression studies. In a simulation, the model intensity μ_{ig} would be hypothesized. In addition, simulation requires values for the measurement error parameters shown in equation 6.2, which are δ^2, $\varphi_{Ag}(l)$, $\varphi_{Bg}(l)$, $\lambda_g(l)$, and $\sigma_g(l)$. This chapter addresses estimation of these parameters. Values of f_i and e_{ig} are obtained from a random number generator. Because there are only three sites for each platform, using estimates of the site dependent measurement error parameters to answer questions about what one might observe at a fourth site is problematic. One could treat the normalization parameters $\eta_{0i}(l)$ and $\eta_i(l)$, which we estimate for the data, as random and thereby simulate un-normalized intensities to which another normalization method could then be applied.

RESULTS FOR CODELINK PLATFORM

The model given in equations 6.1–6.3 provides a basis for summarizing the technical variation exhibited by the MAQC measurements obtained with the CodeLink Human Whole Genome platform. GE Healthcare arranged for these measurements. Applied Microarrays, Inc. currently offers this platform. We summarize these measurements in four steps. First, we normalize the intensities originally given by the platform software, thereby reducing the effects of technical variation but without distorting the relations among the intensities for the different materials. Second, for sites 1 and 3, we show that there are only a few genes that exhibit strong evidence of a nonlinear relation to concentration in the material and that these genes saturate in keeping with the Langmuir isotherm model. Third, we confirm the necessity of all aspects of the measurement error model given in equation 6.2. Fourth, because there are only three sites, we do not develop a probabilistic model of the interlaboratory variation but just display the differences among sites using discriminant coordinates.

Normalization

Our results describe a subset of the genes targeted by the CodeLink platform. We consider only the 12,091 genes selected in the MAQC project, although we could consider more of the genes targeted by this platform because our purpose does not include interplatform comparisons. We make use of two columns in the original data tables, the one labeled "Normalized_intensity" and the other labeled "Quality_flag." Starting with a background corrected spot intensity, the software for this platform applies an array-by-array scaling to produce "Normalized_intensity" values. Of the

12091 genes, we select only those designated "G" or "L" for all 60 assays. This reduces the number of genes in our subset to 9378. The mechanisms responsible for "Quality_flag" values other than "G" or "L" are not included in our characterization of the technical variation.

We rescale the "Normalized_intensity" values on the basis of these 9378 genes. This array-by-array rescaling makes no difference in the analysis except as it governs the proper choice of δ^2 because our final normalization again readjusts the array-by-array scaling. The range of "Normalized_intensity" values is [−0.28, 306.64]. After our rescaling, the range is [−0.075, 65.478]. The rescaled intensities for our subset of genes are denoted $y_{ig}(l)$.

The normalization consists of estimating $\eta_{0i}(l)$ and $\eta_i(l)$ and then computing the normalized intensities as shown on the left-hand side of equation 6.2. If we knew the values of θ_{Ag} and θ_{Bg}, then we could estimate the normalization coefficients using weighted regression with a somewhat simplified version of equation 6.2. (See the statistical analysis section.) Similarly, if we knew the values of $\eta_{0i}(l)$ and $\eta_i(l)$, then we could estimate the model intensities. As detailed in the statistical analysis section, our normalization method consists of alternatively performing these two regressions until the estimated coefficients converge. What can be seen from this brief description is that the normalization accounts for the differences among the four materials.

Our normalization method gives the estimates $\hat{\eta}_{0i}(l)$, $\hat{\eta}_i(l)$, $\hat{\theta}_{Ag}$, and $\hat{\theta}_{Bg}$. The range of $\hat{\eta}_{0i}(l)$ is [−0.092, 0.058], and the range of $\hat{\eta}_i(l)$ is [0.89, 1.20]. The scaling of the model intensities is such that median($\hat{\mu}_{ig}$) = 0.98.

LINEARITY

For each site, the question of linearity involves the combination of equations 6.1 and 6.3

$$\mu_{ig} + \alpha_{ig}(l) = x_{Ai}\left(\theta_{Ag} + \varphi_{Ag}(l)\right) + x_{Bi}\left(\theta_{Bg} + \varphi_{Ag}(l)\right). \tag{6.4}$$

This equation shows that there are effectively two unknown parameters for each site. Because there are two unknown parameters and four materials, each of which was assayed in replicate, we can test the fit of equation 6.4 as a way of testing for linearity. We do this separately for each site and each gene. The result of this is a ratio of variances for each site and each gene.

Under the hypothesis that equation 6.4 holds, each of these ratios can be compared to the F distribution with 2 and 16 degrees of freedom as a way of testing for linearity. We would like to make this comparison for each site separately and all the genes at once. One way to do this is by means of a quantile–quantile plot. For each site, such a plot is shown in Figure 6.1. Points above the reference line indicate lack of fit. For sites 1 and 3, lack of fit is indicated in a relatively small group of genes at the top of the plots. For site 2, there seems to be a more general lack of fit.

Figure 6.2 suggests the cause of lack of fit for some genes. In each panel, Figure 6.2 shows a modification of the volcano plot often used for gene expression measurements. The top three panels portray the genes for which $\hat{\theta}_{Ag} > \hat{\theta}_{Bg}$, and the bottom three portray the genes for which $\hat{\theta}_{Ag} \leq \hat{\theta}_{Bg}$. On the top, the horizontal axis shows the standardized residual for the mean of the assays on material A, and on

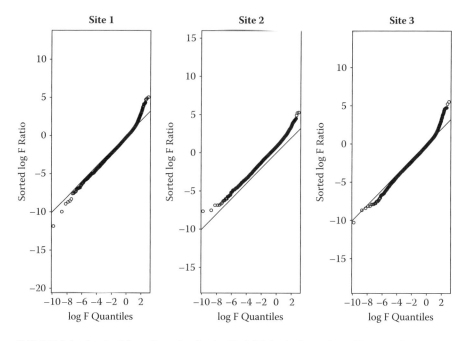

FIGURE 6.1 Lack of fit to linearity for the CodeLink platform. Quantile–quantile plots of the observed F ratio versus the distribution under linearity.

the bottom, the standardized residual for the mean of the assays on material B. The vertical axis shows the p value. For site 1, we see that the genes with lack of fit as indicated by a small p value have standardized residuals that are largely negative. This is what one would expect in the case in which the concentration of the species were so large as to cause saturation as predicted by the Langmuir isotherm (Chudin et al. 2006; Burdin, Pittelkow, and Wilson 2004). Thus, for site 1, we see that saturation largely causes the lack of fit. The same conclusion can be reached for site 3 in Figure 6.2.

On the other hand, there is apparently another cause for lack of fit in the case of site 2. We note that the site 2 measurements considered here are those obtained in a second round, which occurred some months after the site 1 and 3 measurements were made.

We note finally that the saturated genes are mostly the same for sites 1 and 3. This fact provides some confirmation of our observation about saturation causing lack of fit.

MEASUREMENT ERROR

Our analysis of the measurement error is based on the weighted residuals given by

$$z_{ig}(l) = \frac{\left(y_{ig}(l) - \hat{\eta}_{0i}(l)\right)/\hat{\eta}_i(l) - x_{Ai}\hat{\theta}_{Ag} - x_{Bi}\hat{\theta}_{Bg}}{\sqrt{(x_{Ai}\hat{\theta}_{Ag} + x_{Bi}\hat{\theta}_{Bg})^2 + \delta^2}} \tag{6.5}$$

Technical Variation in Modeling the Joint Expression of Several Genes 129

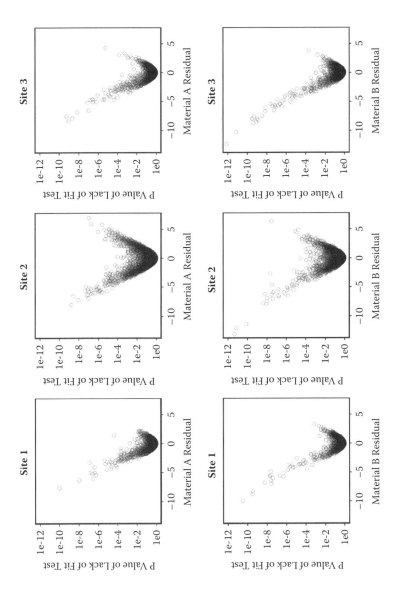

FIGURE 6.2 Volcano plots showing saturation for some genes for the CodeLink platform. Observed F ratio versus standardized residual for material with largest expression.

We analyze these residuals using the model given by

$$z_{ig}(l) = \alpha_{ig}(l)\Big/\sqrt{\hat{\mu}_{ig}^2 + \delta^2} + \lambda_g(l)f_i(l) + \sigma_g(l)e_{ig}(l), \tag{6.6}$$

which is part of the measurement error term in equation 6.2. Note that $\lambda_g(l)$ and $\sigma_g(l)$ depend on site but not assay within site. As discussed in the statistical analysis section, we have an estimate of $\alpha_{ig}(l)$. Using this estimate, we can characterize just the last two terms of equation 6.6 through analysis of

$$z_{ig}(l) - \hat{\alpha}_{ig}(l)\Big/\sqrt{\hat{\mu}_{ig}^2 + \delta^2}.$$

Consider first the fact that the replicates as portrayed in the weighted residuals should have variances that do not depend on the intensity level. For each site, we consider the ratio of the pooled variance for materials A and C to the pooled variance for materials D and B. Were the variances to be independent of the intensity level, these ratios would be F distributed with degrees of freedom eight and eight. For sites 1 and 3, these ratios show little discrepancy from intensity level independence. These ratios for site 2 show considerable discrepancy. This is more evidence that the site 2 measurements are different.

For each site, consider the averages of

$$z_{ig}(l) - \hat{\alpha}_{ig}(l)\Big/\sqrt{\hat{\mu}_{ig}^2 + \delta^2}$$

the materials, which we denote by $\bar{z}_{Ag}(l)$, $\bar{z}_{Bg}(l)$, $\bar{z}_{Cg}(l)$, and $\bar{z}_{Dg}(l)$. Plots of these averages for site 1 are shown in Figure 6.3. To make this figure more easily interpreted, we eliminated the genes that show saturation for site 1 and in addition seven genes that appear to be outliers. Most striking about Figure 6.3 are the high strong negative relationships between materials A and C and between materials B and D. There is also a clear positive relationship between materials A and B. Apparently, there is a dependence of the weighted residuals on the material. We note that this same dependence is evident for other sites and platforms.

Much of what is exhibited in Figure 6.3 is due to the fitting. For each gene, there are four residuals that must satisfy two relations as the result of the fact that two parameters were fit to the original data. Figure 6.3 shows that for materials A and C, the residual patterns (+, −) and (−, +) are favored. Moreover, for materials D and B, the same two residual patterns are favored. On the other hand, for materials C and D, no pattern is favored. Thus, for all the materials in (A, C, D, B) order, the favored patterns are (+, −, −, +), (−, +, +, −), (+, −, +, −), and (−, +, −, +). The other 12 patterns are made less likely by the fitting as can be seen by inspecting them. For example, the pattern (−, −, +, +) is unlikely because this pattern will generally not be consonant with a properly fitting slope.

The term $\lambda_g f_i$ consists of random variation from assay to assay portrayed by f_i that affects genes to a larger or smaller extent depending on the values of λ_g, a property of the measurement system. We want to estimate λ_g in a way that is not

Technical Variation in Modeling the Joint Expression of Several Genes 131

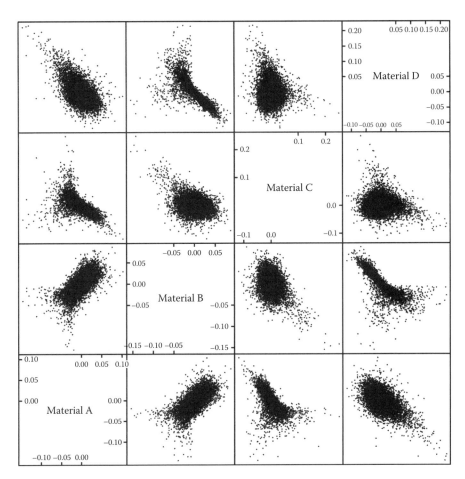

FIGURE 6.3 Relation among materials for the CodeLink platform at site 1. Scatter plot matrix for material means of the weighted and corrected residuals.

influenced by the material-to-material dependence shown in Figure 6.3. The covariance matrix corresponding to the terms $\lambda_g f_i + \sigma_g e_{ig}$ can be estimated from the variation in the replicates for each material. For material A, the weighted residuals z_{ig} centered at their mean give an estimate of this covariance matrix with four degrees of freedom. Pooling this estimate with similar estimates from the replicates of the other materials gives an estimate of the covariance matrix with 16 degrees of freedom. The eigenvector of this matrix corresponding to the largest eigenvalue provides values proportional to λ_g. Using these values to estimate $\lambda_g f_i$ and then changing the scaling so that

$$\sum_i f_i^2 = 16$$

gives λ_g itself.

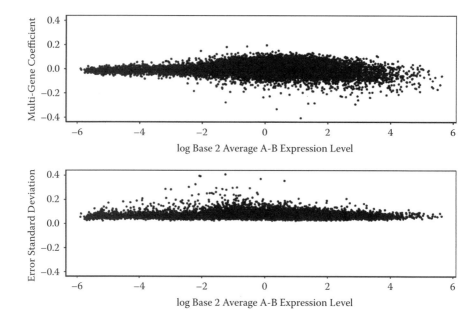

FIGURE 6.4 Repeatability covariance matrix for the CodeLink platform at site 1. The gene-by-gene effects of the multigene source of variation and the standard deviation of the independent error.

Figure 6.4 shows λ_g plotted versus logarithmic values of $(\hat{\theta}_{Ag} + \hat{\theta}_{Bg})/2$. The sizes of the λ_g suggest a source of technical variation that affects many genes substantially and that induces both positive and negative gene-to-gene correlation. This figure shows no dependence of λ_g on expression level. Looking at all the non-zero eigenvalues of the covariace matrix suggests that the term $\lambda_g f_i$ is a necessary part of the model for site 1. The necessity of this term for site 3 is less clear. There are a variety of other methods for judging the importance of the term $\lambda_g f_i$, which also show its necessity for site 1.

The independent error is represented by the term $\sigma_g e_{ig}$. Having corrected z_{ig} for the material means and for $\lambda_g f_i$, the effect of the multigene source, we can estimate σ_g^2. Figure 6.4 also shows σ_g. Again we see no dependence on expression level. Moreover, we see that λ_g and σ_g are comparable in size. Because the estimation error varies from gene to gene, one must be careful in concluding that σ_g^2 itself varies from gene to gene. Were σ_g^2 to be constant, then the estimates would be distributed proportional to the chi-square distribution. That they are not shows that there is real variation in σ_g^2.

INTERLABORATORY VARIATION

We have developed a probabilistic model of the effects of technical variation within a laboratory, but because there are only three sites in the MAQC data set, we cannot develop a probabilistic model of the interlaboratory differences. Rather, we just display the differences among the three sites.

There are two cases: interlaboratory differences between expression level measurements and interlaboratory differences between (within-laboratory) differential expression measurements. Equation 6.2 already provides a model for expression level differences. Consider a differential expression measurement formed from normalized intensities from assays i and j from a particular site. When the expression levels are large enough that the quantity δ^2 is negligible, the following approximation applies:

$$\frac{(y_{ig}(l) - \hat{\eta}_{0i}(l))/\hat{\eta}_i(l)}{(y_{jg}(l) - \hat{\eta}_{0j}(l))/\hat{\eta}_j(l)} \approx \frac{x_{Ai}\hat{\theta}_{Ag} + x_{Bi}\hat{\theta}_{Bg}}{x_{Aj}\hat{\theta}_{Ag} + x_{Bj}\hat{\theta}_{Bg}} \left(1 + z_{ig}(l) - z_{jg}(l)\right) \qquad (6.7)$$

In this subsection, we consider only genes for which $\hat{\theta}_{Ag} > 1$ or $\hat{\theta}_{Bg} > 1$. Equation 6.7 shows that any part of z_{ig} that is constant from material to material within a site influences the interlaboratory differences in the case of expression level measurements but not in the case of differential expression measurements.

Comparison of reproducibility with repeatability consists of comparison of the variability of $z_{ig}(l)$ among sites with its variability within sites. We extend this to the comparison of its variability among sites and materials with its variability among replicates. Consider the following question: Is it possible to discriminate among sites and materials on the basis of the $z_{ig}(l)$? In other words, given a new value of $z_{ig}(l)$ from some site and material, can we tell the site and material to which it corresponds? If such discrimination is possible, then we can conclude that the reproducibility is substantially worse than the repeatability.

Display of the differences among sites requires an approach to dealing with the large number of genes involved in the differences. Our approach is to compute particular linear combinations of the $z_{ig}(l)$ that are called discriminant coordinates. Discriminant coordinates arise from applying linear discriminant analysis to the $z_{ig}(l)$. Our discrimination goal is to separate the 12 combinations of materials and sites from each other in a way that is optimal in light of the replicate-to-replicate variation. Discriminant coordinates are a reduction of the $z_{ig}(l)$ to the combinations on which the best discriminator is based. Using discriminant coordinates to plot the 60 assays provides an image of the clustering by material and site if there is a way to view the data such that clustering exists.

The discriminant coordinates that we derive are based on penalized linear discriminant analysis. The present situation requires penalized linear discriminant analysis because the number of replicates is much smaller than the number of genes. As detailed in the statistical analysis section, such analysis requires specification of a penalty parameter. Depending on the choice of this parameter, the discriminant coordinates can make the reproducibility more or less like the repeatability. We base the choice of this parameter on cross-validation (Hastie, Tibshirani, and Friedman 2001). Cross-validation results in a choice of penalty parameter that optimizes the performance of the classifier in its ability to properly classify an observation that is not used in the derivation of the classifier.

Figure 6.5 shows the first two discriminant coordinates for the z_{ig}. Of the 12 available discriminant coordinates, these two provide the most effective discrimination of the material–site combinations. We see that the sites are well separated and much better separated than the four materials. Liggett (2006) made a similar

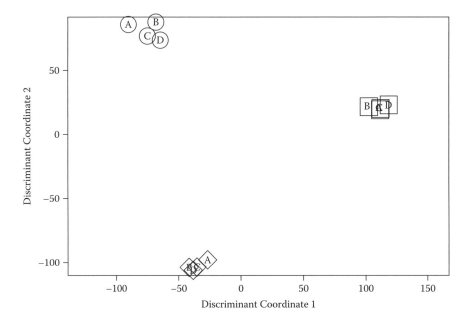

FIGURE 6.5 Comparison of expression level reproducibility with repeatability for the CodeLink platform. Site 1: square; site 2: circle; site 3: diamond. Penalty parameter = 0.25.

observation. This reflects the sizes of the $\alpha_{ig}(l)$ in equation 6.6. This figure shows that the comparison of expression levels among sites has interlaboratory differences as an obstacle.

To display the interlaboratory differences for differential expression measurement, we again compute discriminant coordinates, but this time from the result of centering the $z_{ig}(l)$ for each site. Figure 6.6 shows the first two discriminant coordinates for the centered $z_{ig}(l)$. The differences between the materials show the effect of the interlaboratory differences on differential expression measurement. Consider differential expression between materials A and B. For site 1 (square), B is greater than A in both coordinates. For site 2 (circle), Figure 6.6 shows that A is greater than B in coordinate 1. For site 3 (diamond), A is greater than B in coordinate 2. Apparently, the sites give differential expression results that differ in ways that are not particularly simple. Moreover, we see that the replicates, which are almost indistinguishable, do not account for the site-to-site differences between the material differences.

Consider, finally, discriminant coordinates for $z_{ig}(l)$ corrected for $\alpha_{ig}(l)$. Figure 6.7 shows the first two discriminant coordinates. Although this figure shows some material-to-material differences, it shows almost no site-to-site differences. This leads us to the conclusion that the differences among sites are largely contained in the $\alpha_{ig}(l)$. In other words, for each gene, the sites differ largely in linear calibration curves that connect the species concentrations and the observed intensities. For differential expression, the implication is that the sites could be brought into substantial agreement if an approach to adjustment of the gene-specific calibration curves could be found. The material-to-material differences are largely due to the dependence among residuals induced by the fitting as discussed in connection with Figure 6.3.

Technical Variation in Modeling the Joint Expression of Several Genes 135

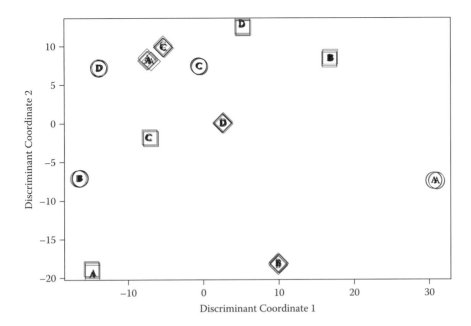

FIGURE 6.6 Comparison of differential expression reproducibility with repeatability for the CodeLink platform. Site 1: square; site 2: circle; site 3: diamond. Penalty parameter = 1.0.

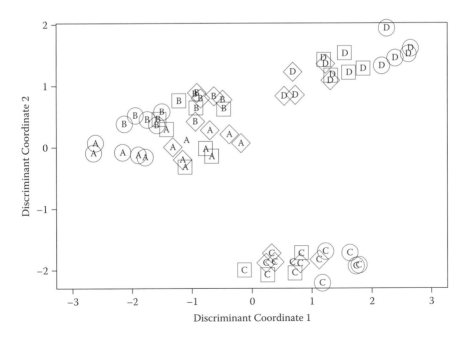

FIGURE 6.7 Comparison of corrected differential expression reproducibility with repeatability for the CodeLink platform. Site 1: square; site 2: circle; site 3: diamond. Penalty parameter = 64.0.

Note that the material-to-material differences in Figure 6.7 are larger than the replicate-to-replicate differences, but not much larger.

RESULTS FOR AFFYMETRIX PLATFORM

The model fitting approach presented in the previous section can be adapted to the case of a multiple probe platform. In this section, we apply such an adaptation to MAQC measurements made with the Affymetrix HG-U133 Plus 2.0 GeneChip. We confine our attention to the perfect match probes. Obviously, these probes alone provide insight into the technical variation. Our choices in this section should not be interpreted as an opinion on the usefulness of the mismatch probes. See Wu and Irizarry (2005) for further discussion of the mismatch probes.

The Affymetrix platform contains 133,839 perfect match probes corresponding to the 12,091 genes selected in the MAQC project. Just as with the 9378 genes of the previous section, we can, in principle, estimate the normalization parameters and model intensities for these Affymetrix probes. To ease the computational burden, we estimate the normalization parameters using only a sample of the probes—namely, 14,871 probes. Having estimated the normalization parameters, we can normalize all the observed intensities and then, from these normalized intensities, estimate the model intensities using interative reweighted least squares. Thus, as in the previous section, we obtain the estimates $\hat{\eta}_{0i}(l)$, $\hat{\eta}_i(l)$, $\hat{\theta}_{Agp}$, and $\hat{\theta}_{Bgp}$, where the subscript p indicates probe p of gene g.

To obtain gene indices from the probe values, we use weighted means, which is possible because we have the estimates $\hat{\theta}_{Agp}$ and $\hat{\theta}_{Bgp}$ from which we can compute weights. This approach is related to the one suggested by Zhou and Rocke (2005). As weights for finding gene indices, we adopt

$$w_{gp} = 1 \bigg/ \left[\left(\frac{1}{20} \sum_{i=1}^{20} x_{Ai} \hat{\theta}_{Agp} + x_{Bi} \hat{\theta}_{Bgp} \right)^2 + \delta^2 \right],$$

where $\delta^2 = 0.25$. Because these weights are independent of the assay, their use does not disturb the linearity. In place of the normalized intensities of the previous section, we use

$$u_{ig}(l) = \frac{\sum_p w_{gp} \left((y_{igp}(l) - \hat{\eta}_{0i}(l)) / \hat{\eta}_i(l) \right)}{\sum_p w_{gp}}.$$

The corresponding model intensity is given by

$$\hat{\mu}_{ig} = \frac{\sum_p w_{gp} \left(x_{Ai} \hat{\theta}_{Agp} + x_{Bi} \hat{\theta}_{Bgp} \right)}{\sum_p w_{gp}}.$$

As regression weights for estimation based on the $u_{ig}(l)$, we use

$$w_{ig} = \left(\sum_p w_{gp} \right)^2 \bigg/ \left(\sum_p w_{gp}^2 \left((x_{Ai} \hat{\theta}_{Agp} + x_{Bi} \hat{\theta}_{Bgp})^2 + \delta^2 \right) \right).$$

Technical Variation in Modeling the Joint Expression of Several Genes

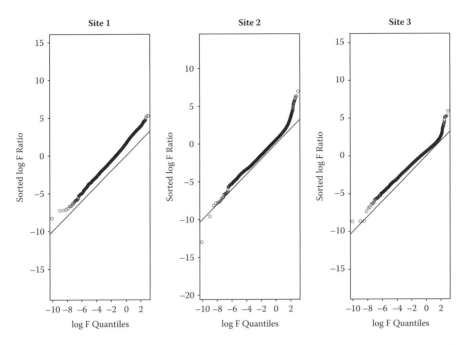

FIGURE 6.8 Lack of fit to linearity for the Affymetrix platform. Quantile–quantile plots of the observed F ratio versus the distribution under linearity.

In equation 6.2, the factor that portrays the increase in the error variance with expression level is given by $1/\sqrt{w_{ig}}$. It replaces $\sqrt{\mu_{ig}^2 + \delta^2}$. We now proceed as in the previous section with $u_{ig}(l)$ instead of the normalized observed intensities, with $\hat{\mu}_{ig}$ computed differently as the estimated model intensities, and with w_{ig} as the weights. These weights do not have the same relation to the estimated model intensities as in the previous section.

We check the linearity for each site individually as in the previous section. Note that in computing $u_{ig}(l)$, we choose weights that are independent of the array-site index so that the linearity is not corrupted by the weighting. Figure 6.8 is comparable to Figure 6.1. There are two aspects to deviations from the reference line. One is general separation of the points from the line, and the other is deviation from the reference line for high lack-of-fit ratios. In terms of general separation, site 1 is the most removed. Sites 2 and 3 exhibit noticeable deviation at the upper end. Note that Figures 6.1 and 6.8 do not provide a legitimate platform comparison. The reason is that separation from the reference line increases both with lack of linearity, which increases the numerator, and with better repeatability, which decreases the denominator.

Figure 6.9 shows volcano plots comparable to those shown in Figure 6.2. We see that, for sites 2 and 3, most of the genes that exhibit lack of fit are genes that seem to exhibit saturation. On the other hand, in the case of site 1, many genes exhibit lack of linearity that cannot be ascribed to saturation.

As in the previous section, we estimate λ_g and σ_g on the basis of the z_{ig} centered at the material means. For site 2, Figure 6.10 shows λ_g plotted versus logarithmic values of $(\hat{\theta}_{Ag} + \hat{\theta}_{Bg})/2$. This figure shows no dependence of λ_g on expression level.

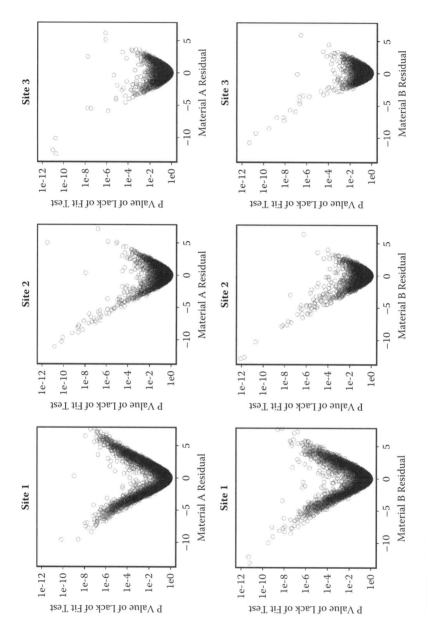

FIGURE 6.9 Volcano plots showing saturation for some genes for the Affymetrix platform. Observed F ratio versus standardized residual for material with largest expression.

Technical Variation in Modeling the Joint Expression of Several Genes

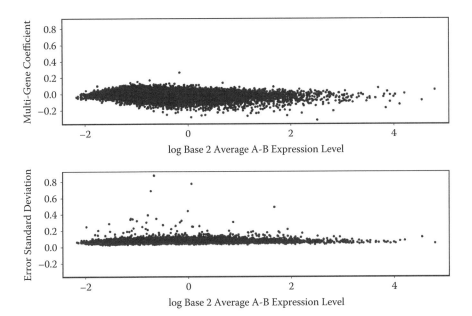

FIGURE 6.10 Repeatability covariance matrix for the Affymetrix platform at site 2. The gene-by-gene effects of the multigene source of variation and the standard deviation of the independent error.

Looking at all the nonzero eigenvalues of the covariance matrix suggests that the term $\lambda_g f_i$ is a necessary part of the model for site 2. Figure 6.10 also shows σ_g. Again we see no dependence on expression level. Moreover, we see that λ_g and σ_g are comparable in size.

Figure 6.11, which is comparable to Figure 6.6, shows the lack of reproducibility in within-site differential expression measurement. As in the case of Figure 6.6, Figure 6.11 is based on discriminant analysis of the z_{ig} centered for each site. The error for a particular site in the differential expression of material A with respect to material B is given by the position of the A points with respect to the B points. We see that the relative position of the A points with respect to the B points is different for each site. This implies the complexity of the interlaboratory variation in differential expression measurement. Figure 6.11 largely reflects the term $\alpha_{ig}(l)$ in equation 6.2. One question about the mismatch probes is whether they can be used to reduce the size of this term.

Removing $\hat{\alpha}_{ig}(l)$ from z_{ig} and performing discriminant analysis leads to Figure 6.12, which is comparable to Figure 6.7. In this figure, the replicates are easily distinguished, which suggests that the material–site differences are not extreme relative to the variation upon replication. The four materials from site 1 (square) have the largest separation. Note that the differences between site 1 and the other two sites are not eliminated by removing $\hat{\alpha}_{ig}(l)$. Apparently, the interlaboratory variability in differential expression measurement cannot be characterized strictly by the size of $\hat{\alpha}_{ig}(l)$. Something else is involved.

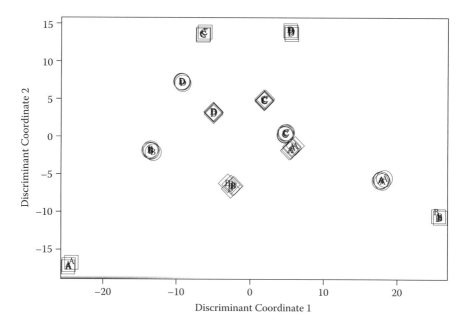

FIGURE 6.11 Comparison of differential expression reproducibility with repeatability for the Affymetrix platform. Site 1: square; site 2: circle; site 3: diamond. Penalty parameter = 3.0.

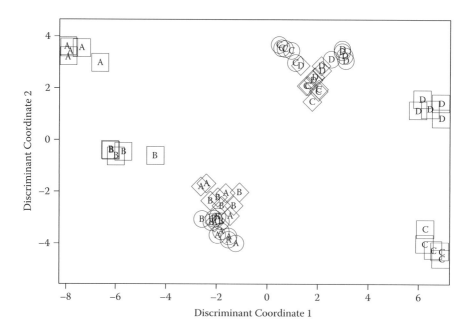

FIGURE 6.12 Comparison of corrected differential expression reproducibility with repeatability for the Affymetrix platform. Site 1: square; site 2: circle; site 3: diamond. Penalty parameter = 16.0.

DISCUSSION

How to model data from single-platform MAQC-like experiments is the subject of this chapter. One aspect of the modeling proposed is integration in the sense that the modeling involves various ways of looking at the data that can be compared and reconciled. The modeling incorporates a normalization method designed to avoid distorting the relations among the different materials and to be simple enough that its effects at later stages of the analysis can be judged. The modeling separates assessment of the mean of the measurements from assessment of their variability. Assessment of the mean involves measurement linearity, and assessment of the variability involves modeling the measurement covariance matrix. The modeling is based on viewing the data on a gene-by-gene basis and in terms of the most remarkable perspective. Gene-by-gene analysis leads to plots such as Figures 6.1–6.4 and Figures 6.8–6.10. Notable perspectives on the data are provided by discriminant coordinates as shown in Figures 6.5–6.7 and Figures 6.11 and 6.12. The measurement error model given in equation 6.2 serves to integrate all these ways of looking at the data.

Another aspect of the modeling proposed is its usefulness in the context of a biological experiment. Think about the usefulness of a MAQC-like study for which the reproducibility dimension is relevant to the biological experiment. It might be useful to perform a MAQC-like study in preparation for a biological experiment because such an experiment can be seriously compromised by unforeseen measurement system variation that occurs during the data collection. Say that the biological experiment depends on adequate reproducibility and that a MAQC-like study shows that the reproducibility is poor. The next step might be to tighten up the measurement protocols to improve the reproducibility. Consider the illustration provided in this chapter. Apparently, the problem is not deviations of the microarray responses from linearity but that the microarray responses differ from site to site. In our modeling, these differences are captured in the term denoted $\alpha_{ig}(l)$. Thus, the next step would be adjustments to reduce the size of $\alpha_{ig}(l)$.

There is another aspect to the usefulness of a MAQC-like study. It might be true that even though the reproducibility is poor relative to the repeatability—as shown, for example, in discriminant coordinate plots, the response to the factors in the biological experiment might be large compared to the lack of reproducibility. See Larkin et al. (2005) for an example of this. In this case, improving the reproducibility might be beneficial but not necessary. However, taking the lack of reproducibility into account may still be necessary. Say that the goal of the biological experiment is specification of a classifier—for example, a classifier intended to distinguish cases from controls. Specifying such a classifier depends not only on the differences among the classes but also on the variation within the classes. Lack of reproducibility might lead to a false picture of variation within the classes. As an extreme example, consider the possibility that all the cases are measured at one site and all the controls at another site. Were this to be the case, the lack of reproducibility would not be reflected in the within-class variability. However, in the application of the classifier, the lack of reproducibility would affect classifier performance. As a partial remedy to this, one might design the biological experiment so that specimens in each class are measured under each set of measurement conditions. In the MAQC case, cases and controls should be measured at each site.

Another use of a MAQC-like study is simulation on the basis of the resulting measurement error model. Think first about simulation of the intrasite variation. Equation 6.2 contains two sources of random variation, f_i and e_{ig}. Estimation of the various parameters in equation 6.2, which is possible for some but not all sites, is discussed in the next section. One would like to perform a simulation for hypothesized values of the model intensities μ_g. If one were to assume that the measurement error parameters do not depend on the model intensities, then one could substitute μ_g for $\hat{\mu}_{ig}$, substitute estimates of the measurement error parameters obtained from the MAQC-like study, and perform the simulation.

Simulation of the effect of different measurement conditions, such as the effect of site in the illustration in this chapter, requires some guess work. Consider the site-to-site variation in differential expression shown in Figures 6.6 and 6.11. Except for the case of Affymetrix site 1, this variation seems largely due to the measurement error term $\alpha_{ig}(l)$. This term portrays site-to-site differences in the linear calibration curves that relate concentrations to intensities. Such a linear calibration curve can be characterized by its intercept, which is related to the non-specific hybridization, and by its slope, which is related to the specific hybridization. Variation from site-to-site in the non-specific hybridization affects the differential expression by shifting both the numerator and denominator. To simulate site-to-site variation in differential expresison, one would have to characterize site-to-site variation in non-specific hybridization as well as its relation to specific hybridization. This characterization is likely complicated by dependence of $\alpha_{ig}(l)$ on the material.

STATISTICAL ANALYSIS

This section details the four steps in the decomposition of single-platform MAQC measurements. The first step is normalization, which involves weighted regression applied alternately in two different ways. The second step is investigation of linearity, which involves testing for lack of fit of a regression model and computation of standardized residuals. The third step is characterizaton of the measurement error through application of multivariate analysis to vectors, each of which contains single-assay results on all the targeted genes. The fourth step is summarization of the additional contribution of inter-laboratory variation through regularized linear discriminant analysis with cross validation.

NORMALIZATION

There are sources of technical variation in microarray expression measurement that cause superficial differences between assays. The first step in the statistical analysis is removal of these differences without affecting actual differences due to differences between the materials. This step is called normalization, and our approach to normalization we call parametric normalization because it depends on a parametric model linking the model intensities for the different materials.

We begin with estimation of the quantity δ^2 shown in equation 6.2. We follow the procedure given by Rocke and Durbin (2001). For each site, material, and gene we can compute the variance of y_{ig} or $\log(y_{ig})$ from the five replicates. According to equation 6.2, such variance estimates reflect variation in

$$\eta_{0i}(l) + \eta_i(l)\sqrt{\mu_{ig}^2 + \delta^2}\left(\lambda_g(l)f_i(l) + \sigma_g(l)e_{ig}(l)\right).$$

If we let the normalization parameters satisfy $\eta_{0i} = 0$ and $\eta_i = 1$ and we ignore the dependence of $\lambda_g^2(l) + \sigma_g^2(l)$ on the gene, then we can assume that the y_{ig} are described by the model given by Rocke and Durbin (2001), $\mu e^H + E$, where H and E are normally distributed with mean 0 and variances σ_H^2 and σ_E^2, respectively. On the basis of a group of genes that are low expressing ($\mu \approx 0$) for all four materials, we can estimate σ_E^2, and on the basis of a group of genes that are high expressing ($\mu \gg 1$) for all four materials, we can estimate σ_H^2 through use of a logarithmic transformation. Rocke and Durbin (2001) show that

$$\delta^2 = \sigma_E^2 / \left(\exp(\sigma_H^2)(\exp(\sigma_H^2) - 1)\right).$$

For the Codelink platform, we did this computation separately for each site and then chose $\delta^2 = 0.1$ as suitable for all three sites. The CodeLink platform section provides a check on this value of δ^2 and thereby shows that the dubious assumptions behind our adoption of the Rocke and Durbin (2001) model did not create a problem. For the Affymetrix platform, we estimated δ^2 at the probe level and chose $\delta^2 = 0.25$ as suitable for all three sites.

Our normalization algorithm is based on the model

$$\frac{y_{ig}(l) - \eta_{0i}(l)}{\eta_i(l)} = \mu_{ig} + \sqrt{\mu_{ig}^2 + \delta^2}\left(\sigma\, e_{ig}(l)\right),$$

which is a simplification of equation 6.2. We use this model to estimate $\eta_{0i}(l)$, $\eta_i(l)$, θ_{Ag}, and θ_{Bg}. Comparing this equation to equation 6.2, we see that some terms in the measurement error model have been omitted. By the use of this model, the term $\alpha_g(l)$ influences the estimates of θ_{Ag} and θ_{Bg}. These estimates become estimates of $\theta_{Ag} + (1/3)\sum_l \varphi_{Ag}(l)$ and $\theta_{Bg} + (1/3)\sum_l \varphi_{Bg}(l)$. Otherwise, we depend on the fact that small deviations from this simplified error model do not affect regression estimates very much.

Our algorithm for parametric normalization is a combination of iterative reweighted least squares (Carroll and Ruppert 1988) and criss-cross regression (Gabriel and Zamir 1979; Liu et al. 2003). To obtain initial values for the weights, we estimate the model intensities θ_{Ag} and θ_{Bg} by regressing x_{Ai} and x_{Bi} on the unnormalized intensities $y_{ig}(l)$ using no weights. We let $\hat{\mu}_{ig} = \max(x_{Ai}\hat{\theta}_{Ag} + x_{Bi}\hat{\theta}_{Bg}, 0)$ and compute as initial weights $(\hat{\mu}_{ig}^2 + \delta^2)^{-1}$. Using these weights, we re-estimate the model intensities to obtain refined values for weights.

The main iteration in our algorithm follows: First, we use the current estimate of the model intensities to estimate the normalization coefficients η_{0i} and η_i. These estimates are obtained by fitting the model

$$\eta_{0i} + \eta_i(x_{Ai}\hat{\theta}_{Ag} + x_{Bi}\hat{\theta}_{Bg})$$

to the un-normalized intensities $y_{ig}(l)$ with weighted least squares based in the weights $\hat{\eta}_i^{-2}(\hat{\mu}_{ig}^2 + \delta^2)^{-1}$ computed from the previous values of η_i, θ_{Ag}, and θ_{Bg}. Second, we renormalize the intensities

$$\frac{y_{ig} - \hat{\eta}_{0i}}{\hat{\eta}_i}$$

and use these to re-estimate the model intensities. We re-estimate the model intensities θ_{Ag} and θ_{Bg} by regressing x_{Ai} and x_{Bi} on the normalized intensities using the weights $(\hat{\mu}_{ig}^2 + \delta^2)^{-1}$, where $\hat{\mu}_{ig} = \max(x_{Ai}\hat{\theta}_{Ag} + x_{Bi}\hat{\theta}_{Bg}, 0)$ is computed from the previous values of θ_{Ag} and θ_{Bg}. We then begin the next iteration by re-estimating the normalization.

The remainder of our analysis is based on the observed intensities normalized with the parameters $\hat{\eta}_{0i}(l)$ and $\hat{\eta}_i(l)$. In addition, we use the estimates $\hat{\theta}_{Ag}$ and $\hat{\theta}_{Bg}$ to provide evaluation of

$$\sqrt{\mu_{ig}^2 + \delta^2}$$

so that we can apply weighted least squares.

LINEARITY

For each site, we evaluate the fit of equation 6.4. With the dependence on site suppressed, the equation that is the basis of this analysis is

$$\frac{y_{ig} - \hat{\eta}_{0i}}{\hat{\eta}_i} = \mu_{ig} + \alpha_{ig} + \sqrt{\hat{\mu}_{ig}^2 + \delta^2}\left(\sigma_g e_{ig}\right),$$

where $\mu_{ig} + \alpha_{ig}$ is given by equation 6.4. The quantity $\sqrt{\hat{\mu}_{ig}^2 + \delta^2}$, which leads to the regression weights, is evaluated at $\hat{\theta}_{Ag}$, $\hat{\theta}_{Bg}$ and is therefore known. All that is missing from this model is the term $\lambda_g f_i$. For each individual gene, inclusion of this term would have no effect beyond substitution of $\sqrt{\lambda_g^2 + \sigma_g^2}$ for σ_g and therefore would not change the validity of weighted least squares. However, the term $\lambda_g f_i$ recognizes a gene-to-gene dependence that might be of concern in the interpretation of the lack-of-fit tests for many genes.

We regress x_{Ai} and x_{Bi} on the normalized intensities using the weights $(\hat{\mu}_{ig}^2 + \delta^2)^{-1}$, where $\hat{\mu}_{ig} = \max(x_{Ai}\hat{\theta}_{Ag} + x_{Bi}\hat{\theta}_{Bg}, 0)$ is computed from the established estimates of θ_{Ag} and θ_{Bg}. This gives estimates of $\theta_{Ag} + \varphi_{Ag}$ and $\theta_{Bg} + \varphi_{Bg}$. Subtracting the established estimates of θ_{Ag} and θ_{Bg} and $\hat{\theta}_{Ag}$ and $\hat{\theta}_{Bg}$ from these estimates gives estimates of φ_{Ag} and φ_{Bg}, which we denote by $\hat{\varphi}_{Ag}$ and $\hat{\varphi}_{Bg}$. Corresponding to these estimates is the estimate $\hat{\alpha}_{ig}$.

To assess the linearity of the probe responses after parametric normalization, we determine how well the mixture model fits the normalized measurements. Our measure of lack of fit is a weighted sum of squared deviations, which we compute for each gene. For a particular laboratory, let the normalized intensities for gene g be denoted by u_{ig}, $i = 1, \ldots, 20$. The averages of the five normalized intensities for each material deviate from the values computed with the mixture model in accordance with

Technical Variation in Modeling the Joint Expression of Several Genes

$$d_{Ag} = \bar{u}_{Ag} - \hat{\theta}_{Ag} - \hat{\varphi}_{Ag}$$
$$d_{Bg} = \bar{u}_{Bg} - \hat{\theta}_{Bg} - \hat{\varphi}_{Bg}$$
$$d_{Cg} = \bar{u}_{Cg} - \gamma_C\left(\hat{\theta}_{Ag} + \hat{\varphi}_{Ag}\right) - (1-\gamma_C)\left(\hat{\theta}_{Bg} + \hat{\varphi}_{Bg}\right)$$
$$d_{Dg} = \bar{u}_{Dg} - (1-\gamma_D)\left(\hat{\theta}_{Ag} + \hat{\varphi}_{Ag}\right) - \gamma_D\left(\hat{\theta}_{Bg} + \hat{\varphi}_{Bg}\right).$$

Note that these deviations reflect departure of the normalized probe responses from linearity. Lack of fit for each probe is indicated by the weighted sum of squares of these deviations

$$w_{Ag}d_{Ag}^2 + w_{Bg}d_{Bg}^2 + w_{Cg}d_{Cg}^2 + w_{Dg}d_{Dg}^2,$$

where the weights w_{Ag}, w_{Bg}, w_{Cg}, and w_{Dg} are inversely proportional to $\hat{\mu}_{ig}^2 + \delta^2$ for the appropriate value of i. (Recall that μ_{ig} depends on the material but not the particular replicate for a material.)

One way to interpret our deviations is to compare them with what would be expected were the mixture model to fit perfectly (Draper and Smith 1981). We obtain a standard deviation from the five normalized intensities for a material and denote these by s_{Ag}, s_{Bg}, s_{Cg}, and s_{Dg}. The comparison involves the ratio of the sum of squared deviations with $2\hat{\sigma}_g^2/5$, where

$$\hat{\sigma}_g^2 = \frac{1}{4}\left[w_{Ag}s_{Ag}^2 + w_{Bg}s_{Bg}^2 + w_{Cg}s_{Cg}^2 + w_{Dg}s_{Dg}^2\right].$$

The scale factor 2/5 depends on specifics of the MAQC design and the number of unknown parameters in the mixture model. If the mixture model fits, this ratio is F distributed—that is, distributed as the ratio of two independent variances. The numerator and denominator degrees of freedom are 2 and 16.

To complete this interpretation of lack of fit, we combine the F values from all the genes into a distribution and compare this observed distribution with the F distribution for 2 and 16 degrees of freedom. We make this comparison with a quantile–quantile plot (Venables and Ripley 2002). Each point on this plot corresponds to a gene. The y value is the natural logarithm of the F value for this gene. Thus, a y value corresponds to an observed quantile. The corresponding x value is obtained from the ranking of this F value among all the F values. The x value is obtained as follows: First, find the proportion of all the F values that are less than the particular F value. Let this proportion be m/n, where n is the total number of genes. Second, find the value such that the cumulative F distribution for this value is $(m + 0.5)/n$. The logarithm of this value is the x value for the gene. Thus, an x value corresponds to a quantile hypothesized on the basis of perfect fit to the F distribution. In addition to the points, there is the x = y line. If there were no lack of fit, the points would lie on this line.

We would like to assess the individual values of d_{Ag} for genes that have material A expression level greater than material B expression level. The reason is that if such a gene exhibits a negative value of d_{Ag}, then we can hypothesize that the lack of fit is due to saturation. Similar reasoning applies to genes that have material B expression level greater than material A expression level.

To compare the values of d_{Ag} and d_{Bg}, we standardize them—that is, divide each by its standard deviation. The formulas for standardization are best presented in matrix notation. Let W_g be the 20 by 20 matrix

$$W_g = diag\left(\left(\hat{\mu}_{ig}^2 + \delta^2\right)^{-1}\right),$$

and let X be the 20 by 2 matrix

$$X = \begin{bmatrix} x_{Ai} & x_{Bi} \end{bmatrix}.$$

From these two matrices, we can compute

$$H_g = X(X'WX)^{-1}X^TW.$$

Let the diagonal element of H_g corresponding to material A be denoted by h_{Ag}. We obtain for material A the standardized deviations

$$d_{Ag}\Big/\sqrt{(1-h_{Ag})\hat{\sigma}_g^2/(5w_{Ag})}.$$

Standardized deviations for the other materials can be obtained similarly.

Measurement Error

Properties of the measurement error we characterize by analyzing the weighted residuals $z_{ig}(l)$, which are computed using equation 6.5. As in the previous subsection, we consider each site separately. With the dependence on the site suppressed, the model for the weighted residuals is given by

$$z_{ig} = \frac{\alpha_{ig}}{\sqrt{\hat{\mu}_{ig}^2 + \delta^2}} + \lambda_g f_i + \sigma_g e_{ig}.$$

We treat the first term as fixed and the other two terms as random. We defer discussion of the first term until the next subsection and consider here the last two terms.

For a particular gene, f_i and e_{ig} are independent random variables with mean 0 and variance 1. Because of this, we have

$$\text{Var}(z_{ig}) = \lambda_g^2 + \sigma_g^2.$$

We can obtain two independent estimates of this variance, one from materials A and C and the other from materials D and B. Comparing these two estimates provides a test of an aspect of the measurement error model, the inclusion of the factor

$$\sqrt{\hat{\mu}_{ig}^2 + \delta^2}.$$

Technical Variation in Modeling the Joint Expression of Several Genes

Under the null hypothesis, the ratio of these two estimates is F distributed with eight degrees of freedom in numerator and denominator. These F ratios can be inspected by means of a quantile-quantile plot.

One might expect the α_{ig} term in the model for z_{ig} in the next section. One might expect that the values of z_{ig} corrected for this term would not depend on the material. To check this, we compute for material A:

$$\bar{z}_{Ag} = (1/5) \sum_{i=1}^{5} \left(z_{ig} - \hat{\alpha}_{ig} \Big/ \sqrt{\hat{\mu}_{ig}^2 + \delta^2} \right)$$

and, similarly, $\bar{z}_{Bg}(l)$, $\bar{z}_{Cg}(l)$, and $\bar{z}_{Dg}(l)$. Plotting these pair by pair shows an apparent dependence on material that has already been discussed.

To estimate the terms $\lambda_g f_i + \sigma_g e_{ig}$, we center the replicates for each material and form an estimate of the covariance matrix. Let $\mathbf{Z}_A$ be the 5 row matrix with terms

$$z_{ig} - \bar{z}_{Ag} - \hat{\alpha}_{ig} \Big/ \sqrt{\hat{\mu}_{ig}^2 + \delta^2}$$

and similarly $\mathbf{Z}_B$, $\mathbf{Z}_C$, and $\mathbf{Z}_D$. The pooled covariance matrix

$$\hat{\Sigma} = \left(\mathbf{Z}_A^T \mathbf{Z}_A + \mathbf{Z}_B^T \mathbf{Z}_B + \mathbf{Z}_C^T \mathbf{Z}_C + \mathbf{Z}_D^T \mathbf{Z}_D \right) / 16,$$

which has 16 degrees of freedom, has, as its first eigenvector, a vector proportional to λ_g. To find this eigenvector, one has to work around the fact that this covariance matrix is large. Let

$$\mathbf{Z} = \left(\mathbf{Z}_A^T \quad \mathbf{Z}_B^T \quad \mathbf{Z}_C^T \quad \mathbf{Z}_D^T \right)^T,$$

and let $\mathbf{UDV}^T$ be its singular value decomposition. Let $\mathbf{R}_A$ be rows 1–5 of $\mathbf{UD}$, let $\mathbf{R}_B$ be rows 6–10, let $\mathbf{R}_C$ be rows 11–15, and $\mathbf{R}_D$ be rows 16–20. To compute λ_g, we find the first eigenvalue and eigenvector of $(\mathbf{R}_A^T \mathbf{R}_A + \mathbf{R}_B^T \mathbf{R}_B + \mathbf{R}_C^T \mathbf{R}_C + \mathbf{R}_D^T \mathbf{R}_D)/16$, premultiply the eigenvector by $\mathbf{V}$ and multiply each component of the result by the square root of the eigenvector. To compute σ_g^2, we subtract λ_g^2 from diagonal element g of $\hat{\Sigma}$.

INTERLABORATORY VARIATION

Characterization of the measurement error $z_{ig}(l)$ is one way to gain insight into interlaboratory variation. Rewriting equation 6.3 leads to the following relation between the normalized intensities and the measurement error:

$$\frac{(y_{ig}(l) - \hat{\eta}_{0i}(l))}{\hat{\eta}_i(l)} = \hat{\mu}_{ig} \left(1 + \frac{\sqrt{\hat{\mu}_{ig}^2 + \delta^2}}{\hat{\mu}_{ig}} z_{ig}(l) \right) \approx \hat{\mu}_{ig} \left(1 + z_{ig}(l) \right).$$

The approximation holds if δ^2 is small in comparison to $\hat{\mu}^2_{ig}$. We choose genes for our characterization with this in mind. We choose genes for which $\hat{\theta}_{Ag} > 1$ and $\hat{\theta}_{Bg} > 1$.

A characterization of interest is a comparison of the interlaboratory (and intermaterial) variation in $z_{ig}(l)$ with replicate single-laboratory (and single-material) variation. Ideally, the site-to-site and material-to-material variations are not greater than the variation exhibited by replicates. This comparison is not easily made because of the high dimension of assays, the range of the index g. We can make this comparison by posing the following classification problem: How well can the 12 site-material groups be distinguished on the basis of the $z_{ig}(l)$?

The characterization can be displayed graphically through the use of discriminant coordinates. It turns out that the contribution of each assay to the classification problem posed can be reduced from the number of original values to 12 values; the number of values equals the number of groups. This transformation defines a 12-dimensional discriminant space. In this space, the Euclidean difference between two assays gives the difference between the assays. If distances between replicate assays in the same site-material group are as large as distances between assays in different groups, then we can conclude that technical variation associated with different sites or materials does not affect the assays very much. Looking at points in 12 dimensions is still something of a challenge. Fortunately, we can start by looking at the points in the first two dimensions because discriminant coordinates are ordered according to their contribution to classification performance.

An algorithm for computing the discriminant coordinates that are part of penalized discriminant analysis is given by Hastie, Buja, and Tibshirani (1995) and Hastie and Tibshirani (2004). We present here the equations that specify the computation. An explanation of the mathematics behind the algorithm can be found in the references. The computation for the case of expression level begins with the application of the singular value decomposition to

$$z_{ig} - \frac{1}{60}\sum_{i=1}^{60} z_{ig}.$$

Combining the first two factors of the singular value decomposition gives a 60 by 60 matrix, which we denote $\mathbf{R}$. The entire decomposition is given by $\mathbf{RV}^T$. Let $\mathbf{Y}$ be the 60 by 12 indicator matrix with each row an indicator of the group to which the corresponding assay belongs. We carry out the computations in the equation

$$\hat{\mathbf{Y}} = \mathbf{R}(\mathbf{R}^T\mathbf{R} + \lambda\mathbf{I})^{-1}\mathbf{R}^T\mathbf{Y},$$

where λ is the penalty parameter. Applying the eigenvalue decomposition, we obtain

$$\mathbf{Y}^T\hat{\mathbf{Y}} = \mathbf{\Theta}^T\mathbf{A}\mathbf{\Theta}.$$

By rescaling $\mathbf{\Theta}$, we obtain $\tilde{\mathbf{\Theta}}$ that satisfies

$$(1/12)\tilde{\mathbf{\Theta}}^T\tilde{\mathbf{\Theta}} = \mathbf{I}.$$

Denote the diagonal elements of $12\mathbf{A}/60$ by a_{ii}. Each row of

$$\text{diag}\left(1/\sqrt{a_{ii}(1-a_{ii})}\right)\tilde{\Theta}^{\mathrm{T}}\hat{\mathbf{Y}}^{\mathrm{T}}$$

contains one of the discriminant coordinates for the 60 assays—the first row the first discriminant coordinate, the second row, etc. We use the first two rows for our figures.

To choose the penalty parameter λ, we use leave-out-one-at-a-time cross-validation (Hastie, Tibshirani, and Friedman, 2001). We use the deviance as the basis of our loss criterion.

DISCLAIMER

Certain commercial equipment, instruments, or materials are identified in this chapter to foster understanding. Such identification does not imply recommendation or endorsement by the National Institute of Standards and Technology, nor does it imply that the materials or equipment identified is necessarily the best available for the purpose.

REFERENCES

Burdick, R. A., Borror, C. M., and Montgomery, D. C., *Design and analysis of gauge R&R studies: Making decisions with confidence intervals in random and mixed ANOVA models.* Society for Industrial and Applied Mathematics, Philadephia, PA, 2005.

Burdin, C. J., Pittelkow, Y. E., and Wilson, S. R., Statistical analysis of adsorption models for oligonucleotide microarrays, *Statistical Applications in Genetics and Molecular Biology*, 3, Art. No. 35, 2004.

Carroll, R. J. and Ruppert, D., *Transformation and weighting in regression.* Chapman and Hall, New York, 1988.

Chudin, E., Kruglyak, S., Baker, S. C., Oeser, S., Barker, D., and McDaniel, T. K., A model of technical variation of microarray signals, *Journal of Computational Biology*, 13, 996–1003, 2006.

Cook, D., Hofmann, H., Lee, E-K., Yang, H., Nikolau, B., and Wurtele, E., Exploring gene expression data, using plots, *Journal of Data Science*, 5, 151–182, 2007.

Draper, N. R. and Smith, H., *Applied regression analysis,* 2nd ed., John Wiley & Sons, New York, 1981.

Gabriel, K. R. and Zamir, S., Lower rank approximation of matrices by least squares with any choice of weights, *Technometrics,* 21, 489–498, 1979.

Gnanadesikan, R., *Methods for statistical data analysis of multivariate observations,* 2nd ed., John Wiley & Sons, New York, 1997.

Hahn, G. J., Hill, W. J., Hoerl, R. W., and Zingraf, S. A., The impact of six sigma improvement—A glimpse in the future of statistics, *The American Statistician,* 53, 208–215, 1999.

Hastie, T., Buja, A., and Tibshirani, R., Penalized discriminant analysis, *The Annals of Statistics,* 23, 73–102, 1995.

Hastie, T. and Tibshirani, R., Efficient quadratic regularization for expression arrays, *Biostatistics,* 5, 329–340, 2004.

Hastie, T., Tibshirani, R., and Friedman, J., *The elements of statistical learning, data mining, inference, and prediction,* Springer–Verlag, New York, 2001.

ISO, *International vocabulary of basic and general terms in metrology*, 2nd ed., International Organization for Standardization (ISO), Geneva, Switzerland, 1993.

Irizarry, R. A. et al., Multiple-laboratory comparison of microarray platforms, *Nature Methods*, 2, 345–349, 2005.

Irizarry, R. A., Wu, Z., and Jaffee, H. A., Comparison of Affymetrix GeneChip expression measures, *Bioinformatics*, 22, 789–794, 2006.

Larkin, J. E., Frank, B. C., Gavras, H., Sultana, R., and Quackenbush, J., Independence and reproducibility across microarray platforms, *Nature Methods*, 2, 337–343, 2005.

Liggett, W., Normalization and technical variation in gene expression measurements, *Journal of Research of the National Institute of Standards and Technology*, 111, 361–372, 2006.

Liu, L., Hawkins, D. M., Ghosh, S., and Young, S. S., Robust singular value decomposition analysis of microarray data, *PNAS*, 100, 13167–13172, 2003.

MAQC Consortium, The MicroArray Quality Control (MAQC) project shows inter- and intraplatform reproducibility of gene expression measurements, *Nature Biotechnology*, 24, 1151–1161, 2006.

Qiu, X., Xiao, Y., Gordon, A., and Yakovlev, A., Assessing stability of gene selection in microarray data analysis, *BMC Bioinformatics*, 7, Art. No. 50, 2006.

Rocke, D. M., Design and analysis of experiments with high throughput biological assay data, *Seminars in Cell & Development Biology*, 15, 703–713, 2004.

Rocke, D. M. and Durbin, B., A model for measurement error for gene expression arrays, *Journal of Computational Biology*, 8, 557–569, 2001.

Shippy, R. et al., Using RNA sample titrations to assess microarray platform performance and normalization techniques, *Nature Biotechnology*, 24, 1123–1131, 2006.

Venables, W. N. and Ripley, B. D., *Modern applied statistics with S*, 4th ed., Springer–Verlag, New York, 2002.

Wolfinger, R. D., Gibson, G., Wolfinger, E. D., Bennett, L., Hamadeh, H., Bushel, P., Afshari, C., and Paules, R., Assessing gene significance from cDNA microarray expression data via mixed models, *Journal of Computational Biology*, 8, 625–637, 2001.

Wu, W., Dave, N., Tseng, G. C., Richards, T., Xing, E. P., and Kaminski, N., Comparison of normalization methods for CodeLink Bioarray data, *BMC Bioinformatics*, 6, Art. No. 309, 2005.

Wu, Z. and Irizarry, R. A., Stochastic models inspired by hybridization theory for short oligonucleotide arrays, *Journal of Computational Biology*, 12, 882–893, 2005.

Zhou, L. and Rocke, D. M., An expression index for Affymetrix GeneChips based on the generalized logarithm, *Bioinformatics*, 21, 3983–3989, 2005.

7 Biological Interpretation for Microarray Normalization Selection

Phillip Stafford and Younghee Tak

CONTENTS

Abstract 151
Introduction 152
Background 153
Technical Quality of Commercial Arrays 153
Expression Platform Types 154
Alternative Expression Technologies 156
Normalization Methods: History 157
Normalization Methods for Commercial Arrays 159
Causes and Consequences of Imprecise Measurements 160
Cross-Normalization Comparisons—Experimental Design 161
Results 163
Discussion 166
Conclusion 169
References 170

ABSTRACT

Molecular profiling has become a fundamental part of biomedical research, but much more importantly, it is now becoming a component of health care [1–3]. New clinical trials are identifying the benefits of treatments supported by microarray data. This device for highly parallel measurements of gene expression was once restricted to pure research laboratories. It has now taken its place next to pathology reports and imaging devices as a tool to elaborate alterations in genetic pathways in order to provide clues to disease status and progression. It is this new paradigm that puts pressure on the mathematical algorithms and data manipulations that normalize molecular profiling data. In order to obtain high precision, we must correct for inherent, repeatable biases in raw expression data. It is incumbent upon the biologist to understand, even anecdotally, the underlying principals of the profiling devices themselves, from image acquisition to processing and data normalization. With this understanding, one knows whether data are being appropriately processed. In this chapter we describe several Affymetrix normalization methods along with a way to compare normalization methods using biologically interpretable analyses. We provide data that illuminates not only how normalization affects array-to-array precision, but also how one can use the gene ontology and gene regulatory and metabolic pathway tools to quickly interpret how normalization can affect the results.

INTRODUCTION

Since the advent of the expression microarray, increasing attention has been given to normalization of raw data. Essentially three commercial platforms are available today: Illumina [4], Agilent [5], and Affymetrix [6]. Each uses unique techniques both to create gene probes and to read out the amount of mRNA that binds to those probes. Each technology has different strengths, weaknesses, and normalization techniques that accommodate particular biases. The National Institute of Standards and Technology's MAQC (Microarray Quality Control Consortium) recently published several very thorough papers on the sources of variance that arise between samples, between laboratories, and across array types, with the consensus being that array results themselves can be very correlative in the proper context [7–12].

Biological interpretation of microarray data can be quite revealing under the right circumstances. Genes that alter their expression due to physical perturbations or genotype changes (such as developing cancer cells) can lend sufficient information to appropriately classify medically relevant and clinically distinct samples. However, the expression data must accurately reflect the state of the transcriptome in order for this method to work. Accuracy is difficult to define, because there is no single gold standard for measuring mRNA abundance, and the task of encapsulating a "gene" as a single defined transcript can be quite contentious. Splice variants, orthologous gene families, and single nucleotide polymorphisms can have an effect on the apparent copy number of a particular gene. Quantitative real-time PCR (qRT-PCR) often provides a good estimate of mRNA copy number, but it suffers from probe sequence selection, starting material (sensitivity), and gene family similarity (selectivity). Indirect methods exist, such as using expression data to measure the error rate of a classifier. By testing the classifier on diverse data sets that span platforms, laboratories, samples, and more, one can iteratively minimize the error of the classifier and determine the biological information contained in the features. Uncharacterized genes can be winnowed, leaving only a core of genes that have some bearing on the biomedical samples *and* that perform well in a classifier.

There are problems inherent in each of these methods; we attempt to provide a way to compare biologically meaningful results statistically using the gene ontology and genetic regulatory networks. We examined Affymetrix data simply because the variance imposed on 25 mers from nuisance factors is profound, and because there are numerous competing normalization methods available for normalization of Affymetrix expression data. We used GeneSpring GX 7.2 (Agilent Technologies) for the expression analysis and database functions, and www.biorag.org (University of Arizona, Tucson) for access to Biocarta, GenMapp, and KeGG pathways.

BACKGROUND

Expression arrays are by far the most developed of the wide assortment of high-throughput profiling devices that work at the whole-genome level. Physically, these devices most often consist of a high-quality glass slide or other nonreactive substrate, made suitable for gene transcript hybridization by eliminating surface imperfections, reactive groups, trace oxidants, fluorescent residuals, and microscopic fractures. Some competing technologies rely on fluorescent beads (Luminex [13,14]), gel-based separation (ABI [15]), electronic detection of DNA resistance (GeneOhm), sequencing-based methods such as SAGE (serial analysis of gene expression) [16,17], or MPSS (massively parallel signature sequencing) [18] molecular calipers [19] (Calipertech)—even a CMOS-like device (Combimatrix [20]). In this discussion, we will focus mainly on the typical glass-surface, oligomer-based expression microarray with some attention paid to bead-based probes. The three main competitors in this marketplace are Affymetrix [21], Agilent [5], and Illumina [4].

TECHNICAL QUALITY OF COMMERCIAL ARRAYS

The need for precision manufacturing becomes apparent when one looks at the steps necessary for detecting mRNA molecules. An optical system that counts the photons per unit area is the primary interface between the microarray and the user. Often the area on the slide that is being measured is very small—a single feature (probe) is often less than 10 μm across in order to accommodate potentially millions of features per slide. As technology moves the field forward, arrays will accommodate even more features and the corresponding probe size will continue to shrink. The new Affymetrix exon array will exceed 5,000,000 features per slide at 5 μm per feature (http://www.affymetrix.com/products/arrays/exon_application.affx). In order to obtain meaningful data, one needs highly precise and sensitive optics, reproducible enzymatic reactions, and robust fluorophores.

Slight variations in the distance from the feature to the confocal scanner caused by minor warps or imperfections will drastically change the readings. Autofluorescence changes the local and global background, adding random and systematic noise to the overall signal [22–24]. The surface of the slide is also exposed to myriad reactive chemicals during processing. The coating on the slide surface must not interfere with hybridization or binding of probe and target, yet it must be stable enough to prevent removal of the feature during stringent washing. All of these and more play a role in the resulting technical variance of the array. The goal is to create an environment where the optics are unobstructed, electronic and thermal noise are mitigated, the surface is chemically inert, and the array must be chemically stable during storage. The situation array that manufacturers strive for occurs when biological variance overwhelms all possible technical nuisance factors. This is arguably the current state of expression array technology; however, the Achilles' heel of commercial arrays remains the quality of starting mRNA. However, even with degraded RNA, some computational methods exist for predicting signal deviation due to degraded RNA.

TABLE 7.1
Product Specifications or Measurements of the Three Main Expression Manufacturers

	Reproducibility	Precision	Sensitivity	Dynamic Range
Illumina Human WG-6 (47,000 probes)	$R2 = 0.990$ to 0.997	1.3-fold	0.25 pM	2.8 logs
Agilent Whole Human Genome (56,000 probes)	$R2 = 0.997$ to 0.999	1.2-fold	1 transcript/cell/ million cells	3 logs
Affymetrix Plus2 Human (44,000 probes)	$R2 = 0.890$ to 0.990	1.5-fold	1 transcript/cell	2.9 logs

Notes: Reproducibility is the r-squared value, precision is the minimum detectable fold-change, sensitivity is the detection of a target against a background of typical noise, and dynamic range is the range of fluorescence across which reproducible signal is detected.

In 2006, the major expression platforms were manufactured by Agilent (Palo Alto, California), Affymetrix (Palo Alto, California), and Illumina (San Diego, California) as seen in Table 7.1. All have made substantial gains in performance, price, and market share enabling them to dominate the field. Custom cDNA arrays still maintain their position as the most prevalent expression technology in academia due to price, increased spotting precision, and data quality, but are not standardized in any way [25,26]. Public data repositories have paralleled the market prevalence, reflecting a near 50:50 split between custom cDNA arrays and the two major commercial arrays, Agilent and Affymetrix [27]. Due to this valuable resource of historical data, the impetus for better normalization never ends. Older data are often supplied with the original scanned TIFF image, so the best normalization algorithm at the date of the experiment may not be the best available today. Normalization of older data has proven useful for reanalysis of classic experiments.

EXPRESSION PLATFORM TYPES

Agilent, Affymetrix, and Illumina arrays have always used some sort of mathematical normalization to correct for systematic biases and error, although Illumina has arguably the most on-array replication and power from averaging. However, it has been the prevalence of cDNA microarrays that forced the development of advanced normalization methods that improved precision in self-spotted long cDNA arrays. These arrays are ubiquitous due to their relatively low cost, but when one compares cDNA array data between labs or even across arrays, the lack of concordance becomes apparent. In early 2001, Agilent leveraged their expertise in ink-jet technology to build a robust cDNA array with over 22,000 features. These arrays were very reproducible but presented Agilent with a normalization challenge, which they met by adopting two-channel loess normalization. This type of normalization has become very sophisticated in the two-color field. Currently, Agilent provides this type of normalization plus a sophisticated error model as part of their Feature Extraction analysis software. Currently, there is little use of loess in single-color express arrays, but for multichannel arrays this technique is ideal.

Within a few years Agilent had introduced a method of applying nucleosides via ink-jet applicators to enable high-quality *in situ* synthesis of 60-mer oligonucleotides on the surface of the array. The ink-jet technology uses an electronic charge to direct individual phosphoramadite bases onto a prepared glass surface. The synthesis is >99.5% complete through 60 rounds of synthesis, yielding ~75% full-length 60-mer probes, with most probes exceeding 50 bp. Spots range in size from ~120 μm to less than 65 μm and much smaller in future releases. Current spot size is limited to enable back-compatibility with less expensive scanners. The physical characteristics of each array are listed in Table 7.1 and represent the current product line. Agilent currently offers single- and dual-color modes for their arrays. For the two-channel mode it utilizes both CY3 and CY5 as the labels in a competitive hybridization and an advanced error model that incorporates loess normalization, spatial detrending, and background estimation [28]. Single-channel mode uses only CY3.

For any two channel array, several physical characteristics play a role in causing an intensity-based deviation from perfect correlation between the channels. The differential dye incorporation and energy output cause severe biases. Although hybridizing two species simultaneously in one solution is the most accurate way to measure the relative differences between two samples since the competition occurs near equilibrium, the uncorrected color bias tends to ameliorate this benefit. Fortunately, scatter plot smoothing, or loess (aka lowess, although the algorithms are not technically identical), reduces the effect of differential dye incorporation and fluorescent energy by using a locally weighted linear regression to accommodate this effect. The cause of dye bias is predictable and therefore well suited for this type of regression approach. A first-degree polynomial least squares regression (loess) or a second-degree polynomial fit (lowess) closely approximates the highly correlated nature of the relationship between fluorescence intensity and the deviation of the color channels from one another, with loess being less susceptible to overfitting. Often the channels use CY5 and CY3 dyes, but other fluorescent dyes, including the Alexafluors from Invitrogen, are also useable with even greater robustness.

Currently, the Agilent expression products use loess and a custom error model within their Feature Extraction software to produce a robust set of measurements. Agilent currently recommends no background subtraction because the wash steps are so stringent that background fluorescence is negligible while positive signals remain very high. The software does provide a measure of the raw, mean, median, subtracted, processed, and error estimates for each calculation, so the normalization effect can be calculated per spot if desired.

Affymetrix has been the leader in shadow-masking technology. Affymetrix's technology relies on photolithography to create an in situ synthesized probe 25 bases long. Recent shadow-masking technology is approximately 90% efficient at each step, yielding approximately 7% full-length probe. New technologies have increased this to ~95% efficiency, yielding ~27% full-length probe. This basic synthesis efficiency has led to hybridization issues that Affymetrix has accommodated by including mismatch (MM) probes. The MM probe is designed to provide an estimate of ectopic and mishybridization—basically, the background hybridization for that probe sequence. Any normalization method that uses MM probes is said to utilize background estimates. Perfect match minus mismatch (PM–MM) was an early method of removing background signal from an ideal match and is the method used in Microarray Suite 5 (aka MAS5). Advancements in normalization have occurred since, and many of the more popular methods actually discard mismatch data. As Affymetrix pushes the envelope in shrinking feature size to enable high-density exon and single nucleotide polymorphism (SNP) arrays, the issue of normalizing data of 25-mer and shorter probes becomes vital in order to provide sufficient precision. Physical characteristics and performance metrics are listed in Table 7.1.

Illumina markets a bead-based technology that consists of beads with 50-mer probes applied at random locations across a pitted surface. Illumina hybridizes each array in house to provide a map to every probe. The Sentrix® Human-6 Expression BeadChip is a relatively new array from Illumina. It contains at least 30 replicate 50-mer probes for each of 47,296 transcripts including 24,385 RefSeq annotated genes, alternative splice variants, and predicted transcripts for each of the six separate enclosed and independent subarrays. BeadChips are imaged on Illumina's BeadArray Reader and output to third-party software for analysis or can be analyzed using Illumina's BeadStudio software. One feature of Illumina's technology is the cost differential between most commercial arrays and Illumina, with the BeadArray being marketed at one fifth the cost of the GeneChip® array with similar data quality but slightly different gene annotation and content. The BeadArray name encompasses the analysis and quality control (QC) packages in Bioconductor, a biotechnology suite of analysis tools for R. Normalization methods offered in BeadStudio include background subtraction, cubic spline (noted as the best in the MAQC reports), quantile normalization, and variance stabilization. Many users report reproducible and accurate results without any normalization applied.

ALTERNATIVE EXPRESSION TECHNOLOGIES

Several technologies that differ in technology from the industry standard Affymetrix- and Agilent-type arrays are attracting renewed attention. NimbleGen's (Madison, Wisconsin) maskless array synthesis (MAS) technology utilizes photoactivated chemistry for deprotection steps and a DLP-like digital micromirror device developed by Texas Instruments. Currently, a resolution of 1024 × 768 (~15 μm feature size) is used but new resolutions can easily quadruple the density, allowing for millions of features per array. The light projection enables much tighter control over the *in situ* probe synthesis, and is not subject to the problem of repeatedly manufacturing a disposable shadow mask that is subject to photonic degradation [6]. NimbleGen's technology allows probe lengths up to 70 bases, allowing for varying-length probes and an isothermal probeset if desired.

Combimatrix [20] (Seattle, Washington) utilizes CMOS-like technology to create a dense surface of microelectrodes that are independently addressable. Each electrode directs the chemical synthesis of biomolecules in a contained microenvironment. The customization of this design makes it possible to create a catalog of timely focused arrays such as the Cancer Genome, or Inflammation or Stress.

Combimatrix also makes the array manufacturing equipment available as a desktop device for researchers who require proprietary sequences to be generated. Massively parallel signature sequencing (MPSS) [18,29] is based on Lynx Technologies' Megaclone and uses a large number of identifiable tabs and fluorescent beads coupled with FACS (fluorescently activated cell sorting) to quantitatively identify all available mRNA species. This technique is remarkably precise and accurate; nearly all misidentifications are caused by problems in the RNA sample preparation.

Solexa, which acquired Lynx Technologies (and itself was subsequently purchased by Illumina), did not pursue the MPSS technology and it and the Megaclone system were abandoned. Applied Biosystems, an Applera unit, has a unique sequencer-derived system that utilizes ABI's experience in capillary gel electrophoresis, chemiluminescent detection, and special software to enable high sensitivity and reproducibility with minimal sample that leverages the ABI sequencers that many laboratories have in abundance [15]. Special probes that bind RNA sequences are utilized to modulate the rate of migration rate in an electric field, and utilize in-line detectors to identify targets. ABI has claimed to detect thousands of novel transcripts not in public genomic databases obtained through their close association with Celera. Since ABI also owns the TaqMan® technology, there is a relatively straightforward path from transcript detection to qRT-PCR validation.

NORMALIZATION METHODS: HISTORY

Benefiting from the past 10 years of research efforts, a set of rigorous measurement, quality control, and normalization steps have been developed for expression profiling. Numerous methods allow the user to account for and correct the intrinsic biases in the fluorescent signals that make up expression array data. Much of the credit must be laid at the feet of researchers who constantly battled to improve the quality of self-spotted cDNA data during the time when high imprecision was routine (Quackenbush [30], Eisen [31], Brown and Botstein [32], Bittner [33], Chen [34], and others). Variance continues to stem from sources that range from differences in dye incorporation, differential energy output from the fluorophores, scanner variation, background fluorescence from different array surfaces and substrates, probe design, and increasingly from biological sources: ectopic or failed hybridization, differential mRNA degradation, RNA integrity, and biological variability. For both cDNA and oligo-based expression arrays, quality control is a major issue, although the details differ. Normalization will correct for technical biases, but in some cases, the adjustment is so extreme that the normalized data bear little resemblance to raw data, and interpretation of the results relies on the quality of the normalization technique.

The choice of using either intensities or ratios to represent expression levels, as well as which quality control measure is most appropriate, can be approached by understanding the full range of the peculiarities and performance parameters of each array platform. Ratio measurements are commonly used in two-color cDNA arrays [34,35] as well as the Agilent expression system. Ratio measurements have been shown to be very precise and accurate because of the kinetics of competitive hybridization. Measuring two mRNA species at the same time in the same hybridization solution reduces interarray variability and provides a robust comparison of treatment versus control. Expression intensities from single-color data provide a semidirect assay of transcript abundance, which is common practice for experimental designs that look at hundreds or thousands of conditions. Agilent now offers protocols for both one and two channels to accommodate highly parallel large-scale experiments.

The External RNA Controls Consortium (ERCC, founded in 2003) [36] is addressing the problem that the FDA's Critical Path and MAQC groups have identified as a quality control issue; that is, in order to make expression microarrays meet the standards imposed by the Clinical and Laboratory Standards Institute and Clinical Laboratory Best Practices, the microarray device must provide cross-laboratory reproducibility. Throughout the history of expression arrays, external standards have been used to validate and normalize data. The choice of spike-ins was at the discretion of the scientist or the manufacturer, and has at various times variously included bacterial, yeast, fly, human, and nonsense transcripts. The ERCC is making a case for spike-in transcripts that prove the most useful information across the widest assortment of platforms. At the most basic level, this could be the construction of a limited library of cloned human transcripts that perform in predictable ways across a wide variety of platforms, conditions, laboratories, and samples.

This is one of the stated goals of the ERCC, but to take their idea one step further—the concept of a *full* library of clones—one could spike in a series of gene-by-gene dilutions into complex mixtures of mRNA. This method would allow one to measure the endpoints in sensitivity for every one of 40,000+ probes, and would help identify sequence-specific factors that cause seemingly random inaccuracy. At the very least, a resource like this would enable the creation of a hybridization response curve for every individual probe. Nonlinearity in probe response could be modeled directly and empirically, rather than using a global error model. Model-based probe-by-probe normalization would be the most efficient way to compensate for variant probes, and would reward the expression platform that is currently the most precise. In this scenario, high reproducibility equates to high accuracy. Although this solution is unworkably expensive, other indirect measures of counting mRNA molecules can be possible through titration experiments of RNA from various cell lines and tissues [37,38].

In a more realistic scenario, genomic DNA has often been used as a reference to measure the fluorescence that results from the binding of a known concentration of target; in this case the known concentration of target would be from the single pair of molecules from a particular allele. Even though gene copy number is variant for more than 10% of human genes and complexities arise when considering gene families, duplications or deletions, the genomic DNA for a particular gene is usually stable at two copies. This has in fact been attempted and works moderately well for cDNA arrays [39]. However, several problems arise: The complexity of the genome is several orders of magnitude greater than the complexity of the transcriptome; 25 or even 60 mers designed to bind mRNA sequences may lack the specificity to uniquely bind single-copy genomic DNA without high imprecision. Also, this method provides a single endpoint only. In order to estimate a nonlinear response such as fluorescence-based mRNA copy number, one needs low, middle, and high copy observations in order to estimate the regression curve properly.

NORMALIZATION METHODS FOR COMMERCIAL ARRAYS

MAS5 [40,41]: Affymetrix uses a short 25-mer probe designed for maximum specificity. However, short probes are prone to mishybridization; in an attempt to rectify this Affymetrix added a centered mismatch probe that in theory accounts for ectopic hybridization. However, typically 20–40% of mismatch probes per array show a signal higher than the matched PM probes, leading to the development of the IM, or ideal match. The IM calculation now used in MAS5 no longer yields negative values, and has been shown to provide higher accuracy in spike-in studies. Microarray Suite 5 normalization with its present, marginal, and absent flags is the most commonly reported data normalization in public databases. Microarray Suite 5 does suffer from a relatively high degree of false positives and there is a noticeable exaggeration, or expansion, of ratios and a concomitant compression of signal, which makes it difficult to estimate true copy number.

RMA and GC-RMA [42,43]: Robust multiarray averaging is a method that normalizes data towards a common distribution under the assumption that very few genes are differentially expressed across two or more conditions. This method tends toward false negatives but has among the highest precision of any of the popular GeneChip (GC) normalizations. Robust multiarray averaging differs from GC-RMA in that the latter utilizes the sequence information of the MM probes in order to calculate probe affinity values during background correction. In this way, although a little general, at least some of the thermodynamic qualities of the probe can be incorporated into the normalization.

dChip [44]: Li and Wang developed a model-based approach for estimating unusual but predictable behavior in Affymetrix probe set data. This method greatly reduces the effect of outlier probes, and the model for the most part predicts the majority of the variance from individual probe sets. The model uses an "explained energy" term that predicts deviance from an expected norm and, much like the GC modification to RMA, is probe specific. This method can use or ignore mismatch data, but does a much better job predicting expression values from probes that are close to background, probes that would be called "absent" in the MAS5 system. dChip also automatically identifies and corrects for slide imperfections, image artifacts and a general class of probe outliers. dChip has been shown to be quite robust and compares favorably to the other common Affy normalization methods.

Quantile normalization [42]: Specifically targeted for early two-channel cDNA arrays, the quantile normalization method assumes that the distribution of gene abundance is almost identical for all samples and takes the pooled distribution of probes on all arrays (the so-called "target distribution"). The algorithm subsequently normalizes each array by transforming the original value to that of the target distribution, quantile by quantile. The target distribution is obtained by averaging the sorted expression values across all arrays and reassigning a transformed value to each array, then re-sorting back to the original order. This method has been shown to work best for samples from similar tissues or conditions with little large-scale transcriptional changes, and it is one of the better performing methods for Illumina expression data. As with RMA, this method suffers from false negatives and false positives that are caused by inappropriately low intrasample variance and compressed expression ratios.

Loess and cyclic loess normalization [45–47]: The most common normalization for two-channel cDNA arrays is loess, in which a piecewise polynomial regression is performed. For two-channel array data, loess records the changes needed to achieve a convergence of center-weighted means followed by a calculation of the distance of each point to the best-fit curve through the data. The distance is reduced iteratively until no further changes in data will affect the magnitude of the residuals. Loess is highly dependent on the span of points that will be used in calculating each local estimate, and selecting a span is not automatic, but must be carefully chosen to prevent overcorrection. Cyclic loess is based upon the concept of the M/A plot, where M is the difference in log expression and A is the average of log expression values. Although often used for two-channel cDNA arrays, this method can also be used for pairs of single-channel data, or by using all possible pairwise combinations, for multiple arrays at once. Through each iteration, the curve from the M/A plot is redrawn and adjustments are made to flatten this curve. Overfitting with this method simply results in pairs of arrays being highly similar while still allowing extreme outliers to remain far from the majority of the data.

Others: dChip has been shown to possess qualities that tend to balance false positive and false negative errors when making estimates for the amount of correction necessary to account for systematic biases in probe-level expression values; however, like RMA it requires many arrays to perform well. Position-dependent nearest-neighbor has been shown to have similar performance to dChip PM, but achieves this level of performance by doing a chip-by-chip normalization. In this case, Li and Wong [44] use the tendency of a probe to have similar sequence or to physically overlap the same sequence when probing the same gene. Only Affymetrix currently uses an abundance of different probes. Most companies have tested many different probes and retain only those that provide the proper estimate of mRNA copy number at any concentration range of that particular mRNA.

CAUSES AND CONSEQUENCES OF IMPRECISE MEASUREMENTS

Both glaring and subtle problems arise because of improper microarray normalization. In the simplest case, a simple lack of precision or correlation to raw values exists, indicating a large bias in the distribution of values. In a more complex case, certain genes change a great deal based on a particular normalization while others do not. In many cases, the rank of differentially expressed genes changes substantially based on the normalization method. Each probe is designed to detect mRNA molecules above background (minimum detectable fold change) all the way to thousands of copies per cell (saturation). There are physical constraints that are imposed upon the probe design process: sufficient sensitivity to detect the low end of mRNA copy number and adequate selectivity to distinguish related or spliced versions of the same gene or gene family.

The dynamic range within a cell may be quite large, from less than one mRNA to thousands. Typical dynamic range for commercial arrays must therefore exceed three orders of magnitude minimally; most commercial arrays meet or exceed this but the problems lie in the nonlinearity of probe response to mRNA copy number and intensity-dependent bias. A small percentage of probes on an array will randomly fail or perform below specifications due to unexpected behavior in certain mRNA samples, most notably in degraded samples [48]. Some probes, however, will fail for complex thermodynamic reasons. Even Universal RNA, a broad collection of pooled RNA from cell lines, may not contain the mRNA concentration that can identify poorly behaving probes. Normalization methods usually do not help in situations like this; no amount of coercing data into a desired data distribution would work for nonresponsive or promiscuous probes. However, the knowledge we obtain from cross-platform and qRT-PCR experiments brings us closer to creating the ideal probe for each gene in the human genome. Other failure modes to note include mishybridization, excessive competition for a target, crossing splice sites, or missing splice variants.

CROSS-NORMALIZATION COMPARISONS— EXPERIMENTAL DESIGN

In order to fully understand the complex interactions between normalization methods and the results of expression analysis in a biological context, we felt that an actual controlled biological example would speak volumes. We present data from pooled commercial RNA extracted from several healthy human tissues on the Affymetrix platform. We have used the most popular normalization techniques to create distinct data sets for biological analysis. We favor experiments that utilize healthy tissues so that one can examine the cause-and-effect relationship of normalization, array, tissue type, and biological interpretation without having to deal with any underlying alteration in the cell's regulatory mechanisms that results from, or causes, the disease phenotype, such as cancer. The complexity of working with the transcriptome is sufficiently difficult without additional nuisance factors. Most importantly, specialized tissues have such carefully regulated processes that the expression profile tends to be very stable, especially when examining pooled samples.

Using the Affymetrix U133Av2 arrays, 24 Affymetrix expression arrays were run using commercially available (Stratagene, La Jolla, California) human spleen, liver, and lung RNA (Table 7.1). Affymetrix CEL files were normalized using RMA, GC-RMA, probe logarithmic intensity error (PLIER), dChip PM–MM, and RAW. Robust multiarray averaging and GC-RMA [43] utilize a variance-stabilization algorithm with GC-RMA correcting for GC content. PLIER (probe logarithmic intensity error; http://www.affymetrix.com/support/developer/plier_sdk/index.affx) is a controversial method proposed by Affymetrix with typically good results, but based on a flawed assumption of mismatch destabilization energy.

dChip PM–MM [44] has been shown to correct for some of the variant behavior seen in Affymetrix's MAS5 background subtraction algorithm, and is a model-based approach that also uses a strong background subtraction component. dChip's PM-only was also used to compare the model-based approach without the strong background subtraction component. RAW data were generated by averaging all PM probes per probe set and represent no background subtraction or normalization at all. We performed all possible pairwise t-tests among the three tissues and selected 1000 of the most significant genes differentially expressed between liver and spleen, liver and lung, and lung and spleen (Table 7.2). Since the tissues are very different,

TABLE 7.2
T-Test Results from Three Separate Tissue Comparisons

Sample	RAW	RMA	GCRMA	PM	PM–MM	MAS5	PLIER
Liver:lung	3.22×10^{-15}	4.55×10^{-18}	3.37×10^{-20}	9.54×10^{-15}	7.34×10^{-17}	1.22×10^{-15}	2.38×10^{-26}
	2.18×10^{-06}	1.91×10^{-07}	1.78×10^{-08}	2.05×10^{-06}	2.31×10^{-07}	2.33×10^{-07}	3.66×10^{-08}
Liver:spleen	1.77×10^{-14}	1.75×10^{-17}	3.95×10^{-03}	2.62×10^{-19}	4.34×10^{-15}	1.20×10^{-13}	2.81×10^{-26}
	9.04×10^{-06}	5.18×10^{-06}	1.86×10^{-14}	5.02×10^{-08}	5.28×10^{-06}	8.51×10^{-05}	2.65×10^{-05}
Lung:spleen	1.64×10^{-13}	1.34×10^{-14}	4.0×10^{-17}	1.17×10^{-12}	1.06×10^{-13}	7.07×10^{-12}	3.70×10^{-21}
	1.36×10^{-05}	1.60×10^{-05}	7.74×10^{-11}	1.52×10^{-05}	1.81×10^{-05}	5.43×10^{-04}	3.71×10^{-05}

Note: For each tissue comparison, the *p*-values for the 1000 most differentially expressed genes are listed in the corresponding box.

the p-values were sufficiently low that no false discovery correction is necessary. We computed the common gene ontology functional categories and the gene regulatory pathways from the 1000 significant genes for each comparison. T-test figures, pathways, and gene ontology (GO) assignments are seen in Tables 7.2, 7.3, and 7.4.

RESULTS

Figure 7.1 is an overview of technical reproducibility. Scatter plots graphically show the relationship between intensity and precision. For MAS5, one immediately notices a large scatter near the low intensity values—a direct result of mismatch subtraction. RAW data, although rife with uncorrected bias, still show apparent technical precision. Robust multiarray averaging and GC-RMA were so similar that RMA was excluded. The model-based methods (dChip PM, PM–MM) show more apparent scatter than the variance stabilization methods (GC-RMA) but as seen in later results, precision does not fully encompass expression array quality.

Figure 7.2 illustrates commonality between gene lists for different normalization methods and tissue comparisons. Several patterns of generalizable behavior were noted. First, dChip PM–MM and dChip PM normalize data such that significant genes appear in many of the other normalization methods. No other normalization yields significant genes that fall within the other methods. It is known that the model-based dChip methods balance false positives and negatives quite well. As seen in Tables 7.4 and 7.5, except for the liver:spleen comparison (where PLIER and MAS5 showed the highest commonality), the dChip methods tended to yield the highest overlap in the GO terms and pathways identified. This has been seen for data outside of these three normal tissue samples (specifically, the expO data set from GEO, GSE21090; data not shown).

It is easy to see that MAS5 has much more sample-to-sample variance than GC-RMA, the normalization with the least sample-to-sample variance. The dChip and PLIER methods were precisely between these two normalization methods in terms of sample variance. How does this affect the biological integrity of the data? In one sense, we can see a tremendous impact on the ratios between housekeeping genes (Table 7.6). We see ratio compression with GC-RMA and ratio expansion with MAS5, but we also see changes in the direction of those ratios. As well, the absolute p-values for the 1000 most differentially expressed genes across all of the tissue comparisons were highest for MAS5 (less significance) and lowest for GC-RMA (most significance). This intuitively makes sense: p-values are affected not only by the magnitude of the gene expression difference between two tissues, but also by within-sample precision. In parallel, we see the most false positives in the MAS5 data and the most false negatives in the GC-RMA data—again, a consequence of technical precision, but not of biological accuracy.

The other methods fall between MAS5 and GC-RMA. dChip PM and PM–MM showed the best balance between false positives and false negatives in this experiment. This again suggests that the model-based dChip algorithms produce data that behave like a compromise between classic background oversubtraction (MAS5) and overstabilization of variance (GC-RMA).

TABLE 7.3
Pathway Consensus Data

	RAW	RMA	GCRMA	PM	PM–MM	MAS5	PLIER
Liver:lung (636 pathways)	364 (57%)	429 (67%)	111 (17%)	**437 (70%)**	**437 (69%)**	409 (64%)	308 (48%)
Liver:spleen (695 pathways)	431 (62%)	443 (64%)	452 (65%)	450 (60%)	441 (63%)	**503 (72%)**	**503 (72%)**
Lung:spleen (632 pathways)	281 (44%)	355 (56%)	176 (28%)	**375 (59%)**	**382 (60%)**	295 (47%)	287 (45%)

Note: Each tissue comparison and the number of relevant pathways determined from 1000 significant differentially expressed genes is listed along with the number of times each pathway was identified with BioRag.org from each normalization method.

TABLE 7.4
Gene Ontology Biological Function Consensus Data

	RAW	RMA	GCRMA	PM	PM–MM	MAS5	PLIER
Liver:lung (4405 terms)	4101 (93%)	**4131 (94%)**	3798 (86%)	4118 (94%)	**4128 (94%)**	4096 (93%)	4001 (91%)
Liver:spleen (4332 terms)	4051 (94%)	4079 (94%)	4091 (94%)	4076 (94%)	4080 (94%)	**4103 (95%)**	**4093 (94%)**
Lung:spleen (4351 terms)	3987 (92%)	**3988 (92%)**	3931 (90%)	3992 (92%)	**4030 (93%)**	3995 (92%)	3957 (91%)

Note: There were 4405, 4332, and 4351 common GO terms in biological function identified for each of the lists of 1000 differentially expressed genes from the three tissue comparisons. Since the gene ontology is redundant, there were fourfold more terms than genes. Under each normalization name is the number of common terms.

Biological Interpretation for Microarray Normalization Selection 165

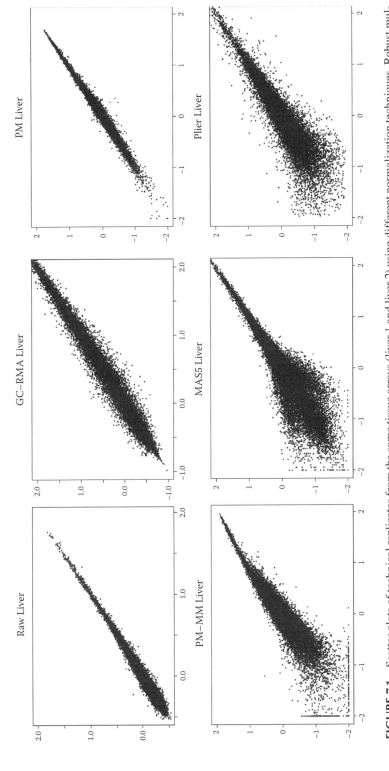

FIGURE 7.1 Scatter plots of technical replicates from the same tissue arrays (liver 1 and liver 2) using different normalization techniques. Robust multiarray averaging was excluded since it appears identical to GC-RMA in this view. Scatter plots indicate apparent precision.

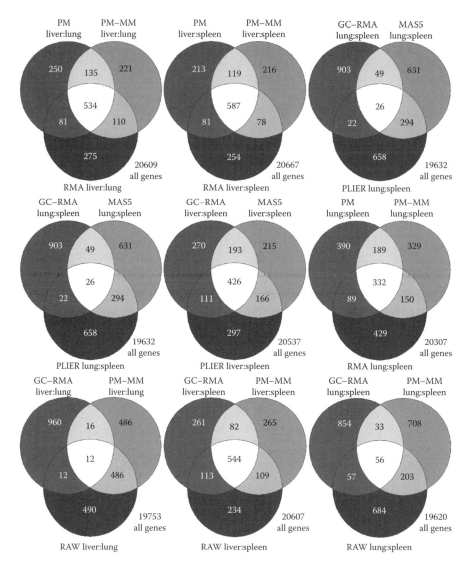

FIGURE 7.2 (See color insert following page 80.) Venn diagrams of 1000 overlapping probes across the seven normalizations. Three-way common genes appear in the white section.

DISCUSSION

Normalization issues have been plaguing microarray analysis for years, particularly for those arrays that absolutely need, and consequently are most affected by, normalization. In our case, we examined normalization effects on the most widely used expression array, the Affymetrix U133Av2. We have created a somewhat artificial experiment using tissues that differ greatly from one another at the transcriptional level. Since we measured changes in gene lists, pathways, and gene ontology

TABLE 7.5
Most Common Pathways and GO Terms across the Seven Normalizations

Sample	Most Common Pathways and Gene Ontology Terms
Liver:lungGO:Pathway:	GO:0046049—UMP metabolismAcetyl-coenzyme A carboxylase beta metabolic pathways
Liver:spleenGO:Pathway:	GO:0006542—glutamine biosynthesisDefensin alpha pathway
Lung:spleenGO:Pathway:	GO:0048227—Plasma membrane to endosome transport

Note: The GO terms and pathways were identified by using the 1000 genes identified as differentially expressed across the three tissues—those that were identified most often in all of the normalization methods are listed here.

TABLE 7.6
$\log_2$ Ratios of Six Common Housekeeping Genes in Liver for Each of the Normalization Methods

Liver Tissue	RAW	RMA	GCRMA	PM	PM–MM	MAS5	PLIER
$\log_2 \dfrac{GAPD}{ACTB}$	−8.6	1.4	1.4	2.6	2.2	−3.6	−1.1
$\log_2 \dfrac{PGK1}{LDHA}$	4.2	−0.7	−0.8	−1.2	−1.5	2.1	1.4
$\log_2 \dfrac{RPSL19}{ILF2}$	3.7	1.8	1.6	1.9	2.3	−3.5	−0.9

Note: 0 is no change.

categories as the calibration tool, we needed large and easily quantifiable differences. For this sort of calibration, one should not use cancer or other tissues that have a nonoptimal transcriptome because (1) changes in gene transcription may be very slight from sample to sample, and (2) the chromosomal alterations that led to the cancer phenotype in the first place have created an uncontrolled transcriptome.

The large changes we see between, say, liver and spleen are ideal for gaining insight into the amount of normalization effect (or bias) that can be tolerated while still maintaining a sensible biological interpretation. Housekeeping genes are a useful readout in this situation. Many constitutively expressed genes are considered stable enough to use for estimating normalization bias (Table 7.6). We looked at the ratios of GAPD (glyceraldehyde-3-phosphate dehydrogenase)/ACTB (beta actin), PGK1 (phosphoglycerate kinase)/LDHA (lactate dehydrogenase), and RPSL19 (ribosomal protein L19)/ILF2 (interleukin enhancer binding factor 2) [37] in order to detect distribution-biased normalization effects. In all cases the ratios differed, in some cases extensively and in opposing directions. The ratio analysis of housekeeping genes supports the results from the analysis of tissue-specific genes and points to the need for setting calibration by using as many biologically relevant genes as possible.

This leads to another important point: We have shown that normalization routines affect the actual values of individual probes. More importantly, they also have an effect on the cumulative analysis of an experiment. Changes in values do not matter as much if the *order* of the values is preserved. In our case, we have shown that the order and content of 1000 of the most significantly expressed genes among three highly variant tissues changes drastically depending on the normalization used. Even a nonparametric rank-sum test, which is not exposed to normality issues like parametric tests (t-test, analysis of variance [ANOVA], etc.), shows enormous changes in the rank of sets of genes (data not shown). This in itself may not be terribly concerning if the biological interpretation remains the same—and it can if the genes that are different across normalizations are simply substituted from the same biological pathways. We have shown that this is not the case. The fact that the ratios between housekeeping genes change based on the normalization method supports the divergence between GO and pathway interpretations.

With several publications espousing the benefits of the model-based dChip methods, we provide additional anecdotal evidence that dChip provides results that are biologically sound and balance potential error and types I and II error. We initially predicted no algorithm would yield values similar to another, and that correlations among them would likewise be low. In fact, only the dChip PM and PM–MM, and RMA and GC-RMA, data were similar. We assumed that if any method had a chance to overlap the others, it would be MAS5 due to its conservative processing. Instead, the two dChip normalization methods yielded statistically significant gene lists that actually have no particular correlation with other lists, but have overlap when viewed in the context of *biological content*. That is, when translating from gene lists to biological pathway or to a set of common biological functions (gene ontology), we find that the two dChip normalization methods yield biological interpretations that are also found in other normalization methods, producing a sort of consensus.

The image processing components of dChip identify and correct for a great deal of the image acquisition error that causes deviant values. Also, dChip can either use or ignore local background information (mismatch data). The ability to include or ignore these controversial probes adds flexibility to dChip's analysis. A model-based approach rather than a variance stabilization or weighted median method prevents biases that plague other normalizations. Microarray Suite 5 would be almost as good as dChip, given sufficient replication to accommodate the imprecision of mid- to low-range signals, but based on the current data, we propose that dChip has achieved adequate technical precision balanced with good apparent accuracy.

Machine learning approaches are popular for classification of expression data sets into biologically relevant groups. They rely on low inherent error in the input data sets in order to distinguish two or more biological classes. Imprecision or bias destroys the ability of classifiers to partition samples correctly, and leads to high levels of overfitting. Feature selection and classification are highly susceptible, often resulting in nonoverlapping sets of genes being selected as the best classifiers without a robust way of accounting for cross-platform and cross-normalization differences, or incorporating prior knowledge as selection criteria. We have shown in a previous study that dChip has the least effect on the error of a classifier. The measured error of a classifier can be a scale by which one can tune the normalization. The error between training and test data sets can be minimized through normalization selection; the lowest classifier error often points to the best normalization method.

CONCLUSION

A direct correlation exists between the particular normalization method of microarray expression data and accuracy and precision. We have discussed expression technologies—cDNA, short and long oligo technologies, two-color and single-color modes—and we have shown the biological interpretations that can be made from simple differential gene lists. We wish to emphasize several important points:

- cDNA arrays have made significant progress in the last few years in terms of accuracy and precision. Most new pin spotters are capable of producing a commercial-grade expression array that can easily cover the annotated human genome on one slide. Precision of cDNA arrays can be exceptional, depending on the quality of the starting EST library. With the proper normalization technique, cDNA expression arrays can be used for identifying the correct pathways and functional gene lists. With adequate technical replication and controls, self-spotted cDNA arrays are a valid alternative to highly precise commercial arrays. However, comparing data from historical cDNA data sets is an almost worthless pursuit.
- Oligo expression products are available in a wide assortment of technologies, from Nimblegen, Combimatrix, ABI, Illumina, Luminex, Clontech, Solexa/Lynxgen, and many others. Some technologies use fluorescent microbeads, and some use microseqencing, gel filtration, microfluidics, or e-detection. With all of these detection methods, it might seem that comparison of samples on a gene-by-gene basis would be fruitless, but most data from the last few years are quite good. Most technologies will report accurate differences from gene to gene, especially when the input RNA is high quality. Biological pathways that identified across most of these technologies are very similar.
- The tests we used to measure the effect of normalization on our results are very simple. The steps are straightforward:
  - Obtain a differential list of genes. Knowing the biological basis behind some of the differences between the tissues will add to the analysis, but is not necessary.
  - Using this differential list of genes between two samples, process your data with other normalization methods. Obtain the GO terms using Go-miner, Onto-Express, or any of a number of gene ontology programs or Websites. Compare the number of genes per functional category between the normalizations. Do the same thing for pathways using www.biorag.org to obtain pathway names from the input gene lists.
  - Analyze the results. Gene lists and pathways that defy conventional biological wisdom are likely incorrect. Look for a consensus among normalization methods. Similar methods like dChip PM and dChip PM–MM, or RMA and GC-RMA, should provide results that are close

to one another. Armed with the knowledge about how different normalization methods behave (for instance, GC-RMA is susceptible to type II error, and MAS5 is more susceptible to type I errors), one can make a choice on a particular normalization that reflects the needs and size of the experiment.

We looked at the Affymetrix U133Av2 array. The biological interpretation techniques described here can be applied to any array type and are so high level that many of the technical differences between platforms disappear. Being able to identify biological processes correctly is a fundamental aspect of expression microarrays that has driven the amazing progress in genomics research over the last 5 years and will continue to do so.

REFERENCES

1. C. Sander, Genomic medicine and the future of health care, *Science*, 287, 1977–1978, 2000.
2. J. D. Brenton, L. A. Carey, A. A. Ahmed, and C. Caldas, Molecular classification and molecular forecasting of breast cancer: Ready for clinical application? *Journal of Clinical Oncology*, 23, 7350–7360, 2005.
3. H. Liu, J. Li, and L. Wong, Use of extreme patient samples for outcome prediction from gene expression data, *Bioinformatics*, 21, 3377–3384, 2005.
4. A. Oliphant, D. L. Barker, J. R. Stuelpnagel, and M. S. Chee, BeadArray technology: Enabling an accurate, cost-effective approach to high-throughput genotyping, *Biotechniques*, Suppl., 56–58, 60–61, 2002.
5. T. R. Hughes, M. Mao, A. R. Jones, J. Burchard, M. J. Marton, K. W. Shannon, S. M. Lefkowitz, et al., Expression profiling using microarrays fabricated by an ink-jet oligonucleotide synthesizer, *Nature Biotechnology*, 19, 342–347, 2001.
6. R. J. Lipshutz, S. P. Fodor, T. R. Gingeras, and D. J. Lockhart, High-density synthetic oligonucleotide arrays, *Nature Genetics*, 21, 20–24, 1999.
7. L. Guo, E. K. Lobenhofer, C. Wang, R. Shippy, S. C. Harris, L. Zhang, N. Mei, et al., Rat toxicogenomic study reveals analytical consistency across microarray platforms, *Nature Biotechnology*, 24, 1162–1169, 2006.
8. L. Shi, W. Tong, H. Fang, U. Scherf, J. Han, R. K. Puri, F. W. Frueh, et al., Cross-platform comparability of microarray technology: Intraplatform consistency and appropriate data analysis procedures are essential, *BMC Bioinformatics*, 6 Suppl 2, S12, 2005.
9. L. Shi, L. H. Reid, W. D. Jones, R. Shippy, J. A. Warrington, S. C. Baker, et al., The MicroArray Quality Control (MAQC) project shows inter- and intraplatform reproducibility of gene expression measurements, *Nature Biotechnology*, 24, 1151–1161, 2006.
10. T. A. Patterson, E. K. Lobenhofer, S. B. Fulmer-Smentek, P. J. Collins, T. M. Chu, W. Bao, H. Fang, et al., Performance comparison of one-color and two-color platforms within the MicroArray Quality Control (MAQC) project, *Nature Biotechnology*, 24, 1140–1150, 2006.
11. W. Tong, A. B. Lucas, R. Shippy, X. Fan, H. Fang, H. Hong, M. S. Orr, et al., Evaluation of external RNA controls for the assessment of microarray performance, *Nature Biotechnology*, 24, 1132–1139, 2006.
12. R. D. Canales, Y. Luo, J. C. Willey, B. Austermiller, C. C. Barbacioru, C. Boysen, K. Hunkapiller, et al., Evaluation of DNA microarray results with quantitative gene expression platforms, *Nature Biotechnology*, 24, 1115–1122, 2006.

13. H. Cai, P. S. White, D. Torney, A. Deshpande, Z. Wang, B. Marrone, and J. P. Nolan, Flow cytometry-based minisequencing: A new platform for high-throughput single-nucleotide polymorphism scoring, *Genomics*, 66, 135–143, 2000.
14. J. Chen, M. A. Iannone, S. Li, J. D. Taylor, P. Rivers, A. J. Nelson, K. A. Slentz-Kesler, et al., A microsphere-based assay for multiplexed single nucleotide polymorphism analysis using single base chain extension, *Genome Research*, 10, 549–557, 2000.
15. Y. Wang, C. Barbacioru, F. Hyland, W. Xiao, K. L. Hunkapiller, J. Blake, F. Chan, et al., Large scale real-time PCR validation on gene expression measurements from two commercial long-oligonucleotide microarrays, *BMC Genomics*, 7, 59, 2006.
16. M. Z. Man, X. Wang, and Y. Wang, POWER_SAGE: Comparing statistical tests for SAGE experiments, *Bioinformatics*, 16, 953–959, 2000.
17. O. L. Griffith, E. D. Pleasance, D. L. Fulton, H. Oveisi, M. Ester, A. S. Siddiqui, and S. J. M. Jones, Assessment and integration of publicly available SAGE, cDNA microarray, and oligonucleotide microarray expression data for global coexpression analyses, *Genomics*, 86, 476–488, 2005.
18. S. Brenner, M. Johnson, J. Bridgham, G. Golda, D. H. Lloyd, D. Johnson, S. Luo, et al., Gene expression analysis by massively parallel signature sequencing (MPSS) on microbead arrays, *Nature Biotechnology*, 18, 630–634, 2000.
19. S. Mouradian, Lab-on-a-chip: Applications in proteomics, *Current Opinion Chemical Biology*, 6, 51–56, 2002.
20. E. Tesfu, K. Maurer, S. R. Ragsdale, and K. F. Moeller, Building addressable libraries: The use of electrochemistry for generating reactive Pd(II) reagents at preselected sites in a ChiB, *Journal of the American Chemical Society*, 126, 6212–6213, 2004.
21. M. Chee, R. Yang, E. Hubbell, A. Berno, X. C. Huang, D. Stern, J. Winkler, et al., Accessing genetic information with high-density DNA arrays, *Science*, 274, 610–614, 1996.
22. D. V. Nicolau, R. Raghavachari, and Society of Photo-optical Instrumentation Engineers, *Microarrays and combinatorial techniques: Design, fabrication, and analysis II*. 25–26 January 2004, San Jose, California, USA. Bellingham, WA: SPIE, 2004.
23. D. V. Nicolau, R. Raghavachari, and Society of Photo-optical Instrumentation Engineers, *Microarrays and combinatorial technologies for biomedical applications: Design, fabrication, and analysis*. 26–28 January 2003, San Jose, California, USA. Bellingham, WA: SPIE, 2003.
24. M. L. Bittner, *Microarrays: Optical technologies and informatics*. Bellingham, WA: SPIE, 2001.
25. S. L. Carter, A. C. Eklund, B. H. Mecham, I. S. Kohane, and Z. Szallasi, Redefinition of Affymetrix probe sets by sequence overlap with cDNA microarray probes reduces cross-platform inconsistencies in cancer-associated gene expression measurements, *BMC Bioinformatics*, 6, 107, 2005.
26. P. Stafford and M. Brun, Three methods for optimization of cross-laboratory and cross-platform microarray expression data, *Nucleic Acids Research*, doic10.1093/nar/gkl1133, 2007.
27. D. L. Wheeler, T. Barrett, D. A. Benson, S. H. Bryant, K. Canese, V. Chetvernin, D. M. Church, et al., Database resources of the National Center for Biotechnology Information, *Nucleic Acids Research*, 35, D5–12, 2007.
28. M. Zahurak, G. Parmigiani, W. Yu, R. Scharpf, D. Berman, E. Schaeffer, S. Shabbeer, and L. Cope, Preprocessing agilent microarray data, *BMC Bioinformatics*, 8(1), doi1:03943, 2007.
29. A. S. Siddiqui, A. D. Delaney, A. Schnerch, O. L. Griffith, S. J. Jones, and M. A. Marra, Sequence biases in large scale gene expression profiling data, *Nucleic Acids Research*, 34, e83, 2006.
30. J. Quackenbush, Microarray data normalization and transformation, *Nature Genetics*, 32 Suppl, 496–501, 2002.

31. M. B. Eisen, P. T. Spellman, P. O. Brown, and D. Botstein, Cluster analysis and display of genome-wide expression patterns, *Proceedings of the National Academy of Sciences, USA*, 95, 14863–14868, 1998.
32. V. R. Iyer, M. B. Eisen, D. T. Ross, G. Schuler, T. Moore, J. C. Lee, J. M. Trent, et al., The transcriptional program in the response of human fibroblasts to serum, *Science*, 283, 83–87, 1999.
33. J. L. DeRisi, L. Penland, P. O. Brown, M. L. Bittner, P. S. Meltzer, M. Ray, Y. Chen, et al., Use of a cDNA microarray to analyze gene expression patterns in human cancer, *Nature Genetics*, 14, 457–460, 1996.
34. Y. Chen, E. R. Dougherty, and M. L. Bittner, Ratio-based decisions and the quantitative analysis of cDNA microarray images, *Journal of Biomedical Optics*, 2, 364–374, 1997.
35. S. Attoor, E. R. Dougherty, Y. Chen, M. L. Bittner, and J. M. Trent, Which is better for cDNA-microarray-based classification: Ratios or direct intensities? *Bioinformatics*, 5, 177, 2004.
36. S. C. Baker, S. R. Bauer, R. P. Beyer, J. D. Brenton, B. Bromley, J. Burrill, H. Causton, et al., The external RNA controls consortium: A progress report, *Nature Methods*, 2, 731–734, 2005.
37. P. D. Lee, R. Sladek, C. M. Greenwood, and T. J. Hudson, Control genes and variability: Absence of ubiquitous reference transcripts in diverse mammalian expression studies, *Genome Research*, 12, 292–297, 2002.
38. N. Novoradovskaya, M. L. Whitfield, L. S. Basehore, A. Novoradovsky, R. Pesich, J. Usary, M. Karaca, et al., Universal reference RNA as a standard for microarray experiments, *BMC Genomics*, 5, 20, 2004.
39. R. Shyamsundar, Y. H. Kim, J. P. Higgins, K. Montgomery, M. J. A. Sethuraman, M. van de Rijn, D. Botstein, et al., A DNA microarray survey of gene expression in normal human tissues, *Genome Biology*, 6, r22.1–r22.9, 2005.
40. J. Seo and E. P. Hoffman, Probe set algorithms: Is there a rational best bet? *BMC Bioinformatics*, 7, 395, 2006.
41. F. F. Millenaar, J. Okyere, S. T. May, M. van Zanten, L. A. Voesenek, and A. J. Peeters, How to decide? Different methods of calculating gene expression from short oligonucleotide array data will give different results, *BMC Bioinformatics*, 7, 137, 2006.
42. B. M. Bolstad, R. A. Irizarry, M. Astrand, and T. P. Speed, A comparison of normalization methods for high-density oligonucleotide array data based on variance and bias, *Bioinformatics*, 19, 185–193, 2003.
43. R. A. Irizarry, B. M. Bolstad, F. Collin, L. M. Cope, B. Hobbs, and T. P. Speed, Summaries of Affymetrix GeneChip probe level data, *Nucleic Acids Research*, 31, e15, 2003.
44. C. Li and W. H. Wong, Model-based analysis of oligonucleotide arrays: Expression index computation and outlier detection, *Proceedings of the National Academy of Sciences, USA*, 98, 31–36, 2001.
45. A. Oshlack, D. Emslie, L. M. Corcoran, and G. K. Smyth, Normalization of boutique two-color microarrays with a high proportion of differentially expressed probes, *Genome Biology*, 8(1), doi19532, 2007.
46. A. Fujita, J. R. Sato, L. O. Rodrigues, C. E. Ferreira, and M. C. Sogayar, Evaluating different methods of microarray data normalization, *BMC Bioinformatics*, 23, 7:469, 2006.
47. K. V. Ballman, D. E. Grill, A. L. Oberg, and T. M. Therneau, Faster cyclic loess: Normalizing RNA arrays via linear models, *Bioinformatics*, 20, 2778–2786, 2004.
48. L. Y. Liu, N. Wang, J. R. Lupton, N. D. Turner, R. S. Chapkin, and L. A. Davidson, A two-stage normalization method for partially degraded mRNA microarray data, *Bioinformatics*, 21, 4000–4006, 2005.

8 Methodology of Functional Analysis for Omics Data Normalization

Ally Perlina, Tatiana Nikolskaya, and Yuri Nikolsky

CONTENTS

Introduction 173
MetaDiscovery Data Analysis Suite 174
Functional Analysis Steps 175
    Distribution by Canonical Pathways Maps 176
    Distribution by GeneGo Processes 177
    Distribution by GO Processes 179
    Distribution by Diseases 179
Network Generation and Analysis 181
References 203

INTRODUCTION

The purpose of global gene expression assays is to help to explain phenotypes such as disease manifestations, drug response, or diet effect. The commonly used methods of statistical analysis allow one to establish a correlation between the expression profile (usually presented as a selected "gene signature") and the phenotype (reviewed in Draghici [1]), but they are insufficient for understanding the mechanism behind the phenotype. Moreover, gene signatures as descriptors are notorious for poor robustness and reproducibility across platform and experimental conditions. It is now widely accepted that the methods of functional, "systems level" analysis such as pathways, networks, and "interactome" modules are needed for interpretation of "Omics" data sets [2]. For years, functional tools such as KEGG maps [3] and GO ontology [4] were mostly used as a follow-up step after statistical analysis as an attempt to understand the underlying biology behind the gene signature. Since most statistics-derived gene lists make little functional sense, the potential of functional analysis in Omics data mining stayed largely unrealized.

    Only recently has functional analysis started to be developed as a stand-alone methodology for high-throughput data analysis in studies on gene set enrichment analysis [5], "signature networks" [6], and interactome [7,8]. Certainly, both statistical and functional tools are needed for interpretation of enormously massive and complex data sets such as microarray experiments. An important initiative by the FDA known as the MAQC (MicroArray Quality Control) project [13] is currently under way for comparison and development of different statistical and functional techniques in controlled experimental conditions on multiple disease and drug response data sets. Here, we present a functional genomic analysis of tissue expression data in lung versus spleen via different platforms and normalization methods described in Stafford and Brun [12].

METADISCOVERY DATA ANALYSIS SUITE

The analysis was conducted in an integrated functional analysis platform, MetaDiscovery, which we have been developing for over 7 years. MetaCore is a comprehensive suite of data bases and tools for annotation of a wide range of published data: network, pathways, gene enrichment, and interactome analysis [10]. The platform, currently installed at all major drug companies and over 150 academic institutions, is based on a comprehensive collection of protein–protein, protein–DNA, protein–RNA, protein–compound and compound–compound interactions manually curated from "small experiments" literature and patents over the past 6 years. The toolbox is divided into two major modules: (1) MetaCore, which is used for visualization and analysis of Omics data, including mRNA and microRNA gene expression, single nucleotide polymorphisms (SNPs) and other DNA-level data, proteomics, and metabolomics; and (2) MetaDrug, which provides analyses and visualization of small molecule compounds structures, interactions and pathways, and screening data. The system enables a variety of functional analyses ranging from gene set enrichment analysis in GO terms (http://www.geneontology.org/), along with six proprietary ontologies (canonical pathways maps, processes, diseases, toxicity categories, etc.), as well as statistical evaluation of the global interactome [11] to local biological networks built by multiple algorithms and prioritization of hubs on local networks (e.g., convergence and divergence hubs, major receptors, ligands, transcription factors).

The editing tools, MetaLink and MapEditor are used for generating networks from user-discovered interactions, drawing custom pathway maps, editing existing maps, and mapping public source and proprietary data onto canonical or custom maps and networks. MetaCore/MetaDrug is also a robust Omics data warehousing system, consisting of warehoused and searchable data from thousands of genome-wide experiments in cancer and other diseases. The platform is integrated with many third-party packages, varying from statistics-centered data analysis tools (Spotfire, GeneSpring from Agilent, Resolver and Syllago from Rosetta, Phylosopher from GeneData) to knowledge databases (Discovery Gate from MDL/Elsevier, Swiss-Prot, Gene Oontology), data integration software (Inforsense, PipelinePilot from Accelrys), translational medicine packages (Xenobase by Transmed/Van Andel Institute, Inforsense's TM module), and modeling software (ChemTree by GoldenHelix). MetaDiscovery is applied in a wide variety of research, clinical, and drug discovery fields ranging from target and biomarkers discovery to compound prioritization based on indications and toxicity assessment, clinical trials, and selection of drug targets by clinical physicians.

FUNCTIONAL ANALYSIS STEPS

First, gene expression data were collected from different tissues using Affy and Agilent microarray platforms, and then subsequently analyzed with different widely used normalization methods for each. Since it has been previously noted that various normalization methods of CY5 and CY3 Agilent data do not yield great disparity in terms of statistically significant gene groups, we would like to use one representative set of Agilent results (MEAN-normalized method) and compare it with four different normalization results from Affy platform for the same tissues (lung and spleen). To do that, we compare the top 100 differentially expressed genes between lung and spleen via Affy-GC-RMA, Affy-Mas5, Affy-dChip-PM-MM, Affy-dChip-PM only, and Agilent-CY5-MEAN normalization results.

The "compare experiments" workflow allows us to profile common, similar, and unique network objects that result from software parsing and interpretation of each 100-gene set (or "experiment") listed in Figure 8.1. Then, it is important to identify the common underlying themes that describe tissue differences between lung and spleen, before focusing on unique normalization-related pathway biases. Interestingly, perfect match (PM)-only normalized Affy data did not produce any significant overlaps with network objects resulting from the top 100 genes produced by the other four methods (Figure 8.2)

We can now explore the enrichment categories corresponding to the intersection of the four experiments that produced some overlap (marked by a rectangle on Figure 8.2). Enrichment analysis consists of matching gene IDs for the common, similar, and unique sets of the uploaded files with gene IDs in functional ontologies in the MetaCore database. The sorting here will be based on the similar network object (i.e., shared between some, but not all, of the samples). The degree of "relevance" to different categories for the uploaded data sets or intersections is defined by p-values, so the lower p-value gets higher priority.

GebeGo enrichment analysis is based on several specific categories or ontologies. We will focus on four specific categories: canonical pathway maps, GeneGo processes, GO processes, and diseases.

1.	☐	Affy-dChip-PM-MM-Lung-Spleen
2.	■	Affy-dChip-PM-Lung-Spleen_
3.	■	Affy-GC-RMA-Lung-Spleen_
4.	■	Affy-Mas5-Lung-Spleen_
5.	☐	Agilent-CY5-MEAN-Lung-Spleen

FIGURE 8.1 The experiments uploaded for comparative analysis

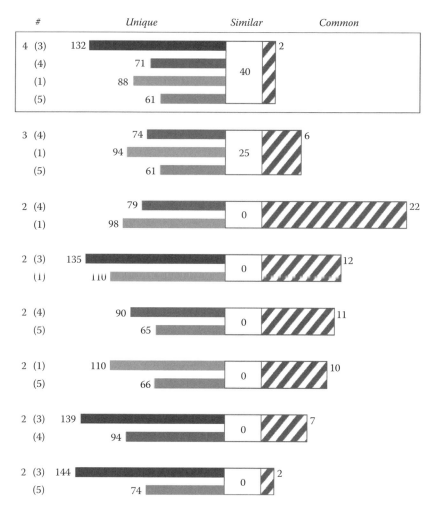

FIGURE 8.2 Experiments' comparison and intersections. The gene content is aligned among all uploaded files. The set of common gene IDs is marked as dark and light stripes. The unique genes for the files are marked in varying shades of gray bars. The genes from the "similar" set are present in all but one (any) file.

DISTRIBUTION BY CANONICAL PATHWAYS MAPS

Canonical pathway maps represent a set of about 500 signaling and metabolic maps covering human biology in a comprehensive way. All maps are drawn from scratch by GeneGo annotators and manually curated. Experimental data are visualized on the maps as blue (for down-regulation) and red (up-regulation) histograms. The height of the histogram corresponds to the relative expression value for a particular gene/protein.

The histogram on Figure 8.3 was generated via similar gene sorting. However, some differences in relative significance are already apparent between different normalization methods. BCR-, PIP3-, and CXCR4-related pathways seem to

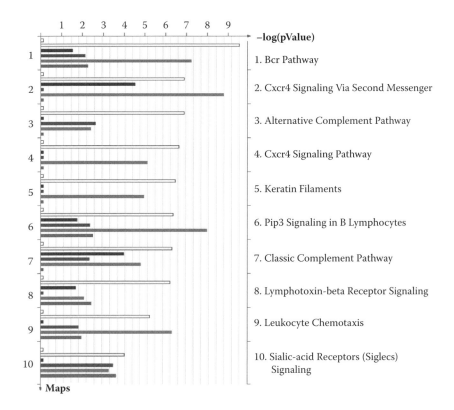

FIGURE 8.3 Distribution by canonical pathways maps. Sorting is done by similar network objects.

be significant with respect to overall similar genes, or more PM-MM (mismatch) normalization-specific set.

The most significant canonical pathway map (Figure 8.4) demonstrates which genes along the pathways are highly significant for only one or more normalization results, designated by red thermometers with the same experiment numbering as on Figure 8.1. Here, we see that CXCR4 itself is highly significant only for the fist experimental set (PM-MM-normalized dChip data), while its ligand, SDF1, is highly significant on other top 100 lists that correspond to MAS5 and Agilent data. This is a clear example of the importance of microarray analysis in the context of functional groups of genes, such as pathway maps, rather than in the context of statistical significance of individual differentially regulated genes.

Distribution by GeneGo Processes

There are about 110 cellular and molecular processes whose content is defined and annotated by GeneGo. Each process represents a prebuilt network of protein interactions characteristic for the process. Experimental data can be mapped on the processes and displayed as red (up-regulation) and blue (down-regulation) circles of different intensity. Relative intensity corresponds to the expression value.

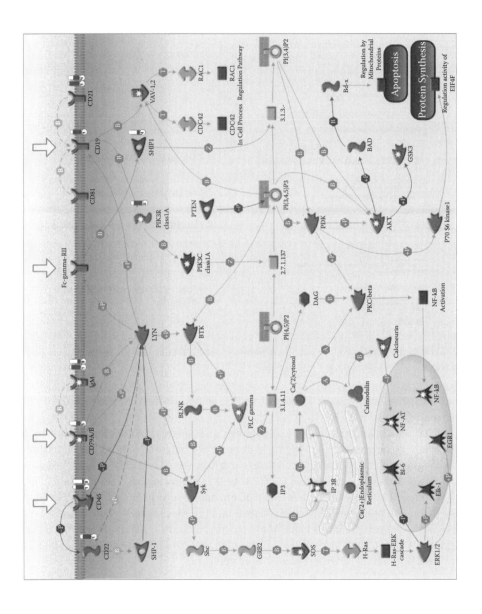

Some of the functional themes (BCR, immune and checmokine-signaling, blood processes/complement activation) that were pronounced in the canonical pathways category can be also seen in the top significant GeneGo processes histogram. Muscle contraction has distinguishable significance in GC-RMA-normalized data and may suggest the potential functional platform bias implicated here.

MetaCore also allows sorting of similar genes based on *differentially affected* ontologies. When we sort by differentially affected GeneGo functional process networks, the differences between normalization methods become even more apparent (Figure 8.8). Adhesion is most significant for Agilent data set, while development and neurogenesis are more highly pronounced in MAS5-normalized data.

DISTRIBUTION BY GO PROCESSES

Although GeneGo processes represent more of a comprehensive, up-to-date overview on gene and interaction levels for each functional process, GO processes are still referred to for ontology analyses. These are the original gene ontology (GO) cellular processes, represented in their own enrichment category within the GeneGo software environment. Since most GO processes have no gene/protein content, the "empty terms" are excluded from p-value calculations.

Due to the redundancy of the hierarchical structure of GO folders, we see mainly immune-response-related categories comprising the top 10 histograms of the graph on Figure 8.9. Gene ontology processes do not present us with prebuilt networks behind each bar, but we can choose to build it for the genes of the significant GO folders, or we can explore individual gene information within some of the representative processes, such as T-cell proliferation, for instance (Table 8.1).

Representative GO process genes were exported to Excel from MetaCore with corresponding values for each evaluated experiment, which represent the genes' order on the top 100 ranked gene list. Here, we see that besides CXCL12, PTPRC, and SFTPD, these genes only appear on one of the top 100 normalization experiment sets each. This again raises the concern of interplatform and internormalization method concordance and, in a way, once again delineates the importance of looking at functional groups of genes for quality evaluation, rather than relying on robustness and reproducibility of individual significant genes.

DISTRIBUTION BY DISEASES

In the same manner, we can evaluate the significant enrichment categories in the context of gene association with various diseases and conditions (Figure 8.10). Disease folders represent over 500 human diseases with gene content annotated by GeneGo. Disease folders are organized into a hierarchical tree. Gene content may

---

FIGURE 8.4 *(facing page)* **(See color insert following page 80.)** BCR pathway. The top scored map (map with the lowest p-value) based on the distribution sorted by similar. Experimental data from all files are linked to and visualized on the maps as thermometer-like figures representing relative p-values assigned during pre-MetaCore data processing and statistical analyses.

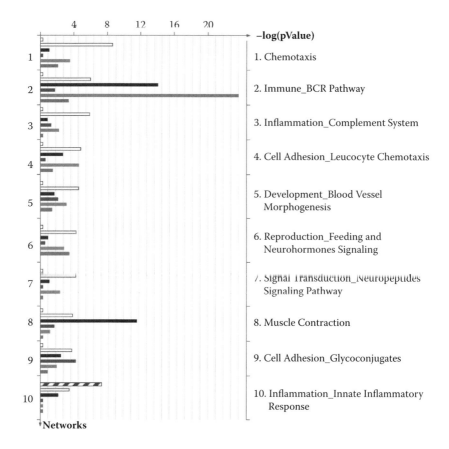

FIGURE 8.5 Distribution by GeneGo processes. Sorting is done by similar network objects.

vary greatly between such complex diseases as cancers and some Mendelian diseases. Also, coverage of different diseases in literature is skewed. These two factors may affect p-value prioritization for diseases.

For more divergent experiment comparisons where no overlap is observed, such as in the case of a PM-normalized dChip data set, one may inquire about the top significant ontologies irrespective of common or similar genes, but taking the entire sample population for distribution analysis as a whole. In MetaCore this can also be achieve by selecting "functional analysis" (histograms not shown). Those distributions display most significant ontologies for solid color bars only, thus allowing us to explore the overall character of the sample set including the PM-normalized gene list information. To see how the overall biological themes change with the inclusion of the PM-normalized sample, we compare the two different types of enrichment analyses across all four categories in Table 8.2. Curiously enough, despite the fact that none of the top 100 genes for the PM-normalized sample set overlapped with other methods, functional analysis reveals that major biological themes distinguishing lung- and tissue-based expression profiles are very similar.

Methodology of Functional Analysis for Omics Data Normalization

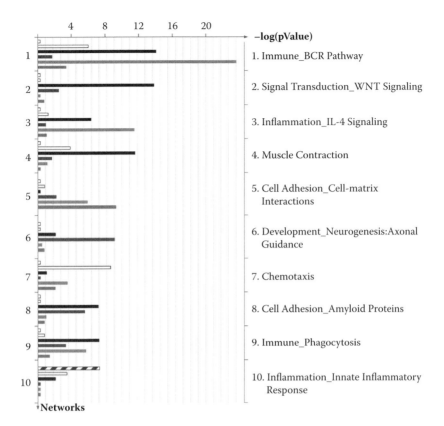

FIGURE 8.6 Distribution by GeneGo processes. Sorting is done by differentially affected networks

This is especially well recapitulated in the "top GeneGo processes" category, whereas the "top canonical pathways" category is more dynamic. It is not unexpected, since only the most well-annotated multistep pathways with numerous literature references make it onto the maps created by GeneGo. Individual gene/protein differences on the level of canonical pathway involvement can be explored and compared in Table 8.5 across all five different microarray platform/normalization methodologies and corresponding gene rankings specific for each.

NETWORK GENERATION AND ANALYSIS

After evaluating main biological themes represented by the most significant functional modules in the enrichment analyses, we can proceed to building networks "on the fly" from the experimental data. The broad functional observations drawn from the analyses of ontologies and prebuilt GeneGo content can serve as a solid base and cross-reference point as we delve into higher levels of gene and interaction data analyzed on the networks. MetaCore allows network construction from common, similar, or unique genes of any data set that was used for comparison. The software also allows leveraging of biological information represented in canonical pathways

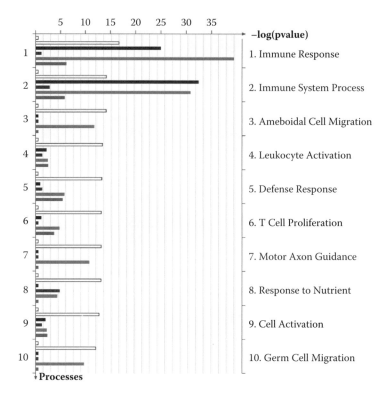

FIGURE 8.7 Distribution by GO processes. Sorting is done by similarly represented processes.

as part of the default setting in the network options. Interactions that are involved in canonical pathways are accounted for by the software in the network building algorithm, which will then highlight and overlay them onto the resulting network.

To explore more details that account for the differences in the five methodologies, we will focus on the networks of genes uniquely represented on individual experimental lists of top 100 most significant genes. We will thereby generate method-specific signature networks all demonstrating differentially affected genes and their interconnectivity between lung and spleen tissue expression data.

The analyze networks (AN) algorithm is used here in accordance with the software default settings (see results in Table 8.3). This is a variant of the shortest paths algorithm with main parameters of (1) relative enrichment with the uploaded data, and (2) relative saturation of networks with canonical pathways. These networks are built on the fly and are unique for the uploaded data. They are subdivided into smaller subnetworks that are prioritized based on the number of fragments of canonical pathways on the network, as well as p-value, z-score, and g-score information. Each subnetwork contains 50 nodes, with root nodes or target genes being those network objects that go into a network building list.

Process distribution is also calculated for the resulting network list on the fly and reported with corresponding percentage representation in each subnetwork. Even

Methodology of Functional Analysis for Omics Data Normalization

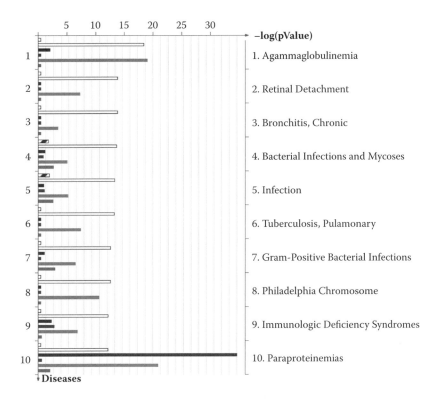

FIGURE 8.8 Distribution by diseases.

without viewing any of the resulting networks, we can see what the main themes are—regulators of which we expect to see. Here, the character and the order of unique signature networks representing each of the five methods are already apparent. Even the recurring themes have different percentage coverage and different prioritization of the corresponding network on the list.

Now we can explore these networks on individual gene and interaction level to learn more about each network specificity, interaction flow, and topology (Figures 8.11–8.13; only top scoring networks are shown for each experiment). All the summary information on main transcriptional regulators, hubs, nodes, and their edges can now be compared using the information provided in the network statistics for the "compare experiments workflow" (Table 8.4). The top three networks are shaded in 3 different colors for each set of unique networks. Although some major regulators occur on signature networks for more than one experiment, their network localization and regulatory prioritization may be strikingly different.

In summary, this sample comparison of only two tissue expression profiles run and normalized on two different platforms and analyzed in MetaCore has allowed us to recognize not only interplatform but also intraplatform differences in the final resulting biological pathway picture. A PM-normalized dChip set, for instance, was more different from other Affy-normalized data than the Agilent-produced results. We have also uncovered the susceptibility of missing important biological pathway,

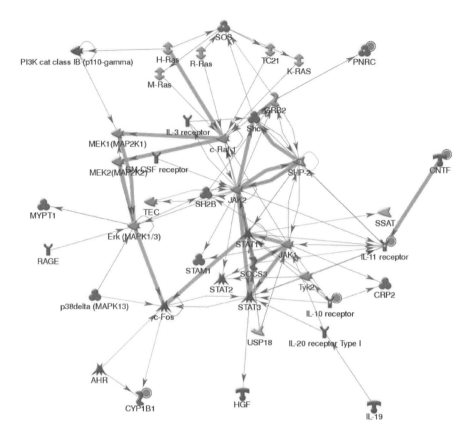

FIGURE 8.9 The top scored (by the number of pathways) AN network from unique Affy-GC-RMA-Lung-Spleen_. Thick lines indicate the fragments of canonical pathways. Up-regulated genes are marked with circles and down-regulated with clustered circles. The "checkerboard" indicates mixed expression for the gene between files or between multiple tags for the same gene.

TABLE 8.1
T-Cell Proliferation

#	Gene Symbol	Protein	Class	Protein name	Affy dChip PM-MM Lung Spleen Gene Rank	Affy GC-RMA Lung Spleen Gene Rank	Affy dChip Lung Spleen Gene Rank	Affy Mas5 Lung Spleen Gene Rank	Agilent-CY5 MEAN Lung Spleen Gene Rank
10	CXCL12	SDF1_HUMAN	T	Stromal cell derived factor 1 precursor				91	17
11	CXCR4	CXCR4_HUMAN	Y	C-X-C chemokine receptor type 4		26			
18	ICOSLG	ICOSL_HUMAN	Y	ICOS ligand precursor			97		
34	PRKCQ	KPCT_HUMAN	◂	Protein kinase C theta type				25	
35	PTPRC	CD45_HUMAN	Y	Leukocyte common antigen precursor		83	16		
36	RIPK2	RIPK2_HUMAN	◂	Receptor-interacting serine/threonine-protein kinase 2					
37	SFTPD	SFTPD_HUMAN	?	Pulmonary surfactant-associated protein D precursor		22		19	

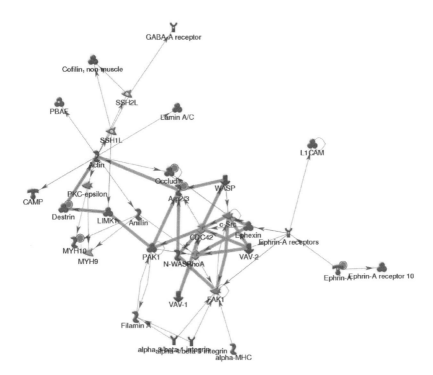

FIGURE 8.10 The top scored (by the number of pathways) AN network from unique Affy-Mas5-Lung-Spleen_. Thick lines indicate the fragments of canonical pathways. Up-regulated genes are marked with circles and down-regulated with clustered circles. The "checkerboard" indicates mixed expression for the gene between files or between multiple tags for the same gene.

process, mechanism, or other ontology data if only individual gene significance were to be taken into account. Here, enrichment analyses helped identify main underlying themes, despite some discrepancies, simply because it was done in a functional grouping-specific way (e.g., a map or network may come up as significant even if completely different genes on it are on the top 100 list for different methods). Hence, relying on individual genes to perform pathway or biomarker analysis would be even less robust or reproducible.

Finally, network construction, uniquely tailored to sample-specific gene lists allowed us to explore the main genes and regulatory steps that may account for some of the observed discrepancies in enrichment analyses, and may be attributable to the normalization method bias. MetaCore helped illustrate that functional module interpretation of microarray data may be more consistent and informative and provide more control while attempting to filter some results of microarray studies.

GeneGo is actively involved with the MAQC project in an effort to help develop more robust and comprehensive, standardized, and more biologically informed ways of assessing the quality and significance of expression data results and choosing the correct unifying path of data processing and normalization.

TABLE 8.2
Enrichment Analyses Summary

Enrichment Analysis Is Method	Top Canonical Pathways Maps	Top GeneGo Processes	Top GO Processes	Top Diseases
Intersections—sorted by similar genes for four experiments[a]: (2) Affy-GC-RM A-Lung-Spleen_; (5) Agilent-CY5-ME AN-Lung-Spleen); (1) Affy-Mas5-Lung-Spleen_; (4) Affy dChip-P M-MM-Lung-Spleen	BCR pathway, CXCR4 signaling via second messenger, alternative complement pathway, CXCR4 signaling pathway, keratin filaments, PIP3 signaling in B lymphocytes, Classic complement pathway, lymphotoxin-beta receptor signaling, leukocyte chemotaxis, sialic-a	Chemotaxis, Immune_BCR pathway, Inflammation_Complement system, Cell adhesion_Leucocyte chemotaxis, Development_Blood vessel morphogenesis, Reproduction_Feeding and Neurohormones signaling, Signal transduction_Neuropeptides signaling pathways, Muscle con	Immune response, immune system process, ameboidal cell migration, leukocyte activation, defense response, T cell proliferation, motor axon guidance, response to nutrient, cell activation, germ cell migration	Agammaglobulinemia, retinal detachment, bronchitis, chronic, bacterial infections and mycoses, infection, tuberculosis, pulmonary, Gram-positive bacterial infections, Philadelphia chromosome, immunologic deficiency syndromes, paraproteinemias
Functional analyses for five experiments (2) Affy-GC-RMA-Lung-Spleen_; (4) Affy-dChip-PM-Lung-Spleen; (5) Agilent-CY5-ME AN-Lung-Spleen); (1) Affy-Mas5-Lung-Spleen_; (4) Affy-dChip-P M-MM-Lung-Spleen	CXCR4 signaling via second messenger, ECM remodeling	Signal Transduction_Cholecystokinin signaling, Proliferation_Positive regulation cell proliferation, Inflammation_Amphoterin signaling, Inflammation_IL-10 anti-inflammatory response, Cell adhesion_Cell junctions, Signal transduction_ESR1-nuclear pathway	T cell proliferation, leukocyte activation, lymphocyte proliferation, cell activation, smooth muscle contraction, regulation of biological quality, lymphocyte activation, T cell activation	Postoperative complications, metabolic diseases, tuberculosis, mycobacterium infections, recurrence, edema, respiratory tract infections, actinomycetales infections, lupus erythematosus, systemic, wounds and injuries

[a] PM method for dChip normalization did not produce any significant network object overlaps with any other lists of top genes generated via other four methods.

TABLE 8.3
Analyze Networks (AN) Summary for All Unique Genes

No	Processes	Size	Target	Pathways	p-Value	zScore
	Unique for Affy-GC-RMA-Lung-Spleen_					
1	Signal transduction (80.0%), regulation of Notch signaling pathway (13.3%), cell communication (80.0%)	50	10	25	1.90e-13	18.39
2	Immune response (88.4%), immune system process (88.4%), response to stimulus (93.0%)	50	17	0	1.02e-23	30.10
3	I-kappaB kinase/NF-kappaB cascade (30.4%), positive regulation of I-kappaB kinase/NF-kappaB cascade (19.6%), regulation of I-kappaB kinase/NF-kappaB cascade (19.6%)	50	6	7	3.67e-07	10.82
	Unique for Affy-Mas5-Lung-Spleen_					
1	Actin cytoskeleton organization and biogenesis (43.5%), actin filament-based process (43.5%), cytoskeleton organization and biogenesis (50.0%)	50	6	24	6.83e-09	15.40
2	Developmental process (78.9%), cell development (55.3%), cell differentiation (60.5%)	50	11	0	1.56e-16	26.19
3	Cell surface receptor linked signal transduction (48.9%), tissue remodeling (20.0%), cell communication (73.3%)	50	7	0	1.13e-10	18.03
	Unique for Affy-dChip-PM-MM-Lung-Spleen					
1	Immune response (69.6%), immune system process (71.7%), response to stimulus (73.9%)	50	17	0	3.62e-18	27.56
2	Localization of cell (40.0%), cell motility (40.0%), cell adhesion (38.0%)	50	8	6	9.83e-12	18.43
3	Rho protein signal transduction (17.0%), calcium ion transport (17.0%), Ras protein signal transduction (17.0%)	50	7	3	5.51e-10	16.07
	Unique for Agilent-CY5-MEAN-Lung-Spleen					
1	Positive thymic T cell selection (8.9%), positive T cell selection (8.9%), thymic T cell selection (8.9%)	50	13	0	8.36e-22	34.72
2	Cell adhesion (54.2%), biological adhesion (54.2%), cell-matrix adhesion (25.0%)	50	9	2	4.55e-15	25.15

(continued on next page)

TABLE 8.3 (continued)
Analyze Networks (AN) Summary for All Unique Genes

No	Processes	Size	Target	Pathways	p-Value	zScore
3	I-kappaB kinase/NF-kappaB cascade (20.5%), positive regulation of biological process (53.8%), signal transduction (69.2%)	50	5	8	1.04e-07	14.43
	Unique for Affy-dChip-PM-Lung-Spleen					
1	Organelle organization and biogenesis (45.8%; 9.984e-10), ruffle organization and biogenesis (8.3%; 3.939e-09), negative regulation of receptor-mediated endocytosis (6.2%; 1.541e-07), small GTPase-mediated signal transduction (18.8%; 2.360e-07), cellular component organization and biogenesis (58.3%; 1.401e-06)	50	10	0	2.64e-15	22.81
2	Immune response (40.9%; 1.098e-10), immune system process (47.7%; 1.160e-10), recombinational repair (9.1%; 6.828e-08), double-strand break repair via homologous recombination (9.1%; 6.828e-08), response to stimulus (61.4%; 1.393e-07)	50	10	0	3.29e-15	22.58
3	Intracellular signaling cascade (55.6%; 8.118e-12), signal transduction (77.8%; 2.250e-11), regulation of nucleobase, nucleoside, nucleotide, and nucleic acid metabolic process (55.6%; 1.744e-10), regulation of transcription (53.3%; 3.239e-10), regulation of cellular metabolic process (57.8%; 8.565e-10)	50	9	0	2.35e-13	20.27

Methodology of Functional Analysis for Omics Data Normalization 189

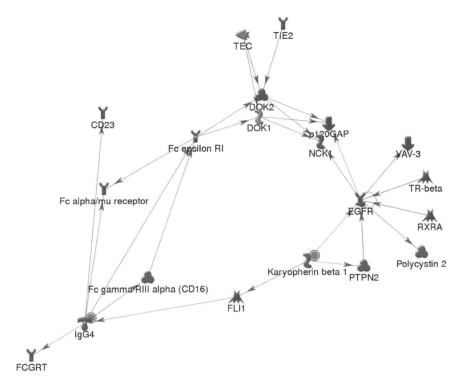

FIGURE 8.11 The top scored (by the number of pathways) AN network from unique Affy-dChip-PM-MM-Lung-Spleen. Thick lines indicate the fragments of canonical pathways. Up-regulated genes are marked with circles and down-regulated with clustered circles. The "checkerboard" indicates mixed expression for the gene between files or between multiple tags for the same gene.

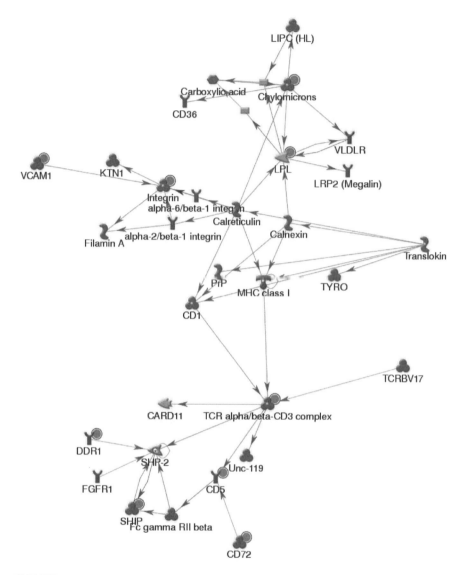

FIGURE 8.12 The top scored (by the number of pathways) AN network from unique Agilent-CY5-MEAN-Lung-Spleen. Thick lines indicate the fragments of canonical pathways. Up-regulated genes are marked with circles and down-regulated with clustered circles. The "checkerboard" indicates mixed expression for the gene between files or between multiple tags for the same gene.

Methodology of Functional Analysis for Omics Data Normalization 191

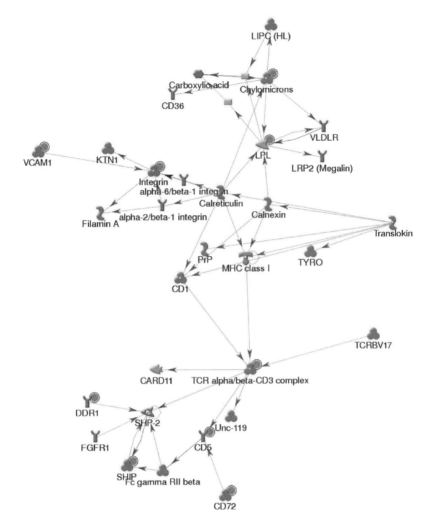

FIGURE 8.13 The top scored (by the number of pathways) AN network from unique Agilent-CY5-MEAN-Lung-Spleen. Thick lines indicate the fragments of canonical pathways. Up-regulated genes are marked with circles and down-regulated with clustered circles. The "checkerboard" indicates mixed expression for the gene between files or between multiple tags for the same gene.

TABLE 8.4
List of Most Relevant Network Objects Based on Network Analysis

	Hubs				Transcription Factors		Receptors		Secreted Proteins	
	Divergence Hubs	Convergence Hubs	Edges IN	Edges OUT	Edges IN	Edges OUT	Edges IN	Edges OUT	Edges IN	Edges OUT
Common nonredundant list in five networks	SHP-1	VAV-1	SHP-2	SHP-2	IRF1	ELF1	PAR4	IL-12 receptor	CCL5	CAMP
	p53	p53	STAT1	STAT1	IRF8	IRF1	TLR2	PAR4	IP10	PSAP
	JAK2	Androgen receptor	SHP-1	SHP-1	STAT1	IRF8	FPRL1	TLR2	CAMP	TRIP6
	STAT3	FAK1	IRF8	IRF8	STAT5	STAT1	PAR4	IL-17 receptor	CDK5	APP
	A2M	STAT3	JAK2	JAK2	Androgen receptor	STAT5	TLR2	IL-18R1	PSAP	CDK5
		APP	IRF1	IRF1	Bcl-6	Androgen receptor	TLR4	PAR4	TGF-beta 2	Conglutinin
		IL-1 beta	Lck	Lck	p53	Bcl-6	Alpha-V/beta-3 integrin	TLR2	TRIP6	PSAP
			Shc	Shc	PML	p53	C1qRp	TLR4	APP	TGF-beta 2
			SHPS-1	SHPS-1	SRF	PML	CD21	Alpha-5/beta-1 integrin	Conglutinin	TRIP6
			Pyk2(FAK2)	Pyk2(FAK2)	Androgen receptor	PU.1	A2M receptor	Alpha-9/beta-1 integrin	IL-18	CDK5
			STAT5	STAT5	c-Fos	SRF	Alpha-5/beta-1 integrin	Alpha-V/beta-3 integrin	CAMP	A2M
			VAV-1	VAV-1	c-Myc	AHR	Alpha-M/beta-2 integrin	C1qRp	CCL5	APP
			CD244	CD244	p53	Androgen receptor	Alpha-V/beta-3 integrin	RAGE	A2M	NGF
			p53	P53	STAT1	c-Fos	Bonzo	A2M receptor	APP	PDGF-B

Androgen receptor	Androgen receptor	STAT3	c-Myc	IGF-2 receptor	L-selectin	IL-1 beta	PSAP
Kallikrein 7	PU.1	AHR	p53	L-selectin	PLAUR (Upar)	IP10	TGF-beta 2
Cathepsin G	Kallikrein 7	Androgen receptor	STAT1	PAR4	TrkA	NGF	
CDK1 (p34)	Cathepsin G	Bcl-6	STAT3	PLAUR (uPAR)		PDGF-B	
Kallikrein 2	CDK1 (P34)	c-Myc	AHR	TrkA			
Kallikrein 3 (PSA)	Kallikrein 2	IRF1	Androgen receptor			PSAP	
MDM2	Kallikrein 3 (PSA)	IRF8	Bcl-6			TGF-beta 2	
MyD88	MDM2	NPAS2	c-Myc				
TRIP6	TRIP6	p53	IRF1				
FAK1	FAK1	PML	IRF8				
JAK2	JAK2	PU.1	p53				
SHP-2	SHP-2	STAT3	PML				
c-Fos	c-Fos	STAT5	PU.1				
STAT3	STAT3	c-Fos	STAT3				
STAT1	STAT1	IRF1	STAT5				
APP	APP	IRF8	c-Fos				
GRB2	GRB2	PU.1	IRF1				
Pyk2(FAK2)	Pyk2(FAK2)	STAT1	IRF8				
c-Myc	c-Myc	STAT3	PU.1				
p53	p53		SRF				
Paxillin	Paxillin		STAT1				
SHPS-1	SHPS-1		STAT3				
STAT3	STAT3						
c-Myc	c-Myc						
IRF8	IRF8						
P53	P53						
IRF1	IRF1						

(cpmtomied pm mext page)

TABLE 8.4 (continued)
List of Most Relevant Network Objects Based on Network Analysis

Divergence Hubs	Convergence Hubs	Hubs		Transcription Factors		Receptors		Secreted Proteins	
		Edges IN	Edges OUT	Edges IN	Edges OUT	Edges IN	Edges OUT	Edges IN	Edges OUT
		SHP-1	SHP-1						
		Erk (MAPK1/3)	Erk (MAPK1/3)						
		Bcl-2	Bcl-2						
		JAK2	JAK2						
		STAT5	STAT5						
		Lck	Lck						
		PU.1	PU.1						
		Bcl-6	Bcl-6						
		A2M	A2M						
		c-Fos	c-Fos						
		MMP-9	MMP-9						
		PLAUR (uPAR)	PLAUR (uPAR)						
		STAT1	STAT1						
		STAT3	STAT3						
		A2M receptor	A2M receptor						
		APP	APP						
		TrkA	TrkA						
		IL-1 beta	IRF8						
		IRF8	Trypsin						
		Trypsin	IRF1						
		iNOS							
		IRF1							

Methodology of Functional Analysis for Omics Data Normalization 195

Experiment Affy (unique) nonredundant list in three networks	Calpactin I light chain	Cholesterol	EGFR	EGFR	ATF-4	ATF-4	CXCR5	CXCR5	HDL	Calpactin I light chain
	c-Src	Ephrin-A receptors	Cholesterol	Calpactin I light chain	ATF-4	NRF2	Galpha(i)-specific peptide GPCRs	MUNC13-2	CAMP	HB-EGF
	Ephrin-A1	SMAD3	HDL	Cholesterol	c-Myc	c-Myc	Galpha(q)-specific peptide GPCRs	NPY1R	Ephrin-A	HDL
	c-Myc		HTR1B	HDL	NRF2	KLF4	SR-BI	SR-BI	APP	Ephrin-A
			c-Src	HTR1B	SMAD2	NRF2	Ephrin-A receptor 1	APOER2	BMP4	Ephrin-A1
			Ephrin-A receptors	c-Src	SMAD3	SMAD3	Ephrin-A receptor 2	Ephrin-A receptor 1	IGF-2	APP
			FAK1	Ephrin-A receptors	SMAD4	SMAD4	Ephrin-A receptor 4	Ephrin-A receptor 2		Calpactin I light chain
			Fyn	FAK1			Ephrin-A receptor 5	Ephrin-A receptor 4		
			CDC42	Fyn			Ephrin-A receptor 6	Ephrin-A receptor 5		
			Actin cytoskeletal	CDC42			Ephrin-A receptor 7	Ephrin-A receptor 6		
			PAK1	Ephrin-A1			Ephrin-A receptor 8	Ephrin-A receptor 7		
			RhoA	Actin cytoskeletal			Ephrin-A receptors	Ephrin-A receptor 8		
			VAV-2	PAK1			GABA-A receptor	Ephrin-A receptors		

(cpmtomied pm mext page)

TABLE 8.4 (continued)
List of Most Relevant Network Objects Based on Network Analysis

Divergence Hubs	Convergence Hubs	Hubs		Transcription Factors		Receptors		Secreted Proteins	
		Edges IN	Edges OUT	Edges IN	Edges OUT	Edges IN	Edges OUT	Edges IN	Edges OUT
		Actin	RhoA			GABA-A receptor alpha-1 subunit	Fc alpha/mu receptor		
		Arp2/3	VAV-2			GABA-A receptor alpha-2 subunit	GABA-A receptor alpha-1 subunit		
		Ephrin-A receptor 4	Actin			Alpha-4/beta-1 integrin	GABA-A receptor alpha-2 subunit		
		DAB1	Arp2/3			Alpha-9/beta-1 integrin	VLDLR		
		c-Myc	Ephrin-A receptor 4			CD23	alpha-4/beta-1 integrin		
		SMAD3	DAB1			ICAM1	alpha-9/beta-1 integrin		
		Actin cytoskeletal	c-Myc			Nod1	ICAM1		
		JNK(MAPK 8–10)	SMAD3						
		PKC-epsilon	Actin cytoskeletal						
		RBP-J kappa (CBF1)	JNK(MAPK8–10)						
		SMAD2	PKC-epsilon						

Experiment Affy (unique) nonredundant list in three networks										
IGH@	IgA1	EGFR	SMAD4 RBP-J kappa (CBF1) SMAD4 TAK1(MAP3K7)	EGFR	SP1	RXRA	ASGR2	Fc alpha receptor	IgA2	IgA2
Lyn	IgG2	IgA1		IGH@	STAT3	TR-beta	CD23	Fc epsilon RI	IgA2 heavy chain	IgE
c-Src	RhoA	Kappa chain (Ig light chain)		IgA1		SMAD4	Fc alpha receptor	Fc epsilon RI gamma	IgE	IgE (heavy chain)
	FAK1	IgA2		Kappa chain (Ig light chain)		SP1	Fc alpha/mu receptor	ITGB1	IgE (heavy chain)	IgG4 (heavy chain)
		Lambda chain (Ig light chain)		IgA2		STAT3	Fc epsilon RI	PGE2R1	IgG	IGH@
		DOK1		Lambda chain (Ig light chain)			Fc epsilon RI gamma	TIE2	IgG2	IGHA1
		DOK2		DOK1			ITGB3	CD22	IgG2 (heavy chain)	IGHE
		IgE		DOK2			CD22	CD45	IgG4	IGHG4
		IgG2		IgE			CD79 complex	CD79 complex	IgG4 (heavy chain)	IgK
		IgM		IgM			CD79A	CD79A	IgM	IGKC
		NCK1		NCK1			TBXA2R	TBXA2R	Kappa chain (Ig light chain)	IGLC2

(continued on next page)

TABLE 8.4 (continued)
List of Most Relevant Network Objects Based on Network Analysis

	Hubs			Transcription Factors		Receptors		Secreted Proteins	
Divergence Hubs	Convergence Hubs	Edges IN	Edges OUT	Edges IN	Edges OUT	Edges IN	Edges OUT	Edges IN	Edges OUT
		Lyn	Lyn			Alpha-10/beta-1 integrin	Alpha-10/beta-1 integrin	Lambda chain (Ig light chain)	IgM
		VAV-1	VAV-1			Alpha-2/beta-1 integrin	Alpha-2/beta-1 integrin	CAMP	Kappa chain (Isg light chain)
		SHP-1	SHP-1			Alpha-3/beta-1 integrin	Alpha-3/beta-1 integrin	Laminin 5	Lambda chain (Ig light chain)
		Rac1	Rac1			Alpha-6/beta-1 integrin	Alpha-6/beta-1 integrin	Laminin 6	LAMA3 (Epiligrin)
		RhoA	SP1			Alpha-V/beta-3 integrin	Alpha-V/beta-3 integrin	Laminin 7	Laminin 5
		SP1	CDC42			C4.4A	CCR2	SDF-1	Laminin 6
		CDC42	RhoGDI alpha			CCR2	CXCR4		SDF-1
		RhoGDI alpha	CD19			CXCR4	DDR2		Thrombin
		CD19	LyGDI			ITGB1	ITGB1		
		LyGDI	CD22			ITGB3	ITGB3		
		CD22	SHIP			PAR1	LTBR2		
		SHIP	SHIP2			PAR4	PAR1		
		SHIP2	TEC			Syndecan-2	PAR4		

Methodology of Functional Analysis for Omics Data Normalization

Network 1[a]	Network 2[b]	Network 5[e]
TEC	c-Src	PGE2R1
c-Src	FAK1	SLC-1
FAK1	CXCR4	Syndecan-2
CXCR4	G-protein alpha-i family	
G-protein alpha-1 family	Alpha-3/beta-1 integrin	
Alpha-3/beta-1 integrin	JAK2	
JAK2	Laminin 5	
Laminin 5	Talin	
Talin	Alpha-2/beta-1 integrin	
Alpha-2/beta-1 integrin	G-protein alpha-q/11	
G-protein alpha-q/11	ITGB1	
ITGB1	PAR1	
PAR1	STAT3	
STAT3	CCR2	
CCR2		

[a] Network 1
[b] Network 2
[c] Network 3
[d] Network 4
[e] Network 5

TABLE 8.5
Canonical Map Genes

Expression data for genes that made it into Canonical Pathway Maps only

#	Gene Symbol	Protein	Class	Protein name	affy-dChip-PM-MM Lung -Spleen Gene Rank	affy-GC-RMA-Lung-Spleen Gene Rank	affy-dChip-Lung -Spleen Gene Rank	affy-Mas5-Lung -Spleen Gene Rank	Agilent-CY5-MEAN Lung -Spleen Gene Rank
1	ACTR2	ARP2_HUMAN		Actin-like protein 2				57	
2	ADRA1B	ADA1B_HUMAN	Y	Alpha-1B adrenergic receptor		77			
3	ADRBK1	ARBK1_HUMAN		Beta-adrenergic receptor kinase 1		80			
4	AKR1C1	AK1C1_HUMAN		Aldo-keto reductase family 1 member C1					20
5	ALDH1A3	AL1A3_HUMAN		Aldehyde dehydrogenase 1A3				85	
6	APOC3	APOC3_HUMAN		Apolipoprotein C-III precursor					36
7	ARF6	ARF6_HUMAN		ADP-ribosylation factor 6			88		
8	AXIN1	AXN1_HUMAN		Axin-1		12			
9	BCDO2	BCDO2_HUMAN		Beta,beta-carotene 9',10'-dioxygenase		26			
10	CACNB1	CACB1_HUMAN	X	Voltage-dependent L-type calcium channel subunit beta-1		61			
11	CD19	CD19_HUMAN		B-lymphocyte antigen CD19 precursor	91				
12	CD22	CD22_HUMAN		B-cell receptor CD22 precursor	65				
13	CD79B	CD79B_HUMAN	Y	B-cell antigen receptor complex-associated protein beta-chain precursor					41
14	CDK5R1	CD5R1_HUMAN		Cyclin-dependent kinase 5 activator 1 precursor		40			
15	CFP	PROP_HUMAN		Properdin precursor	18				30
16	CGB	CGHB_HUMAN		Choriogonadotropin subunit beta precursor			79		
17	CGB7	CGHB_HUMAN		Choriogonadotropin subunit beta precursor			79		
18	CHRNG	ACHG_HUMAN	Y	Acetylcholine receptor protein subunit gamma precursor			92		
19	CLDN5	CLD5_HUMAN		Claudin-5					63
20	CLU	CLUS_HUMAN		Clusterin precursor				72	
21	CNTF	CNTF_HUMAN		Ciliary neurotrophic factor		96			
22	CR1	CR1_HUMAN	Y	Complement receptor type 1 precursor	47				
23	CR2	CR2_HUMAN	Y	Complement receptor type 2 precursor	96				65
24	CSF1R	CSF1R_HUMAN		Macrophage colony-stimulating factor 1 receptor precursor	34				
25	CSNK1G2	KC1G2_HUMAN		Casein kinase I isoform gamma-2		36			
26	CTNNA1	CTNA1_HUMAN		Catenin alpha-1				30	
27	CXCL12	SDF1_HUMAN		Stromal cell-derived factor 1 precursor				91	17
28	CXCL13	SCYB0_HUMAN		Small inducible cytokine B13 precursor	6				
29	CXCR4	CXCR4_HUMAN		C-X-C chemokine receptor type 4	26				
30	CYP1B1	CP1B1_HUMAN		Cytochrome P450 1B1		95			
31	CYP27A1	CP27A_HUMAN		Cytochrome P450 27, mitochondrial precursor				101	
32	CYP2E1	CP2E1_HUMAN		Cytochrome P450 2E1					28
33	DSTN	DEST_HUMAN		Destrin				39	
34	DYNLRB1	DLC2A_HUMAN		Dynein light chain 2A, cytoplasmic				83	
35	EFNA1	EFNA1_HUMAN		Ephrin-A1 precursor				45	
36	ENO1	ENOA_HUMAN		Alpha-enolase				36	
37	EXOSC2	EXOS2_HUMAN		Exosome complex exonuclease RRP4				10	
38	F12	FA12_HUMAN		Coagulation factor XII precursor		16			
39	F2R	PAR1_HUMAN	Y	Proteinase-activated receptor 1 precursor	73				
40	FABP4	FABPA_HUMAN		Fatty acid-binding protein, adipocyte				74	
41	FASLG	TNFL6_HUMAN		Tumor necrosis factor ligand superfamily member 6		74			
42	FBLN1	FBLN1_HUMAN		Fibulin-1 precursor				98	
43	FBP1	F16P1_HUMAN		Fructose-1,6-bisphosphatase 1					8
44	FGFBP1	FGFP1_HUMAN		Fibroblast growth factor-binding protein 1 precursor			12		
45	FIGF	VEGFD_HUMAN		Vascular endothelial growth factor D precursor				41	
46	FN1	FINC_HUMAN		Fibronectin precursor				12	
47	FZD1	FZD1_HUMAN	Y	Frizzled-1 precursor		14			
48	GALK2	GALK2_HUMAN		N-acetylgalactosamine kinase			83		
49	GATM	GATM_HUMAN		Glycine amidinotransferase, mitochondrial precursor					61
50	GK	GLPK_HUMAN		Glycerol kinase		45			
51	GK3P	GLPK3_HUMAN		Glycerol kinase, testis specific 1		45			
52	GNAQ	GNAQ_HUMAN		Guanine nucleotide-binding protein G(q) subunit alpha		53			
53	GNB5	GBB5_HUMAN		Guanine nucleotide-binding protein subunit beta 5		72			

Methodology of Functional Analysis for Omics Data Normalization

TABLE 8.5 (continued)
Canonical Map Genes

#	Gene	UniProt	Description					
				5				
54	GTF2A2	T2AG_HUMAN	Transcription initiation factor IIA gamma chain				37	
55	HBA1	HBA_HUMAN	Hemoglobin subunit alpha					90
56	HBEGF	HBEGF_HUMAN	Heparin-binding EGF-like growth factor precursor				80	
57	HIST1H1T	H1T_HUMAN	Histone H1t			7		
58	HIST1H4C	H4_HUMAN	Histone H4			97		
59	HMOX1	HMOX1_HUMAN	Heme oxygenase 1	14			23	
60	HPGD	PGDH_HUMAN	15-hydroxyprostaglandin dehydrogenase [NAD+]					69
61	HSPA14	HSPA14	heat shock 70kDa protein 14			16		
62	ICOSLG	ICOSL_HUMAN	ICOS ligand precursor			97		
63	IFI6	IN2_HUMAN	Interferon-induced protein 6-16 precursor		9			
64	IFNA10	IFN10_HUMAN	Interferon alpha-10 precursor				89	
65	IFNA17	IFN17_HUMAN	Interferon alpha-17 precursor				89	
66	IFNA4	IFNA4_HUMAN	Interferon alpha-4 precursor				89	
67	IFNA7	IFNA7_HUMAN	Interferon alpha-7 precursor				89	
68	IGF1R	IGF1R_HUMAN	Insulin-like growth factor 1 receptor precursor				53	
69	IGHG1	IGHG1_HUMAN	Ig gamma-1 chain C region		9			
70		MUCM_HUMAN	Ig mu chain C region membrane-bound segment	95	9			
	IGHM	MUC_HUMAN	Ig mu chain C region	76				
71	IGHV1-69			76				
72	IGHV3-30			95				
73	IGLV6-57				30			
74	IL10RA	I10R1_HUMAN	Interleukin-10 receptor alpha chain precursor					94
75	IL6ST	IL6RB_HUMAN	Interleukin-6 receptor subunit beta precursor		11			
76	IL8	IL8_HUMAN	Interleukin-8 precursor		9			
77	INPP5D	INPP5D	inositol polyphosphate-5-phosphatase, 145kDa			25		
78	ITGB7	ITB7_HUMAN	Integrin beta-7 precursor					83
79	KPNB1	IMB1_HUMAN	Importin beta-1 subunit	60				
80	KRT19	K1C19_HUMAN	Keratin, type I cytoskeletal 19	8			13	10
81	KRT7	K2C7_HUMAN	Keratin, type II cytoskeletal 7	101			45	
82	KRT8	K2C8_HUMAN	Keratin, type II cytoskeletal 8	71				
83	LAMA3	LAMA3_HUMAN	Laminin subunit alpha-3 precursor	65				
84	LAMA4	LAMA4_HUMAN	Laminin subunit alpha-4 precursor					95
85	LAT	LAT_HUMAN	Linker for activation of T-cells family member 1		25			
86	LCP2	LCP2_HUMAN	Lymphocyte cytosolic protein 2	67				
87	LPL	LIPL_HUMAN	Lipoprotein lipase precursor	12				23
88	LTB	TNFC_HUMAN	Lymphotoxin-beta					90
89	MAG	MAG_HUMAN	Myelin-associated glycoprotein precursor	85				
90	MAP2K3	MP2K3_HUMAN	Dual specificity mitogen-activated protein kinase kinase 3			54		
91	MAP3K7	M3K7_HUMAN	Mitogen-activated protein kinase kinase kinase 7			93		
92	MGST1	MGST1_HUMAN	Microsomal glutathione S-transferase 1					38
93	MMP7	MMP7_HUMAN	Matrilysin precursor					54
94	MMP9	MMP9_HUMAN	Matrix metalloproteinase-9 precursor	90				
95	MRCL3	MLRM_HUMAN	Myosin regulatory light chain 2, nonsarcomeric	75				
96	MUC1	MUC1_HUMAN	Mucin-1 precursor				5	
97	MYH10	MYH10_HUMAN	Myosin-10				26	
98	NDUFB2	NDUB2_HUMAN	NADH dehydrogenase [ubiquinone] 1 beta subcomplex subunit 2, mitochondrial precursor			67		
99	NPY1R	NPY1R_HUMAN	Neuropeptide Y receptor type 1				52	5
100	NRCAM	NRCAM_HUMAN	Neuronal cell adhesion molecule precursor		47			
101	OCLN	OCLN_HUMAN	Occludin					70
102	PABPC1	PABP1_HUMAN	Polyadenylate-binding protein 1			82		
103	PAPSS2	PAPS2_HUMAN	Bifunctional 3'-phosphoadenosine 5'-phosphosulfate synthetase 2				42	
104	PDE2A	PDE2A_HUMAN	cGMP-dependent 3',5'-cyclic phosphodiesterase	30				
105	PDE4A	PDE4A_HUMAN	cAMP-specific 3',5'-cyclic phosphodiesterase				22	

TABLE 8.5 (continued)
Canonical Map Genes

#	Gene	Protein	Description	C1	C2	C3	C4	C5	C6
88	LTB	TNFC_HUMAN	Lymphotoxin-beta						96
89	MAG	MAG_HUMAN	Myelin-associated glycoprotein precursor	85					
90	MAP2K3	MP2K3_HUMAN	Dual specificity mitogen-activated protein kinase kinase 3		54				
91	MAP3K7	M3K7_HUMAN	Mitogen-activated protein kinase kinase kinase 7		93				
92	MGST1	MGST1_HUMAN	Microsomal glutathione S-transferase 1						38
93	MMP7	MMP7_HUMAN	Matrilysin precursor						54
94	MMP9	MMP9_HUMAN	Matrix metalloproteinase-9 precursor	90					
95	MRCL3	MLRM_HUMAN	Myosin regulatory light chain 2, nonsarcomeric		75				
96	MUC1	MUC1_HUMAN	Mucin-1 precursor					5	
97	MYH10	MYH10_HUMAN	Myosin-10					26	
98	NDUFB2	NDUB2_HUMAN	NADH dehydrogenase [ubiquinone] 1 beta subcomplex subunit 2, mitochondrial precursor		67				
99	NPY1R	NPY1R_HUMAN	Neuropeptide Y receptor type 1					52	5
100	NRCAM	NRCAM_HUMAN	Neuronal cell adhesion molecule precursor		47				
101	OCLN	OCLN_HUMAN	Occludin						70
102	PABPC1	PABP1_HUMAN	Polyadenylate-binding protein 1		62				
103	PAPSS2	PAPS2_HUMAN	Bifunctional 3'-phosphoadenosine 5'-phosphosulfate synthetase 2					42	
104	PDE2A	PDE2A_HUMAN	cGMP-dependent 3',5'-cyclic phosphodiesterase	30					
105	PDE4A	PDE4A_HUMAN	cAMP-specific 3',5'-cyclic phosphodiesterase 4A			22			
106	PDE6C	PDE6C_HUMAN	Cone cGMP-specific 3',5'-cyclic phosphodiesterase alpha'-subunit			55			
107	PDGFA	PDGFA_HUMAN	Platelet-derived growth factor A chain precursor					80	
108	PIK3R1	P85A_HUMAN	Phosphatidylinositol 3-kinase regulatory subunit alpha		81				
109	[illegible]	[illegible]	[illegible]						
110	PPP2R5C	2A5G_HUMAN	Serine/threonine-protein phosphatase 2A 56 kDa regulatory subunit gamma isoform		71				
111	PPP2R5D	2A5D_HUMAN	Serine/threonine-protein phosphatase 2A 56 kDa regulatory subunit delta isoform		47				
112	PRKCQ	KPCT_HUMAN	Protein kinase C theta type			25			
113	PTCH2	PTC2_HUMAN	Protein patched homolog 2		34				
114	PTPRC	CD45_HUMAN	Leukocyte common antigen precursor	83	15				
115	RBBP8	RBBP8_HUMAN	Retinoblastoma-binding protein 8		83				
116	ROCK2	ROCK2_HUMAN	Rho-associated protein kinase 2		34				
117	RTN4	RTN4_HUMAN	Reticulon-4		85				
118	RUNX1	RUNX1_HUMAN	Runt-related transcription factor 1		70				
119	SELL	LYAM1_HUMAN	L-selectin precursor	33				20	
120	SERPINA5	IPSP_HUMAN	Plasma serine protease inhibitor precursor	54					
121	SFTPA1	SFTA1_HUMAN	Pulmonary surfactant-associated protein A1 precursor	2	19			7	
122	SMAD3	SMAD3_HUMAN	Mothers against decapentaplegic homolog 3					60	
123	SPTBN1	SPTB2_HUMAN	Spectrin beta chain, brain 1				84		
124	SREBF1	SRBP1_HUMAN	Sterol regulatory element-binding protein 1	62					
125	TAF9B	TAF9B_HUMAN	Transcription initiation factor TFIID subunit 9B	30					
126	TCF3	TFE2_HUMAN	Transcription factor E2-alpha		72				
127	TF	TRFE_HUMAN	Serotransferrin precursor						19
128	TNFSF8	TNFL8_HUMAN	Tumor necrosis factor ligand superfamily member 8		87				
129	TRA@	TCA_HUMAN	T-cell receptor alpha chain C region						71
130	UGCG	CEGT_HUMAN	Ceramide glucosyltransferase	64					
131	UQCRB	UCR6_HUMAN	Ubiquinol-cytochrome c reductase complex 14 kDa protein		43				
132	UROD	DCUP_HUMAN	Uroporphyrinogen decarboxylase					60	
133	VCAM1	VCAM1_HUMAN	Vascular cell adhesion protein 1 precursor	13					56
134	WIF1	WIF1_HUMAN	Wnt inhibitory factor 1 precursor					16	

REFERENCES

1. Draghici, S. *Data analysis tools for DNA microarrays*. New York: Chapman & Hall, 2003.
2. Hood, L., Perlmutter, R. M. The impact of systems approaches on biological problems in drug discovery. *Nat. Biotech.*, 2(22), 1218–1219.
3. Kanehisa, M. et al. The KEGG resource for dicephering the genome. *Nucl. Acids Res.*, 2004, 32, D277–D280.
4. http://www.geneontology.org/index.shtml
5. Subramanian, A., Tamayo, P., Mootha, V. M., et al. Gene set enrichment analysis: A knowledge-based approach for interpreting genome-wide expression profiles. *Proc. Natl. Acad. Sci. USA*, 2005, 102(43), 15545–15550.
6. Nikolsky, Y., Ekins, S., Nikolskaya, T., Bugrim, A. A novel method for generation of signature networks as biomarkers from complex high-throughput data. *Tox. Lett.*, 2005, 158, 20–29.
7. Warner, G. J., Adeleye, A. E., Ideker, T. Interactome networks: the state of the science. *Genome Biol.*, 2006, 7(1), 301–309.
8. Lim, J. et al. A protein–protein interaction network for human inherited ataxias and disorders of Purkinje cell degeneration, *Cell*, 2006, 125(4), 801–814.
9. Sharan, R., Ideker, T. Modeling cellular machinery through biological network comparison. *Nat. Biotech.*, 2006, 24(4), 427–433.
10. Ekins, S., Nikolsky, Y., Bugrim, A., Kirillov, E., Nikolskaya, T. Pathway mapping tools for analysis of high content data. In *High content screening handbook*, Totowa, NJ: Humana Press, 2006, 319–350.
11. Yu, H., Shu, X., Greenboum, D., Karro, J., Gerstein, M. TopNet: A tool for comparing biological subnetworks, correlating protein properties with topological statistics. *Nucl. Acids Res.*, 2004, 32(1), 328–337.
12. Stafford, P., Brun, M. Three methods for optimization of cross-laboratory and cross-platform microarray expression data. *Nucl. Acids Res.*, 2007, 35(10), e72. Epub 2007 May 3.
13. Leming, S. et al. (MAQC consortium). The MicroArray Quality Control (MAQC) project shows inter- and intraplatform reproducibility of gene expression measurements. *Nat. Biotech.*, 2006, 24(9), 1151–1161.

9 Exon Array Analysis for the Detection of Alternative Splicing

Paul Gardina and Yaron Turpaz

CONTENTS

Introduction and Background 205
    The Biology of Alternative Splicing 205
    Basic Principles of mRNA Detection by Microarrays 207
Affymetrix Arrays 208
Exon Arrays 209
    Array Design 209
    Algorithms for Signal Estimation 213
    Algorithms for Alternative Splice Predictions 214
    Concepts and Caveats in Identifying Splicing Events 214
Workflow for the Analysis of Exon Data 216
    Phase I. Signal Estimation 217
    Phase II. Quality Control 218
    Phase III. Filtering Signal Data 219
        An Ideal Splicing Event 219
        Anomalous Probe Signals 220
        Filtering Strategies 222
    Phase IV. Alternative Splice Prediction 225
    Phase V. Selection of Candidate Splicing Events 226
    Phase VI. Validation of Splicing Variants 227
Biological Interpretation of Alternative Splicing 229
Concluding Remarks 230
References 230

INTRODUCTION AND BACKGROUND

THE BIOLOGY OF ALTERNATIVE SPLICING

Estimates of the number of genes in the human genome currently range between 20,000 and 30,000 [1]. This is a surprisingly low number considering that the simple bacteria *Escherichia coli* has more than 3000 genes, yeast generally contain about 6000, and the roundworm has nearly 20,000 [2–4]. A major mechanism for

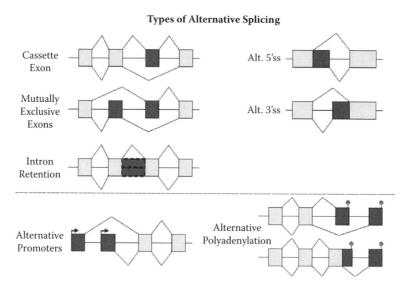

FIGURE 9.1 (See color insert following page 80.) Types of alternative splicing events. Potential exons are represented by boxes. Red boxes are exons that might be included or skipped in any particular transcript. The angled lines above and below the exons represent possible alternatives for joining the exon borders into a mature transcript.

generating genomic complexity from this relative paucity of genes in higher eukaryotes is through alternative splicing of mRNA.

The protein coding regions of human genes are rarely found as contiguous units on the genome; rather, the coding segments ("exons") are scattered over relatively large distances with long "intronic" regions between them. Introns are removed during processing of the mRNA and the exons are spliced together to form the mature coding sequence of the protein. However, the exons are not always stitched together in exactly the same pattern (Figure 9.1); some exons are skipped and others included, the site marking the beginning or end of the exon may be altered ($3'$ or $5'$ splice sites, respectively), and, in some cases, introns may be retained into the final coding sequence. Frequently, the transcriptional start or stop of gene may be drastically shifted by employing alternative promoters or poly-adenylation sites, respectively, leading to the skipping of many consecutive exons. (For a general review of mRNA splicing, see Black [5].)

Because of these splicing variants, exons may be viewed as building blocks that the cell can incorporate in a plethora of different combinations to generate, in effect, many different proteins from the same gene. The resultant proteins may have a stronger or weaker affinity for other proteins, may recognize different ligands, or may include domains that have different functional properties. In extreme cases, two differentially spliced protein products from the same gene may have antagonistic (opposite) functions.

A conservative count based on the strictly curated National Center for Biotechnology Information (NCBI) RefSeq database produces in the neighborhood of 100,000 well annotated exons in the human genome [6,7]. A more liberal estimate including speculative exon annotations suggests that the total number of human exons may be around 250,000. According to this summation, there are approximately nine exons per gene, most of which are less than 200 base pairs (bp) long [8]. However, unbiased assays of the transcriptome, such as with whole-genome tiling arrays, suggest that much transcriptional activity is unaccounted for and clearly there are large gaps in our present annotation of the genome [9,10].

Recent research suggests that a minimum of 75% of human genes are alternatively spliced [10], providing protein diversity for developmental and tissue-specific functions. Some genes have more than 100 exons, generating a theoretically huge number (2^{100} or $\sim 10^{30}$) of possible combinations of protein sequence. In reality, many of the exons appear to be constitutive (usually or always present) and many others are rarely utilized, but the variety of possibilities remains immense. In fact, the complexity of the human brain is partially explained by the high rate of alternative splicing that provides variants of proteins for particular neuronal functions [11,12].

Basic Principles of mRNA Detection by Microarrays

Microarrays and the large-scale sequencing of entire genomes generated a revolution in molecular biology in the 1990s. For the previous four decades, research in gene expression and regulation was conducted one gene at a time, usually requiring much preliminary work to create genetic constructs that could be easily detected in the laboratory. Microarrays altered this paradigm by allowing researchers to follow the expression of thousands, and then tens of thousands, of genes simultaneously.

The essential property of nucleic acids that enables this detection mechanism is the same one that allows DNA to stably exist as a double helix: strand hybridization—the ability of a nucleotide chain to specifically bind to the complementary sequence on the other strand. It was realized that oligonucleotides (short nucleotide chains) attached to a firm matrix, under appropriately stringent conditions, could be used as specific "probes" for complementary nucleotide sequences in solution. If one knew the sequence of the probe and its location on the matrix, one could deduce the sequence *and the quantity* of an unknown nucleotide chain (the "target") that bound to it.

The primary function of expression arrays is to identify the type and amount (either relatively or absolutely) of mRNA in a cell type under particular conditions. The approach is conceptually simple: chemically label the sample's mRNA (or a DNA copy of it, called "complementary DNA" [cDNA]) with a fluorescent dye, then expose it to known probes linked to a solid substrate under conditions that allow specific hybridization between the probe and the target. The intensity of fluorescence at a particular location on the substrate (matrix) should reflect the concentration of a specific mRNA in the original sample.

Early versions of nucleotide arrays evolved through various matrix materials like nitrocellulose paper and glass slides to more sophisticated materials. The probes themselves might be long fragments of DNA derived indirectly from mRNA ("spotted arrays") or produced synthetically as short (25–30) nucleotide sequences ("oligonucleotide arrays") based on known mRNA sequences. Methods of applying the probe oligonucleotides to spotted arrays were analogous to printing methods such as spotting with laboratory pipettes, mechanical pins, and jet printers. Each probe then was a "spot" of identical nucleotide chains fixed onto a known location on the surface of the matrix. Unfortunately, the spots were rarely uniform in size or exactly centered on a known position. This led to a number of difficulties in estimating the probe placement, and therefore determining the intensity of the bound target: Where is the spot? How big is it? Where are its borders? How intense is the background signal (noise due to nonspecific fluorescence)?

Spotted arrays typically use the "two-color" dye method in which mRNA from pairs of samples (a test and a reference) is labeled with two different dyes. One sample is labeled with Cy3, which fluoresces green, and the other is labeled with Cy5, which fluoresces red. The two samples are hybridized to the same array allowing the labeled mRNA derivatives to compete for the probes. If there is more of a particular Cy3 target, the probe spot appears more greenish; if there is more of a particular Cy5 target, the probe spot appears redder, and if the labeled mRNAs from the two samples are present in equal amounts, the spot appears yellow. The advantage of this method is that it allows a direct comparison of gene expression between two samples in one experiment. Comparisons become more cumbersome when there are more than two groups and the samples do not fit into a simple pairwise scheme. Furthermore, the dyes themselves have different properties that may affect the output, such as differential affinities for the target and different inherent intensities. Frequently, these discrepancies are accounted for with "dye-swap" experiments in which each sample pair is tested twice with reciprocal dye assignments. In contrast, in a "single color" array like Affymetrix, one sample is hybridized to each array. In general, differentially expressed genes are identified by the relative changes against the general expression of the mass of genes.

AFFYMETRIX ARRAYS

Affymetrix chips represented a radically different synthetic approach than spotted arrays by combining the photolithography methods developed for semiconductors with large-scale oligonucleotide chemistry (see Figure 9.2) [13,14]. This method of synthesis allowed each probe sequence to be built up *in situ* within a precise microscopic square (a "feature") as part of a solid grid pattern. Each feature is of uniform size and shape, with rigid, square borders, and contains millions of identical copies of that specific probe. This approach not only reduced the computational difficulty of determining the fluorescent intensity of the "spot," but it also enabled the probe patterns to be miniaturized near to the scale of electronic microchips. In fact, each feature of the exon array is approximately the length of an *E. coli* bacterium (~3–4 μm). Special scanners have been developed to detect the fluorescent signals within the grid pattern and produce signal intensities for each feature. The signal intensities for each feature are summarized in "cell intensity files" with the ".CEL" extension.

The binding of two DNA sequences is not absolutely specific; rather, it is a stochastic process that depends on the nucleotide sequence and the hybridization conditions. For this reason, methods of calculating the background signals are often necessary. The traditional Affymetrix arrays contained "mismatch" (MM) probes in tandem with each "perfect match" (PM) probe. The PM probes are exactly complementary to the target sequence and therefore generate a signal representing the expression of a particular mRNA. The MM probes are mismatched to the target

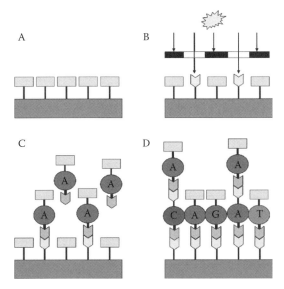

FIGURE 9.2 (See color insert following page 80.) Affymetrix's photolithographic method to synthesize oligonucleotide probes directly on the chip substrate. (A) The surface of the silica-based substrate initially contains chemical moieties that are nonreactive to the addition of new nucleotides, but become reactive upon exposure to UV light. (B) For each place where a new nucleotide should be added to the oligonucleotide sequence, unique masks have been designed with precise windows that allow UV light to penetrate only at predetermined sites. In this example, the mask exposes sites where the next nucleotide will be an adenosine ("A"). (C) Chemically modified adenosine is added and covalently links ("the coupling step") only at the sites where the chemical moiety has been activated by UV light. Each nucleotide itself contains a UV-excitable moiety to allow for extension of the growing oligonucleotide chains. The masking is repeated for specific addition of each of the other three nucleotides to that layer. (D) The oligonucleotides are built up sequentially through repeated cycles of masking/UV exposure and nucleotide coupling. This example shows the situation after the second round of adenosines has been added. Each oligonucleotide chain (a probe) actually represents millions of contiguous identical molecules (a feature).

sequence at the central nucleotide and produce a signal representing the nonspecific binding, or noise, accompanying the PM probe. Some analytical approaches simply subtract the MM value from the PM value to estimate the true signal found at that probe. This is generally most useful in cases where gene expression is expected to be low (i.e., close to the background signal). Other methods may estimate a global background by sampling intensities from other parts of the chip or (as in the exon array) by comparing the PM signal to a pool of background probes with the same quanine-cytosine (GC) content as the target probe.

EXON ARRAYS

Array Design

The U133 Plus 2.0 chip, the most common Affymetrix array for gene-expression experiments, typically employs 11 probes (a "probe set") that primarily target the

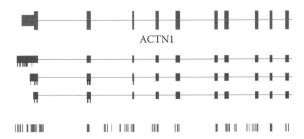

FIGURE 9.3 Comparison of probe set distribution between Affymetrix's U133+2 expression array and the human exon 1.0 ST array. The top row shows part of the RefSeq annotation of the *Actn1* gene with the 3′ end of the gene to the left and known exons shown as gray rectangles. The middle three rows represent consensus transcripts that are targeted by the U133+2 array. Probes present on the array that interrogate these transcripts are shown as small black ticks. The bottom row represents the probe selection regions from the exon array. Note that these are numerous and distributed throughout the gene, rather than confined to the 3′ end. Some probe set regions do not overlap known exon sequences, generally representing more speculative content based on weaker empirical evidence or computational predictions. The view is from Affymetrix's Integrated Genome Browser (IGB).

3′ end of each gene transcript (Figure 9.3). The 3′ end of the transcript is generally more stable and the chemistry used to process the mRNA (termed "3′ IVT" for *In Vitro* Transcription) tends to amplify the signal at this end. The design approach of these expression arrays is largely "one probe set = one gene"; the probes are collectively detecting a relatively limited region of the transcript and the signals from the 11 probes can be statistically combined into a single gene-level signal. Expression arrays are not, in general, designed to detect the various configurations of transcripts that may occur upstream of the probe site and may entirely miss transcripts with alternative poly-A sites, although the U133 frequently has multiple probe sets that target the possible alternative ends of the transcripts (as in Figure 9.3). In total, the Affymetrix U133 Plus 2.0 array contains 1.3 million 11-μm features (i.e., probes) grouped into 54,000 probe sets. The probes constituting a probe set are, in fact, randomly distributed around the chip to suppress the impact of artifacts, like scratches, dust or fingerprints, affecting the surface of the chip.

In contrast, the probe sets on the exon array are designed to detect individual exons. Although there are typically only four probes per probe set, the multiplicity of exons means that each gene is probed by many probes (a median of 30–40 probes for each RefSeq transcript). The analytical philosophy of the exon array is to treat each gene as a cluster of potentially different transcripts that are assembled from various combinations of exons, reflecting the biological reality that one gene actually can produce a myriad of different mRNA products. The GeneChip Human Exon 1.0 ST array contains approximately 5.4 million 5-μm features grouped into 1.4 million probe sets interrogating over one million exon clusters. The exon array simultaneously functions both to detect gene-level expression, as an aggregate signal of all the exons composing the gene, and to assess the contribution of individual exons to the overall expression of the gene. The fact that the probes are distributed throughout the transcript sequence also means that the chemistry to produce the target sequence (termed "whole transcript sense target labeling") is different from that used for expression arrays; the transcripts must be amplified in an unbiased manner, rather than just at the 3' end. It should be noted that Affymetrix also produces the Human Gene 1.0 ST array, which covers only well annotated content. Each gene is represented by 26 probes (a subset of those used on the exon array), which are distributed along its length. The gene array covers approximately 29,000 genes. Many of the gene-level estimation approaches used for the exon array will be applicable to the gene array.

The philosophy of treating exons as building blocks implies two levels of probe set design. The first is the requirement to identify genomic sequences that might serve as exons in real cells. The second is to assemble the putative exons into the relevant biological transcripts at the gene level.

Given that the total gene complement of the human genome (and much less their component exons) is unknown, the Exon 1.0 ST array was designed to be as inclusive as possible, containing probes for both known and predicted exons. Obviously, the strongest evidence for expression of exons is empirical (i.e., the observation that they are contained in transcripts in nature). However, the fact that individual exons, like genes, might or might not be expressed in particular tissues under particular conditions means that exons known from empirical data are only a subset of true biological exons. In general, the exon targets on the exon array are drawn from three levels of annotation:

- core: empirically observed and strongly curated transcripts drawn from RefSeq and full-length Genbank mRNAs
- extended: weaker annotations based on cDNA transcripts, syntenic rat and mouse mRNA, and Ensembl, microRNA, Mitomap, Vegagene, and Vega-Pseudogene annotations
- full: computer predictions by Geneid, GENSCAN, GENSCAN Suboptimal, Exoniphy, RNAgene, SgpGene, and TWINSCAN

Exons present in one of these collections became possible targets for probes on the array by providing the sequence for a probe selection region (PSR). Each PSR represents a region of the genome predicted to act as a coherent unit of transcriptional behavior; that is, it is the minimal region that is either included or left out of a gene product. In many cases, each PSR is an entire exon; in other cases, due to variation in individual exon structures, several PSRs may form contiguous subregions of a true biological exon. The median size of the PSRs is 123 bp with a minimum of 25 bp. About 90% of the PSRs are represented by a probe set of four probes. Some PSRs were too small or contained highly repeated or otherwise problematic sequence and were not included on the array. Altogether, the array contains probe sets that target approximately one million putative exons or subexons.

The second step is to virtually assemble the exons back into gene-level transcripts. One possibility would be simply to reverse the process of the initial exon discovery—that is, to reassemble the exons into the transcripts from which they were originally observed or predicted. However, it is essential at this point to realize that we are typically dealing with *populations* of possible transcripts (depending on

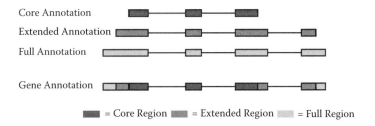

FIGURE 9.4 (See color insert following page 80.) Assembly of exon annotations into a gene-level annotation. Exonic sequences from all three confidence levels (core, extended, or full) are reassembled into transcript level sequences according to a defined set of rules. Each region is annotated according to the highest confidence level that supports it and a gene may contain sequences from all three sets. Genes in the core metaprobe set will contain at least one core exon; conversely, a gene from the full metaprobe set might only contain highly speculative exons from the full set.

the effects of alternative splicing) from a locus rather than the singular "gene" typically associated with gene-level expression arrays. Since any particular gene may produce many different related transcripts (a "transcript cluster"), and since the exon array detects exons rather than transcripts, it has no direct information about which exons are actually physically linked together in the mRNA. There may, in fact, be a number of combinations of exon assemblages (transcripts) that would account for the observed pattern of probe set signals. For this reason, the exon array treats each gene as a summarization of all the exons associated with that locus.

The plethora of exon architectures (e.g., cassette exons, mutually exclusive exons, alternative splice sites, alternative transcriptional starts and stops), variations in confidences of transcript annotations, and the necessity of rapidly incorporating new genomic knowledge have led to a dynamic design for reconstituting exons into genes (see Figure 9.4). A set of rules was created for virtually assembling the probe sets (exon level) into transcript clusters (gene level) based on the confidence level of the supporting evidence and the juxtapositions of the exon borders (see the Affymetrix white paper: "Exon Probe Set Annotations and Transcript Cluster Groupings v1.0"; the white papers can be found at: http://www.affymetrix.com/support/technical/byproduct.affx?product=huexon-st). In general, exons were merged into transcript clusters depending on whether and how many higher confidence clusters or exons they overlap, whether they have common exact splice sites, and whether single orphan exons are bounded within another annotation. The results of these rules are sets of mapping files called "metaprobe sets," which are essentially lists of exons that belong in each transcript cluster.

The metaprobe sets are divided into three levels, depending on the confidence level of the original gene and exon annotations as described earlier. The core set contains 17,800 transcript clusters comprising the "known" genes and their exon components, according to the best available annotation. Of the 1.4 million probe sets on the exon array, approximately 290,000 are supported by the full-length mRNAs in the core set. The extended set contains 129K transcript clusters, which clearly have some empirical support, but still represent far more genes than are currently expected in the human genome. Finally, the full set contains 262K transcript clusters, most of which are solely derived from *ab initio* computer prediction. As stated, the array contains much speculative content but there is evidence to believe that the transcriptome is much more extensive than is currently widely accepted. It should be noted that the confidence level of exons associated with any particular gene may be mixed; that is, a core gene may have exons from all three metaprobe sets.

The exon array does not include a paired mismatch feature for each perfect match feature; rather, surrogate mismatch intensities are derived from 1000 pooled "antigenomic" probes with the same GC content as each perfect match feature (see the Affymetrix white paper: "Exon Array Background Correction v1.0"). Antigenomic sequences are not present in the human, rat, or mouse genomes.

Algorithms for Signal Estimation

Probe set (exon-level) and transcript (gene-level) signals can be estimated by either the robust multichip average (RMA) or PLIER algorithms, both of which are described elsewhere in this book. Robust multichip average (RMA) is a multi-array algorithm that has been widely tested and accepted in the microarray community. Probe logarithmic intensity error (PLIER) is a robust M-estimator that uses a multichip analysis to model feature and target responses for each sample. The target response is the PLIER estimate of signal for a probe set (i.e., exon level). Gene-level PLIER estimates are derived by combining all probe sets that map into the same transcript cluster (according to the metaprobe set list). The model-based algorithms of PLIER tend to down-weight the effect of alternatively spliced exons relative to constitutive exons in predicting the overall gene-level response (see the Affymetrix white paper: "Gene Signal Estimates from Exon Arrays") and may be more robust towards highly variable or weakly annotated transcripts.

A variation of PLIER is "IterPLIER," which iteratively discards features (probes) that do not correlate well with the overall gene level signal. It then recalculates the signal estimate to derive a robust estimation of the gene expression value primarily based on the expression levels of the constitutive exons.

The choice of which method to use depends upon the situation, and standards will evolve as best practices are established. PLIER/IterPLIER is more aggressive in selecting the optimal set of probes to represent the transcript cluster and is probably better suited in discriminating expressed probe sets (exons) within the speculative content of the extended and full metaprobe sets. The trade-off is that IterPLIER runs the risk of choosing a sub-optimal set of representative probes. RMA is generally more conservative and probably a good choice with the core metaprobe set, and it has two further advantages: It is computationally less intensive and generally well known within the microarray community.

The DABG (Detection Above Background) algorithm generates a detection metric by comparing perfect match probes to a distribution of background probes. This comparison yields a p-value that is then combined into a probe set level p-value using the Fischer equation. Since the exon array does not include a paired mismatch feature for each perfect match feature (as in the U133 Plus 2 array), surrogate background intensities are derived from approximately 1000 pooled probes with the same GC content as each perfect match probe.

DABG is appropriate to use for exon-level present/absent calls, but not for gene-level calls. The algorithm makes a very strong assumption that all the probes are detecting the same target, which is likely to be true for individual probe sets (which detect exons) but is not necessarily true at the gene level, due to potential variation in transcript structure.

ALGORITHMS FOR ALTERNATIVE SPLICE PREDICTIONS

A number of algorithms (MIDAS, Robust PAC, etc.) have been developed to detect alternative splicing from exon-level signals. Software from Affymetrix implements an ANOVA (analysis of variance) method (MIDAS [multi-instrument data analysis system]), but all of the algorithms operate on a similar principle: determining the relative contribution of an exon to the overall gene signal in one group of samples versus another group. In each case, the algorithm must account for the possibility that gene expression itself is different between two groups.

The splicing index (SI) is a method to normalize each exon signal to the cognate gene signal in that sample. The signal of exon j is divided by the cognate gene signal in sample i of tissue A to produce a "normalized intensity" (NI_{Aij}), which is then expressed as a log ratio between the two sample groups:

$$SI_j = \log_2 [(\text{median } NI_{Aij})/(\text{median } NI_{Bij})]$$

A graphical representation of SI values for a hypothetical splicing event is shown in Figure 9.5.

A t-test can be applied to the normalized intensities (NI) to produce p-values that estimate the statistical separation of relative exon inclusion in the two sample groups. The SI itself represents the magnitude of the difference for exon inclusion between two tissues. The best candidates for validation are likely to have very small p-values and a large absolute Splicing Index value.

Implemented within the *apt-midas* command line program in Affymetrix Power Tools (APT; www.affymetrix.com/support/developer/powertools/index.affx), MIDAS is conceptually similar to the SI, but allows simultaneous comparisons between multiple sample groups. MIDAS employs the normalized exon intensities in an ANOVA model to test the hypothesis that no alternative splicing occurs for a particular exon. In the case of only two sample groups, ANOVA reduces to a t-test.

CONCEPTS AND CAVEATS IN IDENTIFYING SPLICING EVENTS

There are several concepts and potential pitfalls, some biological and some technical, that should be understood in the analysis of alternative splicing:

1. Alternative variation is typically a subtle event: There is rarely an all-or-nothing change in exon inclusion between one tissue and another. A more likely scenario may be a shift from 50% inclusion of an exon in one tissue to 80% inclusion in a different tissue. Furthermore, alternative splicing is

Exon Array Analysis for the Detection of Alternative Splicing

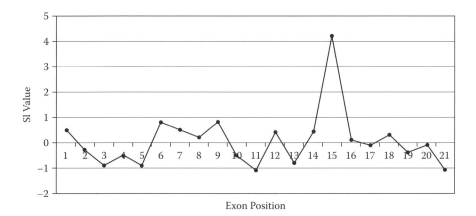

FIGURE 9.5 The Splicing Index. Most of the exons of this gene have SI values close to zero except for exon 15, which is probably included in one sample type and skipped in the other.

a very common phenomenon: ~75% of all genes appear to be alternatively spliced. Between any two tissue types there may be tens of thousands of probe sets that vary in their inclusion into transcripts.

2. Alternative splicing results are highly dependent on the accurate annotation of the transcript clusters and precise quantitation of gene-level estimates. Misannotated transcript clusters or genes with large variations in exon expression may produce less reliable results. Since most algorithms for alternative splice prediction make some comparison of exon expression to gene expression, accurate gene-level estimates are an essential part of the equation. But gene-level estimates are, in turn, based upon the individual signals of exons that might or might not be expressed. Most of the core-level exons are likely to be constitutively present within the transcripts of the respective gene. Therefore, the signal estimate of each core gene is probably well reflected by the signal of the core exons. Speculative metaprobe sets are more likely to include exons that are not expressed in any particular situation and, possibly, are never expressed. This is one of the strengths of the PLIER algorithm: it searches for and evaluates probe signals that are the most consistent within the transcript cluster.

3. Like most statistical tests, these algorithms assume that the splicing pattern is consistent among all of the samples within a group. Since splicing is generally a subtle event, sample quality, homogeneity, and general consistency are important factors in identifying true splicing events. Furthermore, since estimating alternative splicing adds another layer of complexity to gene expression, in general the power needed to detect significant changes in splicing will probably dictate that sample sizes should be larger than for a similar analysis of gene expression.

4. The search for alternative splicing is essentially a search for exon signals that behave differently from the rest of the transcript. Unfortunately, this means that technical artifacts and abnormally behaving probe sets tend to appear significant in the output of most algorithms. For this reason, the

data should be multiply filtered at both the exon level and the gene level to reduce false positives. Visual inspection of the probe set signals from strong splicing candidates is also recommended to screen out artifacts that survive all other filtering. Filtering methods are described in detail later.

WORKFLOW FOR THE ANALYSIS OF EXON DATA

The current standard Affymetrix software for processing exon array CEL files is called the Expression Console, which generates signal estimates, quality control (QC) metrics, and graphical summarizations. Alternatively, APT provides a command-line program called *apt-probeset-summarize* with many options for greater flexibility in the analysis. Some of the workflow described here must be performed with external applications for scripting/filtering (e.g., Perl) or general statistical analysis. Several commercial packages are available for a nearly complete beginning-to-end analysis (see www.affymetrix.com/products/software/compatible/exon_expression. affx for a list of GeneChip-compatible™ software packages). Finally, other commercial packages or public Web-based tools are useful for visualizing probe set signals and transcript structures. The overall workflow is illustrated in Figure 9.6.

This workflow makes the assumption that there are two or more sample groups, and we intend to find differential alternative splicing between the groups. Alternative splicing can occur within one tissue, but we are generally more concerned with the way it differs in different tissues or conditions.

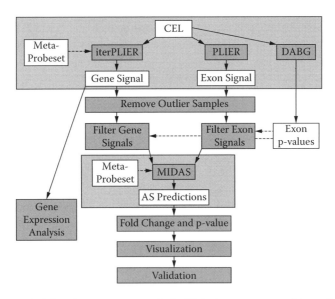

FIGURE 9.6 Workflow for exon array analysis. Light boxes represent files or data sets and dark boxes represent processes or programs. Functions within the large gray rectangles occur within the expression console or APT command line. Gene- and exon-level analyses are performed in parallel, beginning with signal estimation, until they are united at the algorithm (here, MIDAS), which makes the alternative splicing prediction.

Phase I. Signal Estimation

The Expression Console actually performs two parallel analyses, initially generating separate estimates of gene-level and exon-level signals. Eventually, these estimates will be combined within an algorithm that produces alternative splicing predictions, essentially by estimating the relative contributions of each exon signal to its respective gene signal.

As described earlier, genes are reconstructed from exons based on mapping files called metaprobe sets. Depending on the research focus, the analysis can vary from highly supported (core) to highly speculative (full) gene and exon annotations.

Here, we suggest an initial (and fairly conservative) approach to analyze the exon array.

- Gene level:
  - IterPLIER (pm-gcbg): performs a PLIER estimation and iteratively discards noncorrelating probe sets. Background signals are subtracted from the probe signals.
  - antigenomic.bgp: the background pool that is binned by GC content
  - Quantile sketch normalization at 50,000 data points: this option saves memory and is generally sufficient for normalization.
  - Core metaprobe set: this is the set of gene annotations with the highest confidence. The file contains a list of probe sets that are associated with each gene and IterPLIER will use this collection of probe sets to make estimations of the gene signal. In this case, the probe sets represent core exons that are supported by RefSeq annotations of that gene.

Users may prefer to use the RMA method of gene signal estimations. The RMA is generally more conservative than PLIER and may be more appropriate with the conservative gene annotations of the core set. The RMA method is also more generally better known and extensively validated by the microarray community. The optimal approach should become clear as more testing and more exon array data sets become available.

Do not use DABG for gene-level "present/absent" calls. This must be estimated indirectly, as described in the section on filtering.

- Exon level:
  - PLIER (pm-gcbg): probe sets have a maximum of four probes, which is too limited to be useful with IterPLIER at the individual exon level.
  - antigenomic.bgp
  - Quantile sketch normalization at 50,000 data points
  - DABG (pm-only): produces p-values for present calls
  - It is possible to limit the analysis to a subset of exons by providing a list file. In the absence of such a list, signal estimates are generated for all probe sets (exons).

The application produces three signal-specific files: one for gene-level intensity, one for exon-level intensity, and one for exon-level DABG p-values. These files will be manipulated later during the filtering phase to remove problematic probe set signals.

Phase II. Quality Control

As in any experiment, outlier samples should be identified and eliminated or, at minimum, accounted for. This may be particularly true in analysis of exon array data since low sample quality is likely to be highly influential in generating noise, therefore leading to high false positive rates.

The standard QC approaches to microarray data are available for and applicable to the exon arrays. Standard Affymetrix expression control sets such as hybridization controls and poly-A RNA control spikes are included on the arrays. Metric analytics like box plots, MA-plots, and signal histograms can be generated within the Expression Console. While RMA and PLIER will not produce gene-level present/absent calls, these are available at the probe set level as detection p-values.

In addition, exon-based and intron-based probe sets for approximately 100 constitutively expressed genes can serve as positive and negative control sets, respectively. These signals are used to produce an "area under the curve" (AUC) value in a receiver-operator curve (ROC) as a measure of the separation between the positive (exonic) and negative (intronic) controls. An AUC of 1.0 would indicate absolute distinction in the data, while 0.5 would indicate no separation. Note that, in real biological samples, the predicted exonic sequences are not always expressed and the putative intronic sequences may be expressed, so we cannot expect perfect scores for the AUC.

The Expression Console produces two other useful metrics, the mean absolute PLIER deviation (MAD) and the mean absolute relative log expression (RLE). The MAD represents the PLIER residuals, a measure of how well the signals from a particular chip match the overall PLIER model for the entire batch (smaller residuals are therefore better). The RLE measures the aggregate deviation of each probe set signal on the chip from the median signal value of the respective probe set across all chips. The MAD and RLE values are particularly useful since they are relatively robust against experimental conditions.

As yet, there are no absolute cutoff values for these QC metrics. Rather, the metrics should be used to identify potential outliers among the chips. Questionable chips should either be removed (in egregious cases) or treated suspiciously in downstream QC analysis (see later discussion). For a general description of QC for exon arrays, see the Affymetrix white paper: "Quality Assessment of Exon Arrays."

It is frequently useful to analyze the signal estimate data (at both gene and exon levels) with a sensitive multidimensional method like principal components analysis (PCA). In addition to PCA pinpointing possible outlier samples, clustering patterns may also identify systematic factors affecting the data, such as batch or operator effects. In particular, paired samples (e.g., tumor/normal or treated/untreated from the same individual) may indicate tendencies in the data that are not apparent in singular data points; sample pairs that violate the general trend of the relationships also may be outliers. A useful follow-up to PCA is a multiway ANOVA to test the effect of possible outliers on the signal-to-noise ratio. Samples found as extreme outliers should be removed from the analysis, but exercise caution since most studies lack sufficient sample numbers and removing a mild outlier might be costly in terms of power.

Exon Array Analysis for the Detection of Alternative Splicing 219

PHASE III. FILTERING SIGNAL DATA

In order to obtain meaningful splicing information and to decrease the chances of false positives (thereby increasing the verification rate), a number of filtering steps can be performed. Subsequent to producing the CEL files, validation of splicing events in the laboratory (e.g., by real-time polymerase chain reaction [RT-PCR]) is by far the most laborious step. Therefore, additional effort made during computational analysis to reduce false positive rates will ultimately pay off in greatly reducing time spent in the lab. In this section, we present some suggestions for reducing the impact of nuisance factors.

Among the problematic cases are genes and exons with low expression, or genes with highly differential expression. Furthermore, among the 1.4 million probe sets on the exon array, some anomalous probe signals (such as saturated, cross-hybridizing, or nonresponsive probes) are likely to appear as false positives. Given that the detection algorithms are searching for subtle changes against a somewhat noisy background, these artifactual signals tend to score strongly. Fortunately, these anomalous probe sets have particular characteristics that allow for automated identification and removal.

An Ideal Splicing Event

Prior to describing the specifics of filtering, we might conceptualize the ideal scenario for gene and exon expression that would maximize our ability to identify differential splicing. Factors that deviate from this ideal situation tend to lower the probability of correctly identifying splicing events. Characteristics of the ideal alternative splicing event are listed next and illustrated in Figure 9.7, which shows an example of a correctly identified (validated by RT-PCR) alternative splicing event:

1. High gene expression. High gene expression removes the signal from the realm of the background signals and is robust against small changes in apparent signal.
2. Consistent gene expression within the sample group. Consistency across samples is a positive feature among all statistical analysis; the smaller the deviation is within the group, the greater is the power to detect differences. This is especially true when looking at relatively subtle phenomena like splicing.
3. Equal gene expression in both sample groups. As we will see, large discrepancies in relative gene levels combined with several classes of possible anomalous probe signals will lead to false positives.
4. Most probe sets parallel the overall expression of the gene except for a single exon that is alternatively spliced. In other words, if most of the exons belonging to that gene are constitutive, the estimation of the gene-level signal is more reliable.
5. The alternatively spliced exon is always included in one sample group and never included in the other sample group. This is a rare situation, but qualitative all-or-none changes are almost always easier to detect than quantitative changes.

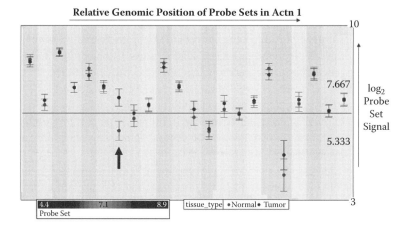

FIGURE 9.7 Characteristics of an ideal splicing event. The average signal and standard error for exon-level (probe set) signals from a single gene are shown for tumor samples (dark) and normal samples (light). The $\log_2$ intensity scale is shown on the right-hand axis. In the validated splicing event (indicated by the arrow), the exon shows low expression relative to the overall expression of the gene in normal tissues. Note there are many characteristics of an ideal scenario: the gene is highly and equally expressed in both groups, and most of the exons behave uniformly. (The data here are visualized with the Partek Genomic Suite.)

Anomalous Probe Signals

Figure 9.8 shows a number of scenarios that might be envisioned to produce false signals in the context of alternative splicing. Panel A shows a true positive in which the same gene in two different sample groups has a strong overall signal and one probe set is differentially expressed relative to the corresponding gene signal. Panel B illustrates a case where the gene is expressed near background in one of the samples. In this case, a generally unexpressed exon might appear to be a relatively strong contributor to the gene with low expression, but a weak contributor to the gene with high expression. Panel C illustrates a probe set that is affected by cross-hybridization, resulting in an artificially high signal in both sample groups. The splicing algorithms will identify this exon as proportionally higher in the sample group with lower gene expression. Panel D shows the effect of the converse situation: unexpressed, poorly hybridizing, or weakly responsive probes that appear to contribute relatively less to the gene with the higher expression. In all of these cases, the false signal is due to a combination of two events: a probe set that does not represent real changes in expression but that is "stuck" at some level in both sample groups; and an overall different gene-level signal between the two groups. The false signal is constant across all samples; it is only the differences in gene signals that lead to different exon/gene ratios.

Figure 9.9 illustrates a gene containing a constitutively high probe set signal. Fortunately, these anomalous probe sets have characteristics that make them relatively easy to identify and filter out:

- approximately equal signal intensity in both sample groups
- low variance relative to other probe set signals
- (possibly) extreme intensities relative to gene expression level

Exon Array Analysis for the Detection of Alternative Splicing 221

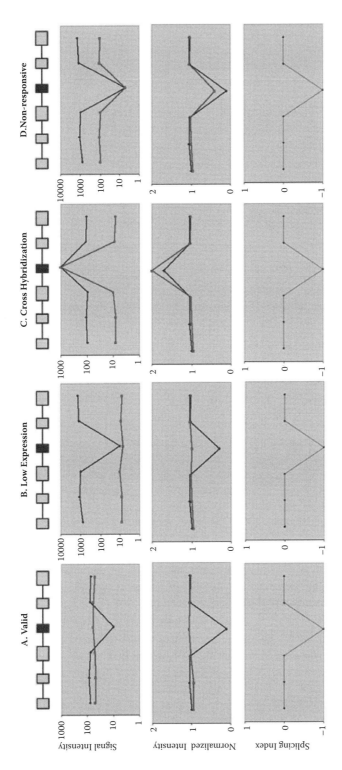

FIGURE 9.8 Several scenarios that might lead to false predictions of alternative splicing events. The top row of panels indicates mean probe set signal intensities from two sample groups for a hypothetical gene with six exons. The resulting gene normalized intensities (exon/gene) for both sample groups are shown in the second row. The resulting SI values are shown in the third row. In cases B–D, it is the combination of misleading probe set results and *differential gene expression* that creates a false prediction of alternative splicing. The filtering steps described in the text seek to eliminate the scenarios illustrated in cases B–D.

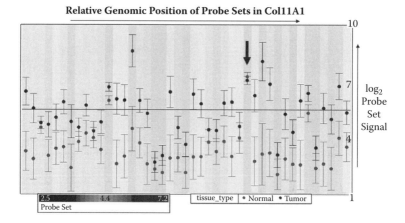

FIGURE 9.9 An anomalous probe set signal. A probable artifact from a probe set with constitutively high signals (indicated by the arrow) may lead to an erroneous splicing prediction. Even though this probe set gives the same signal for both tissue types, it appears to have a relatively high inclusion in normal samples because the gene-level signal is lower in normal than in tumor. This probe set displays three trademark symptoms of this artifact: It is higher than the neighboring probe set signals, it is approximately the same in both tissues, and it shows an extremely low variance across the samples.

Filtering Strategies

Filtering of gene and exon signals can be applied to reduce three classes of problematic cases that might lead to false splice variant predictions:

- low signals
- anomalous signals
- low confidence predictions

The affected genes and exons should be removed from the analysis. Some of these filtering methods have been implemented in third-party software, but they can generally be performed using simple scripts in, for example, Perl.

Low Signal
As is generally the case with microarray data, very low signals frequently reflect the noise in the system. Since alternative splice predictions are usually dealing with ratios between genes and exons, low signal might be interpreted as differential inclusion of the exon when in fact the exon (or gene) may not be expressed at all.

1. Remove exons (probe sets) that are not expressed in at least one sample group. This suppresses the scenarios illustrated in panels B and D of Figure 9.8. It is reasonable that an alternatively spliced exon might be entirely absent from any particular sample group, but it must be expressed in at least one of the groups under comparison to be considered. The following is a good default set of rules as starting points to remove exons that are expressed at low level:

1. If the DABG p-val < 0.0001, the exon is called as "present" in a sample.
2. If the exon is called as present in ≥50% of samples of a group, the exon is expressed in that group.
3. If the exon is expressed in either group, accept it; otherwise, filter it out.

2. Remove genes (transcript clusters) that are not expressed in both sample groups. Differential alternative splicing is meaningless unless the gene is expressed in both sample groups. This eliminates the scenario illustrated in panel B of Figure 9.8, but is more complicated than at the exon level, because there is not currently a direct method for present/absent calls at the gene level. Two possible solutions are presented here: using expression of core exons as a surrogate for gene expression or setting a threshold for gene signals.

The first method uses the aggregate exon-level DABG results to indirectly estimate present/absent for the gene. The rationale is that, even with some alternative splicing, most of the core exons should be constitutively expressed if the gene itself is present. Of course, this method is only valid with core genes.

The following is a good default set of rules to remove genes expressed at low level:

1. If the DABG p-val < 0.0001, the exon is called as "present" in a gene.
2. If >50% of the core exons are present, the gene is called as present.
3. If the gene is called as present in >50% of the samples in a group, the gene is considered expressed in that group. This step may be made more stringent by increasing the percentage of samples in which the gene is called present (e.g., to 75%).
4. If the gene is expressed in *both* groups, accept it; otherwise, filter it out.

The second method is to require a minimum gene signal level. Background gene signal is generally considered to be around 15, while a reasonable, but very conservative, threshold would be >100. If the gene signal exceeds the threshold in the majority of the samples, the gene is considered to be present in that group. As with many statistical strategies, setting a gene-level threshold is a trade-off; genes with higher expression levels are further away from noise and thus are more reliable, but at the cost of missing some potentially interesting events (i.e., more false negatives).

Anomalous Probe Sets

Some probe sets produce signals that are disproportionate to the exon's true level of expression. This class includes probes that cross-hybridize, saturate easily, have high background levels, or are generally nonresponsive. The resulting false relationship between the exon signal and the overall gene signal will tend to erroneously suggest that the exon is alternatively spliced.

1. Remove probe sets with a high likelihood of cross-hybridization. These probe sets are likely to produce in false positives since the signal may originate from an entirely different genomic location. The annotation for the exon array includes a file "HuEx-1_0-st-probeset-annot.csv" with the field "probeset_type" that describes the probe set's potential to cross-hybridize. Probe sets with values other than 1 ("unique") can be removed from the analysis.

2. Remove probes with extremely low variance. This filter seems a bit counterintuitive, but the idea is that probes with very low variance relative to others within the gene are very likely to be either not expressed or saturated in all samples. As illustrated in Figure 9.9, anomalous probe sets tend to generate a narrow range of signals both within and between sample groups.

3. Discard probe sets with a very large ratio to gene intensity (exon/gene > 5.0). Even though probes may show inherently different responses due to varying affinities for target, probe sets with very high signals relative to the cognate gene signal are likely to result from cross-hybridization or high background levels. Exon/gene signal ratios > 5 likely indicate problematic probe sets that should be removed from the analysis.

4. Discard probe sets with very low maximum ratio to gene intensity (exon/gene < 0.20). A very low ratio of the exon signal to the gene signal may indicate that the probe set is nonresponsive or that the exon is simply not expressed. It is quite possible that an exon could be included in 10% of transcripts in group A, but in only 1% of transcripts in group B. This difference still represents a 10-fold higher inclusion rate in group A, and may represent a physiologically meaningful example of alternative splicing. However, if our interest is in reducing the false positive rate and finding suitable validation targets, this class is better off being filtered out.

5. Filter probes with known single nucleotide polymorphisms (SNPs). Single nucleotide polymorphisms represent small genetic differences between otherwise closely related individuals (e.g., members of the same species). Single nucleotide polymorphisms in transcript regions that are targeted by probes are likely to give false signals since they may no longer match the canonical sequence that was the basis of the probe design. This is not likely to be a problem when the SNP is rare or when the alternative alleles representing the SNP are randomly partitioned between the test groupings. However, SNPs may have a significant effect when comparing, for example, two strongly inbred strains of some model organism, where each strain contains a homogeneous complement of SNP variants.

Low Confidence Predictions

Given the speculative content of the exon array, the possibility of anomalous probe set signals, and the statistical difficulties of multiple testing for a large number of probe sets, the researcher may choose to focus his search on putative splice variations with a higher confidence of accurate prediction. This section suggests three approaches that are likely to improve the likelihood of eventual validation of splice prediction.

1. Remove genes that have very large differential expression. Despite the fact that splicing algorithms normalize for gene expression, very large differences in gene expression between the sample groups have a tendency to produce false positives. With large disparities in target level, probe set intensities in the two groups may be disproportionately affected by background noise or saturation. Furthermore, differences in gene expression exacerbate the false signals generated by anomalous probe signals. As noted earlier, it is the combination of false exon signals and differential gene expression that frequently leads to erroneous splice predictions.

A 10-fold difference is an arbitrary cutoff that can be decreased if a more stringent filtering is desired (i.e., filter genes with $|\log2 (gene/gene)| > 3.32$). At the extreme, we might begin by looking only at genes with approximately equal expression.

2. Limit the search to high confidence (core) exons. Focusing on only well annotated exons (the RefSeq-supported exons in the "core" metaprobe set) dramatically reduces the amount of speculative content. Because speculative exons have a much lower level of supporting evidence, they are more likely to result in false positives. Although this eliminates a large number of potential splicing events, it is still possible to uncover novel examples of alternative splicing that involve well annotated exons.

3. Limit the search to known alternative splicing events. Exons that have prior evidence of being involved in an alternative splicing event are more likely to result in positive validations. This filter relies on bioinformatic predictions of alternative splicing based on EST/mRNA sequences or annotations. The UCSC genome browser is a good source of transcript and alternative splicing predictions, and there are several other public databases of known or predicted splice variation. It may be possible for this prediction to be run once to create a list of probe sets that could be used to filter all subsequent analyses. Using this filter does, however, surrender the ability to discover entirely novel alternative splicing events.

PHASE IV. ALTERNATIVE SPLICE PREDICTION

Once the data are filtered, gene-level and exon-level signal intensities can be fed into a splicing algorithm. As noted previously, the general approach of the algorithms is to estimate the contribution of each exon to its respective gene signal and ask whether this contribution is statistically significant between different groups. The APT command line program *apt-midas* implements the ANOVA-based MIDAS algorithm, but other algorithms are available. In particular, the SI takes a similar approach as MIDAS and is simple to implement in Perl or R.

The following is a typical input for alternative splicing analysis in MIDAS:

- Gene signals: the filtered gene-level signal file; the genes included in this set depend on which metaprobe set (core/extended/full) was chosen at the signal estimation stage.
- Exon signals: the filtered exon-level signal file.
- Metaprobe set: core/extended/full; at this point the user can select what confidence level of exons to evaluate. For example, inputting the full metaprobe set means that MIDAS will look at *all* the exons associated with the input genes, even if the user is working with only the core gene set. A more conservative approach would be to look only at core exons by inputting the core metaprobe file here.
- Group file: This file tells MIDAS how samples are partitioned into the groups (e.g., "brain," "lung," "kidney") that should be compared for differential splicing. The ANOVA in MIDAS can compare multiple groups but does not incorporate additional factors like gender, tumor stage, etc.

MIDAS outputs predictions for alternative splicing events as p-values. These should not necessarily be treated as true p-values but rather as scores that reflect relative ranking. The false positive rate is likely to be much higher than indicated by p-values, depending largely on the quality and homogeneity of the samples. High ranking candidates should be further screened as described below and ultimately verified empirically.

PHASE V. SELECTION OF CANDIDATE SPLICING EVENTS

Subsequent to filtering the signals and generating splice predictions, superior candidates for empirical validation need to be identified. The candidates can be initially filtered based on the statistical output of the splicing algorithm and further evaluated by visualizing the probe set signals in a genomic context.

1. Keep probe sets with p-values $< 1 \times 10^{-3}$. Probe sets with the smallest p-values are the most likely to have significant differences in inclusion rate and the cutoff value used can easily be lowered to increase stringency. In addition, multiple testing corrections such as Bonferroni or the Benjamini-Hochberg false discovery rate can determine the p-value at which the differences are considered statistically significant. The optimal cutoff value is really dependent on the number of targets desired for the final validation list.

2. Keep probe sets with an absolute splicing index > 0.5. Recall that the SI represents the ratio of the relative inclusion of an exon in a gene in two sample groups. Whereas p-values measure the ability to statistically separate two groups, the SI gives the magnitude of the difference, represented as a $\log_2$ fold-change. Probe sets with larger magnitudes of predicted changes are more likely to have dramatic splicing changes and result in true splice variants.

3. Visualizing probe set signals in a genomic context. Despite extensive filtering of the data, several classes of false positives are not possible to identify purely based on values from the statistical analysis itself. A simple visual inspection of the data in genomic context can catch many of these potential pitfalls. The probe set intensity data of the target exon and surrounding exons can be visualized in an integrated browser (see Figure 9.10) such as the Affymetrix Integrated Genome Browser (IGB) or BLIS (Biotique Systems, Inc.). Such inspection should confirm whether the probe set signals are consistent with the predicted splicing event. Questions to ask are: Is the signal consistent among the samples in the group? Is the signal identical in both groups? (If so, it may represent an anomalous probe set.) Are the signals from neighboring probe sets consistent? Do neighboring constitutively expressed exons produce signals in line with the overall predicted gene expression? Do several consecutive probe sets appear to have relatively higher or lower inclusion in the gene? (If so, it may suggest that this region of the gene has an alternative start or stop that affects multiple exons.) Finally, if there are multiple probe sets for the exon, it is reassuring that the data for all of them are in agreement.

Visualization and careful observation of the data can often predict the likelihood of a positive result prior to verification. While the process of visual filtering can be time consuming, a well trained eye can reasonably view a hundred or more potential hits in a short time. Our experience suggests that it is well worth the time, and it can be coupled with the design of primer sequences for validation. Several commercially available GeneChip-compatible software packages provide easy click-through visualization of the raw signal data on a gene-by-gene basis, and can be easily used to conduct this step of visual inspection.

Exon Array Analysis for the Detection of Alternative Splicing 227

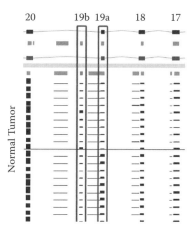

FIGURE 9.10 Probe set signals from two mutually exclusive exons in the *Cttn* gene. The BLIS genome viewer represents the probe set signals in a genomic context while normalizing each signal to the median exon signal within view for that sample. Note that the differential signals are fairly consistent within each group and the neighboring putatively constitutive exons also display a relatively constant signal. This candidate splicing event was confirmed by RT-PCR.

4. Identify known splicing events. While not an absolute requirement for validation, candidates consistent with known examples of alternative splicing are more likely to prove to be true positives. However, this approach also tends to suppress the discovery of novel splicing events.

The sequence of the target exon can be obtained using the probe set ID on the NetAffx Data Analysis Center (www.affymetrix.com/analysis/index.affx) and BLATed against the genome on the UCSC genome browser (genome.ucsc.edu). The BLAT results will also provide information about potential cross-hybridization or if the probe set maps to multiple genomic locations. The genome browser can provide important information about location of the probe set within the gene and whether the exon represents a known splicing event based on EST and mRNA sequences (see Figure 9.11). It may also be possible to observe overlapping, intervening, or independent transcripts (on the same or opposite strand) and potential artifacts of transcript cluster annotations, such as multiple genes being combined into a single cluster. Finally, visualization in a genomic context will aid in identifying appropriate targets, such as neighboring constitutive exons, for primer design in the validation stage.

PHASE VI. VALIDATION OF SPLICING VARIANTS

Ultimately, alternative splicing candidates must be validated experimentally. This is generally accomplished by reverse-transcriptase PCR, which probes the structure of the original RNA samples to distinguish between alternative splicing events. In most instances, a straightforward approach involving PCR primers on constitutive flanking exons will discriminate between splicing events, but in other cases, customized strategies must be employed. The validation rate may vary considerably, with sample quality and heterogeneity likely to be the most critical factors.

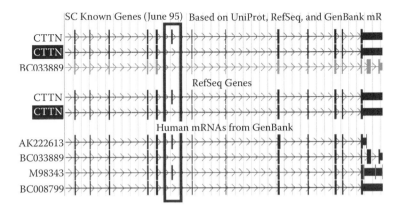

FIGURE 9.11 A candidate splicing event mapped onto the UCSC genome browser. The highlighted region corresponding to the targeted probe set appears to be a cassette exon that is included or skipped in different known transcripts.

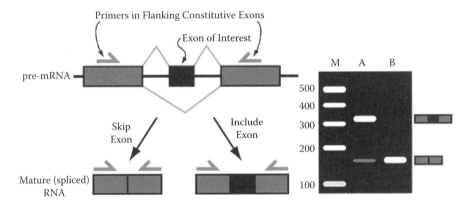

FIGURE 9.12 Primer design for RT-PCR validation of alternative splicing events. Primers (half arrows) are designed on adjacent constitutively expressed exons. The resulting PCR products will differ in size depending on the inclusion or exclusion of the intervening exon. M = marker; A = PCR products of sample A; B = PCR products of sample B.

Validations can be performed with RT-PCR using primers designed on constitutive exons that flank the exon of interest (Figure 9.12). This works well for simple cassette exons and alternative 5′ and 3′ splice sites. Polymerase chain reaction (PCR) products separate by size as they migrate through an agarose gel; the "include" product is larger by the size of the alternative exon and therefore runs more slowly through the gel. It is usually possible to calculate the expected sizes of the alternative PCR products ahead of time.

While carefully designed quantitative (Q) PCR could provide more accurate estimates of the absolute levels of each isoform, Q-PCR is typically not necessary to confirm splice variation. Changes in the relative intensity of the alternative bands are frequently apparent between samples. Amplification efficiencies may differ between

Exon Array Analysis for the Detection of Alternative Splicing

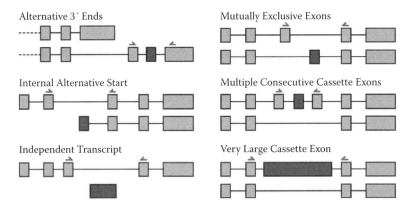

FIGURE 9.13 (See color insert following page 80.) Examples of challenging cases for RT-PCR validation. In these cases, the standard approach to designing primers (half arrows) will not produce PCR products that represent both alternatives of the splice variation. Therefore, an alternative validation strategy must be employed.

the two products, but this bias will be the same for all samples. By starting with equal amounts of input cDNA, simple RT-PCR can be considered semiquantitative.

Each validation case is unique and different types of alternative splicing events require different strategies. For example, detection of alternative starts and stops requires placement of a primer within the exon of interest since the target exon is not flanked by exons on *both* sides. In this case, two separate PCR reactions should be run: an internal one that determines presence/absence of the exon and a reference reaction that measures the expression level of the gene. A similar design strategy can be used for mutually exclusive exons of the similar sizes that cannot be resolved by separation on a gel. A slightly more sophisticated approach is to identify a restriction site that is unique to only one of the alternatives and cleave the PCR products with the appropriate enzyme. The loss of a band and the appearance of fragment bands identify the transcript containing the target sequence. Clearly, design of primers for validation is not a "one size fits all" situation and some problematic cases are illustrated in the Figure 9.13.

BIOLOGICAL INTERPRETATION OF ALTERNATIVE SPLICING

At this point, the researcher should have a list of validated splicing events in hand. Subsequent biological interpretation will depend greatly on the focus of the research, but some general suggestions are given here.

Look for connections within networks that might suggest functional relationships between the alternatively spliced genes. Many public or commercial databases are available to map biochemical or ontological associations. Particularly interesting splice variants may disproportionately concentrate into an area of physiological effect that is relevant to differentiating the groups under study. Splicing studies have revealed modular networks that are altered as a functional unit, such as a set of genes involved in morphogenesis of neurons (Ule). In some cases, the splice variants may be targets of a single splicing factor, suggesting that altered regulation of that factor is responsible for a unified set of coherent changes in protein function.

Particular proteins affected by the splice variation may be known to be important in the groups under study (e.g., a cell cycle protein or an apoptotic factor in tumor cells). In this case, a domain-level analysis may provide clues to the effect of splice variation on protein function. It is important to keep in mind that particular splice variants could be either active participants or passive byproducts of the processes that distinguish the cell types. Determining the actual role of the splice variation will require more focused research. However, even passive byproducts may be useful—for example, as targets for antibody therapy or as diagnostic markers.

CONCLUDING REMARKS

Approaches to the analysis of exon arrays will depend greatly on the goal of the research and the quality of the samples. One possibility is to begin with a very conservative approach by stringent filtering and concentrating on genes that have approximately equal expression. The latter condition will tend to suppress the misclassification of anomalous probe sets and produce a high proportion of true positives in validation. This will initially bias the findings, but the search can be gradually expanded to test the limits of the array and algorithms to identify splicing variants in the particular sample set. As experimental validation of the splicing candidates is by far more laborious than the statistical analysis, the suggested gradual approach can save significant time and resources in the lab.

The speculative content on the arrays allows great flexibility to focus on known transcripts or on a more exploratory search at both the gene and exon level. Whether the goal of the study is to discover novel genes and their splice variants or to focus on well annotated genes and pathways of interest, the organization of the metaprobe sets provides a dynamic analytic mechanism to accomplish either from a single experiment.

Analytical approaches to exon arrays are rapidly evolving as more academic and commercial groups utilize the arrays, refine techniques, and expand the repertoire of data sets. Several commercial software products have already incorporated early lessons in filtering and analysis, providing simple off-the-shelf solutions with flexible analysis options and elegant visualization capabilities. Exon analysis adds an extra layer of complexity to microarray-based investigation, but the return is robust gene-level data and a greater understanding of the true biological picture. Initial results of biological experiments with the exon array are revealing the immense complexity of the transcriptome. It becomes clear that in gene expression, the common question of "at what expression level?" should also be accompanied by the essential question of "of which variant?"

REFERENCES

1. Finishing the euchromatic sequence of the human genome, *Nature*, 431, 931–945, 2004.
2. F. R. Blattner, G. Plunkett, 3rd, C. A. Bloch, N. T. Perna, V. Burland, M. Riley, J. Collado-Vides, et al., The complete genome sequence of *Escherichia coli* K-12, *Science*, 277, 1453–1474, 1997.

3. A. Goffeau, B. G. Barrell, H. Bussey, R. W. Davis, B. Dujon, H. Feldmann, F. Galibert, et al., Life with 6000 genes, *Science*, 274, 546, 563–567, 1996.
4. Genome sequence of the nematode *C. elegans*: A platform for investigating biology, *Science*, 282, 2012–2018, 1998.
5. D. L. Black, Mechanisms of alternative premessenger RNA splicing, *Annu Rev Biochem*, 72, 291–336, 2003.
6. J. Wu and D. Haussler, Coding exon detection using comparative sequences, *J Comput Biol*, 13, 1148–1164, 2006.
7. X. H. Zhang and L. A. Chasin, Comparison of multiple vertebrate genomes reveals the birth and evolution of human exons, *Proc Natl Acad Sci USA*, 103, 13427–13432, 2006.
8. M. K. Sakharkar, V. T. Chow, and P. Kangueane, Distributions of exons and introns in the human genome, *In Silico Biol*, 4, 387–393, 2004.
9. P. Kapranov, A. T. Willingham, and T. R. Gingeras, Genome-wide transcription and the implications for genomic organization, *Nat Rev Genet*, 8, 413–423, 2007.
10. T. A. Clark, A. C. Schweitzer, T. X. Chen, M. K. Staples, G. Lu, H. Wang, A. Williams, and J. E. Blume, Discovery of tissue-specific exons using comprehensive human exon microarrays, *Genome Biol*, 8, R64, 2007.
11. C. J. Lee and K. Irizarry, Alternative splicing in the nervous system: An emerging source of diversity and regulation, *Biol Psychiatry*, 54, 771–776, 2003.
12. P. J. Grabowski and D. L. Black, Alternative RNA splicing in the nervous system, *Prog Neurobiol*, 65, 289–308, 2001.
13. S. P. Fodor, J. L. Read, M. C. Pirrung, L. Stryer, A. T. Lu, and D. Solas, Light-directed, spatially addressable parallel chemical synthesis, *Science*, 251, 767–773, 1991.
14. A. C. Pease, D. Solas, E. J. Sullivan, M. T. Cronin, C. P. Holmes, and S. P. Fodor, Light-generated oligonucleotide arrays for rapid DNA sequence analysis, *Proc Natl Acad Sci USA*, 91, 5022–5026, 1994.
15. J. Ule et al., *Nat. Gen.*, 37(8), 844–852, 2005.

10 Normalization of Array CGH Data

Bo Curry, Jayati Ghosh, and Charles Troup

CONTENTS

CGH Background 233
CGH Data Analysis 234
    Workup Common to Other Array Applications 234
    Error Model Specific to aCGH Arrays 236
    Normalization Specific to CGH Arrays 237
Examples of Centralization 240
Complications for Polyclonal Samples 240
Conclusions 240
Glossary of Terms 241
References 243

CGH BACKGROUND

Since the sequencing of the genomes of humans and other species, research has focused on differences between the "normal" genome of a species, and the genomes of cells from particular individuals. Different individuals differ in their DNA sequences. In particular, they differ in the number of copies of parts of their genomes, ranging in size from the short "tandem repeat" (aka microsatellite) sequences used for forensic genotyping to entire chromosomes. Some of the copy number variation among humans, such as the trisomy 21 responsible for Down syndrome, is associated with specific heritable disorders. Other variations may also be clinically significant because they predispose to particular diseases.

Most cancers arise as a result of an acquired genomic instability and the subsequent evolution of clonal populations of cells with accumulated genetic errors. Accordingly, most cancers and some premalignant tissues contain multiple genomic abnormalities not present in cells in normal tissues of the same individual. These abnormalities include gains and losses of chromosomal regions that vary extensively in their sizes, from microdeletions and amplifications of ~20 base pairs up to and including whole chromosomes. Comparative genomic hybridization (CGH) is a powerful technique for measuring differences in the copy number of regions of genomic DNA from different cell populations. In the last few years, microarray-based CGH (aCGH) has largely displaced classical CGH because of its high *precision* and high *resolution* (see glossary for working definitions of italicized terms.) [1]. Most simple-linkage genetic diseases have already been described, and further progress is focused on the identification of multiloci genetic abnormalities. Array CGH analytical methods have enabled increasingly high resolution measurement of chromosomal alterations in cancer cells [2] and in *somatic* cells of different individuals.

In array CGH, DNA is extracted from test and reference samples, labeled with differentiable fluorophores, and hybridized to a microarray comprising thousands of DNA probes complementary to specific targeted genomic regions. The labeled target DNA hybridizes to the complementary sequences printed on the array, and the relative fluorescence intensity of the test and reference fluorophores hybridized to each probe indicates the relative abundance of the target sequences in the test and reference samples. Oligonucleotide aCGH, developed at Agilent, allows the determination of DNA copy number changes over relatively small chromosomal regions, with resolution that can be much finer than single genes.

An aCGH assay measures the ratio of the test to reference signals at each of the targeted genomic locations, producing a vector of log ratios (see Figure 10.1). Log ratios are, by intention, zero when there is no copy number variation between the test and reference samples, positive in regions of amplification, and negative in regions of deletion. A common first step in analyzing DNA copy number data consists of identifying aberrant (amplified or deleted) regions in a sample. Proper aCGH normalization must correctly identify the regions of log ratio zero, and the error model must correctly evaluate the significance of deviations from log ratio zero. The data analysis algorithms described in this chapter are designed to achieve these goals.

CGH DATA ANALYSIS

WORKUP COMMON TO OTHER ARRAY APPLICATIONS

Image processing of the raw array data from CGH assays is similar to that used for gene expression assays, micro-RNA assays, and other array applications. Agilent CGH assays are two-color assays, though similar processing steps are recommended for one-color assays. Primary analysis of CGH data extracted from the image processing step is also very similar to other microarray analysis. Here we briefly describe the data processing common to all two-color applications. The rationale and the details of the algorithms are described in Agilent's Feature Extraction (FE) manual [3]. Parenthesized terms are the column headings of the Agilent FE9.5 output columns reporting the results of each processing step.

1. Compute the robust average signal per pixel of each feature spot in each channel (rMeanSignal, gMeanSignal).
2. Compute the background subtracted average signal and error bars for each feature (rBGSubSignal, gBGSubSignal, rBGSubSigError, gBGSubSigError). Accurate background subtraction is essential to avoid compression of ratios. The background-subtracted average signal is obtained by subtracting the background signal from the MeanSignal of each feature. The background signal is estimated by fitting a piecewise linear surface (using loess [3,4]) to the *negative control probes*. Interpolating point by point into the surface

Normalization of Array CGH Data

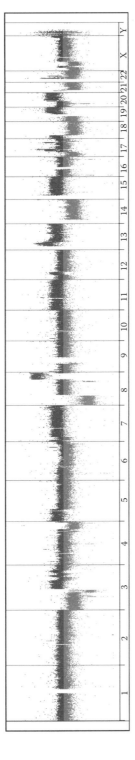

FIGURE 10.1 (**See color insert following page 80.**) Log$_2$ ratios of an HT29 cell line versus normal female DNA, after primary normalization and centralization to the dominant (triploid) peak.

gives a background signal value for each feature. Negative control probes measure the sum of intrinsic fluorescence of the printed features and the scanner dark offset. Hundreds of these probes are scattered randomly about the array to measure the background in all regions. The root mean square (RMS) of the residuals from the background fit is a measure of the background noise in the array. It is reported as a critical quality control (QC) metric for the array, and is also used as the AdditiveError term in the error model. Significant spatial trends in the intrinsic background fluorescence are not common, but can occur due to, for example, spatial asymmetries in washing and drying the slides.

3. Compute the processed signal by performing dye normalization (rProcessedSignal, gProcessedSignal, rProcessedSignalError, gProcessedSignalError). This preliminary normalization, which is uninformed by any presumptions about the biology of the test and reference samples, will be termed the "naïve" normalization. The two channels are first linearly scaled by setting the median signal of the *rank consistent noncontrol* probes that are well above the background noise levels to equal 1000. In some cases a nonlinear (e.g., loess [4]) normalization of log signals is preferable to the linear scaling. Such a nonlinear scaling is available as an option in FE software, but is not recommended in general. Routine application of nonlinear normalization can make the data significantly less accurate.

4. Compute log ratios (stored in LogRatio and LogRatioError, respectively). Log ratios are the $\log_{10}$ (these are converted to $\log_2$ ratios for subsequent CGH analysis) of the ratio of the rProcessedSignal to the gProcessedSignal. To avoid division by 0, processed signals are (by default) *surrogated* before the log ratios are calculated. The LogRatioError is the estimated one-sigma error in the LogRatio, computed by propagation of the estimated errors in the two channels.

The error on each signal is estimated using a variation on the Rosetta error model [5], which includes an additive error term and a multiplicative error term. The additive error is calculated from the background noise level as described before. The multiplicative error term cannot be robustly estimated for a single array; we recommend using 0.1 (10% error), estimated from multiple replicate experiments conducted in different laboratories. The error model includes no probe-specific or sequence-specific corrections.

Error Model Specific to aCGH Arrays

The error model described previously purports to estimate the standard deviation that would be observed for the signal and the log ratio of a single probe in a collection of replicate experiments. Special characteristics unique to genomic data allow a more precise functional error model to be parameterized entirely from internal data.

The aCGH data differ essentially from RNA expression data in that CGH probes can be ordered according to their position along the genome. Under the reasonable assumption that most pairs of consecutive probes are probing regions having the same copy number (i.e., that the number of probes exceeds the number of *breakpoints*), the true log ratios of the targets of most pairs of consecutive probes are expected to be strictly equal. Therefore, variations in the measured log ratios of consecutive probes result from noise, and a good estimate of the probe-to-probe noise can be computed from the robust standard deviation of differences between log ratios of consecutive probes. We report this error estimate as the *dLRsd* for an array. The dLRsd is reported as a critical QC metric for the array, and is also used in estimating the significance of observed aberrations [6].

NORMALIZATION SPECIFIC TO CGH ARRAYS

CGH data also differ from RNA expression data in that it is presumed that measured signal ratios correspond to discrete (i.e., rational) *copy-number* ratios. Sample genomes, however aberrant, are unlikely to be viable if they have fewer than two copies of most genes. These assumptions can be used to refine the naïve normalization described earlier.

The basis for assigning a statistical aberration score to an interval involves the assumption of a zero value for the data—the mean around which we expect the data to be distributed when no aberration is present. Naive normalization of aCGH data sets the average log ratio for all (or for *rank-consistent*) probes to zero. For illustration, consider a male versus female aCGH assay. As all probes residing on chromosome X are expected to yield a nonzero log ratio and all other probes are expected to yield signals centered on zero, we do not, in fact, expect the naïve normalization to yield a value that represents the real center of the diploid part of the sample. In this case, it is easy to overcome the problem by excluding the X-chromosome probes from the normalization process, but in most cases the aberrant regions of the genome are not known a priori.

CGH-specific normalization adds a constant to the $\log_2$ ratios such that the median reported log ratio of probes targeting diploid (or otherwise "normal") regions of the sample, after correction, is exactly zero.* This normalization correction step is termed *centralization*, to distinguish it from the normalization, which is naïve with regard to the chromosome positions probed.

Centralization of aCGH data proceeds in two steps. First, we generate a *centralization curve*, which shows the distribution of $\log_2$ ratios of regions of the genome. A typical centralization curve, computed from the data of Figure 10.1, is shown in Figure 10.2a. A centralization curve can be generated in various ways, the simplest of which is to replace the log ratio of each probe with the average of the log ratios of a number of adjacent probes (typically 50–100 adjacent probes). A histogram of the resulting local averages constitutes a centralization curve. More sophisticated and robust methods for generating a centralization curve have been developed and implemented in Agilent's CGHAnalytics software [7,8].

Ideally, a centralization curve consists of one or more well-separated peaks, each of which corresponds to a distinct integral *ploidy* of one or more regions of the test

---

* If the reference is male, then the true reference copy number of the X chromosome is one, not two. Properly centralized probes to X-chromosome regions thus report a log ratio of zero for regions in which the sample is haploid, and a log ratio of one for diploid regions. In the rest of this chapter, we will implicitly discuss only autosomal probes, or X-chromosome probes versus a female reference.

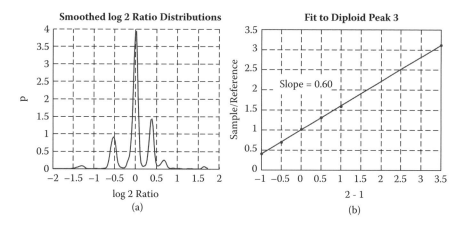

FIGURE 10.2 (a) The centering curve from the HT29 assay of Figure 10.1, centered on the highest peak. (b) A graph of the test/reference ratio versus $c/2-1$ with c as the assigned copy number (0, 1, 2, 3, 4, and 9 for the identified peaks). The linearity of the fit confirms that the copy number assignments are indeed sequential. However, the slope of 0.60 suggests that the diploid peak is misassigned.

sample. (If this is not the case, it is likely that the sample has been compromised, and further normalization may not be possible.). The log ratios of these peaks are determined by using any competent peak-picking algorithm (see Figure 10.2). In general, unless the test and reference samples have equal copy numbers for most regions of their genome, naïve normalization will not suffice to center any of these peaks at exactly zero.

The second step of centralization is to identify the peak corresponding to the "normal" ploidy. The log ratio of the "normal" peak is then subtracted from the log ratios of all the probes, so that "normal" regions of the genome report a log ratio of zero. The remaining problem is to decide which peak of the centralization curve is the "normal" peak.

Depending on the goals of the experiment, it may be sufficient to call as aberrant those regions that differ from the dominant ploidy of the test sample, whatever that is. In this case, it is sufficient to assign the highest peak of the centralization curve to "normal" without actually determining its true ploidy.* In most cases, though, it is desirable to flag as aberrant any nondiploid regions. So we need to assign a true copy number to each of the centralization peaks.

If the experimenter has prior knowledge of true ploidy of any genomic region (e.g., from FISH or SKY data), then the copy numbers of the peaks in the centralization

---

* Even when the experimenter does not care about absolute copy number, this scheme can run into difficulties. Some samples have nearly equal numbers of probes at two different ploidies (e.g., diploid and triploid). Small variations between nearly identical replicate experiments can cause one or another of these peaks to be (marginally) higher in the centralization curve, so that the "normal" peak is assigned differently in the two replicates, and they appear (after centralization) to be very different. As a general design principle, decision rules that can produce discontinuous outputs from arbitrarily small variations in their inputs should be avoided when possible.

Normalization of Array CGH Data

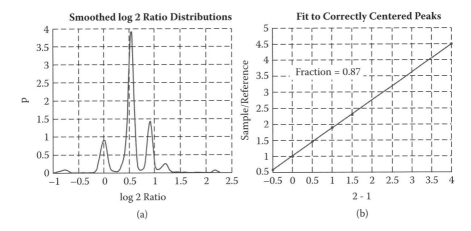

FIGURE 10.3 (a) The same centering curve of Figure 10.2a, recentered to the first significant peak. (b) The assigned copy numbers are updated from Figure 10.2b and are now 1, 2, 3, 4, 5, and 10. The slope of 0.87 is consistent with the correct assignment of the diploid peak. The slope is less than the theoretical value of 1.0 because of log ratio compression. Slopes of 0.9–0.95 are typical for the Agilent assay. SKY data confirm that HT29 is predominantly triploid.

curve can be confidently assigned. If not, then heuristics must be used to choose the most likely copy number for each peak:

H1. Significant peaks represent successively increasing copy numbers.
H2. Any large peak is not less than diploid.*

Heuristic H1 can be tested relatively easily. First, integer copy numbers are tentatively assigned to the peaks of the centralization curve. Since the reference sample is diploid, the centralization curve is tentatively shifted so that the log ratio of the selected "diploid" peak is at zero. If c is the true copy number of a region of the test sample, the ratio (not the log ratio) of test to reference sample will be $c/2$, since the copy number of the reference is two. A plot of ratio versus $(c/2-1)$ should have a slope of 1.0 and an intercept of 1.0. Such copy number plots are shown in Figure 10.2 and Figure 10.3. If the copy number intervals are not correctly assigned (i.e., assumption H1 is not valid), then the points plotted will not fall on a line. If the copy number intervals between the peaks are correctly assigned, but the absolute copy numbers are incorrectly assigned, then the plot will be linear, but the slope will not be 1.0.†

Once the log ratio of the diploid peak has been determined, the data are centered by subtracting the log ratio of the diploid peak in the original centering curve from the reported log ratios of all probes.

---

* Again, any male test sample versus a female reference will have a relatively large haploid peak that includes most X-chromosome probes. This heuristic assumes a sex-matched reference, or a workup including only autosomal probes.
† In practice, the slopes when copy numbers are correctly assigned are somewhat less than 1.0, due to log ratio compression in the assay. For the Agilent CGH assay, typical slopes are 0.90–0.95.

EXAMPLES OF CENTRALIZATION

A simple example of the application of these techniques is a normal male versus normal female. Naive normalization fails to correctly center the data, since the median log ratio lies between the autosomal peak and the X-chromosome peak. Centralization to the dominant autosomal peak shifts the zero to the center of the autosomal peak, so only X-chromosome probes will be identified as "aberrant" (data not shown).

A more interesting example is the HT-29 cell line shown in Figure 10.1. This cell line is known from SKY data to be predominantly triploid [9]. The centralization curve is shown in Figure 10.2, along with a copy number plot made under the assumption that the dominant peak is diploid. The copy number plot is linear, confirming that the major peaks in fact have sequential copy numbers, but its slope is only 0.60. Reassignment of the dominant peak to triploid increases the slope to 0.87, which is acceptably close to 1. An assignment of the dominant peak as tetraploid would have a slope of 1.54, which is clearly wrong.

COMPLICATIONS FOR POLYCLONAL SAMPLES

The normalization methods described before are appropriate for monoclonal test samples. But what if the test sample is polyclonal? In this case, the true copy number of each target region of the sample will be the average of the copy numbers of the different clones, weighted by their relative abundance. Thus, the assumption that copy numbers are integers is no longer valid. Samples from cell lines and normal cells from individuals are unlikely to be polyclonal, but tumor tissue samples are nearly always polyclonal.

Consider a biclonal sample, in which a single aberrant clone is mixed with diploid normal cells. Let a be the fraction of tumor cells (aberrant clone) in the sample, and c be the integral copy number in a genomic region of the tumor clone. Then the copy number of the mixed sample in this region is given by $ac + 2(1 - a)$, and the observed ratio of test/reference signal in a peak of the centralization curve will be $ratio = (c/2 - 1)*a + 1$. A copy number plot is expected to have a slope of a, rather than 1.0 as for a monoclonal sample (see Figure 10.4). When analyzing a polyclonal sample, there is a bit more ambiguity about copy number assignments than for a monoclonal sample, since the slope of the copy number plot cannot be used to verify the peak assignments. However, if the predominant ploidy or the fraction of normal cells is known from other evidence (e.g., cytogenetics or cytological data), a robust centralization and copy number assignment can be made.

When the sample contains more than one aberrant clone, as well as normal cells, it is generally not possible to assign particular aberrations to one or the other clone. However, in favorable cases copy numbers can still be assigned to aberrant regions, and the log ratios can be correctly normalized.

CONCLUSIONS

Array CGH is used to measure, at high resolution, the differences between a normal reference genome and a variant or aberrant genome. Calling aberrations in DNA

Normalization of Array CGH Data

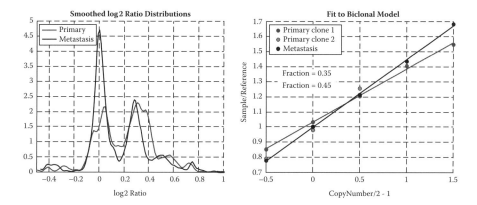

FIGURE 10.4 (See color insert following page 80.) Centralization curves and copy number plots of a primary (blue) and metastatic (black) tumor sample from the same individual versus a matched normal reference. The best fit to the metastatic tissue is consistent with a predominantly diploid genome, with successive peaks assigned copy numbers of 1, 2, 3, 4, and 5, but the copy number slope is only 0.45. We interpret this to mean that the malignant cells are about 50% diluted with normal cells. The primary tumor (blue) is more interesting, as each of the major peaks is doubled. This indicates the presence of more than one aberrant clone in the sample. One peak of each pair can be assigned to the dominant clone, so as to produce a linear copy number curve, which suggests that about 35% of the sample consists of the dominant aberrant clone. These interpretations are consistent with cytological data.

copy number data are based on deviations from a baseline representing the "normal" copy number. Little attention has been given in the literature to the problem of automatically setting this baseline. Traditional methods, such as using the average or median of the data, fail not only for the highly aberrant genomes of cancer cells, but also for individuals exhibiting normal copy number variation [10]. The normalization methods described here should provide reliable copy number estimates for most samples from aCGH data.

GLOSSARY OF TERMS

Breakpoint: In the context of genomic copy number measurements, a breakpoint is the nucleotide position at which the copy number changes. That is, it is the position of the boundary of an aberrant region.

Centralization: In the context of genomic copy number measurements, the estimation of the true copy number of regions of the sample genome, and shifting reported log ratios such that diploid regions, on average, report a log ratio of zero with respect to the (diploid) reference.

Centralization curve: A histogram of the $\log_2$ ratios of genomic regions. Such a curve is expected to show peaks at integral copy numbers of the sample. It can be generated by computing a moving average of successive probe log ratios, or by more sophisticated methods.

Copy number: The number of copies of a particular genomic region in the genotype of an individual.

dLRsd: The derivative log ratio standard deviation, a measure of the 1 sigma error in the log ratios on an array. It is computed from the interquartile range (iqr) of differences of log ratios reported for probes sequentially ordered along the genome: dLRsd == iqr ($\{LR_{i+1} - LR_i\}$)/1.908. In practice, observed dLR distributions are reliably Gaussian between the 25th and 75th percentiles.

Log ratio: The log base 10 or base 2 of the ratio of the normalized red signal to the normalized green signal. Generally, one channel measures the reference genome (whose true copy number is presumed to be known), and the other channel measures the unknown sample genome. If the reference is in the red channel, reported log ratios are usually multiplied by –1, so as to represent copy number changes in the sample relative to the reference. Log_{10} ratios are typically converted to $\log_2$ before CGH analysis and visualization.

Negative control probes: Probes that have been designed not to hybridize to the labeled sample in a microarray experiment. These probes are placed around the array to monitor possible experimental contamination as well as to measure background. There are two types of negative control probes: "hairpin" negative controls, which have sufficiently stable internal structure that they are unlikely to bind to anything else, and "hybridization" negative controls, which are random sequences having no homologies to any known targets existing in the input samples and confirmed not to hybridize significantly to a wide variety of sample types.

Noncontrol probes: Experimental probes to biological target sequences.

Ploidy: The number of copies of each chromosome in the genome of an individual. Ploidy is typically two (diploid) for normal somatic cells of animals and one for normal germ cells, though it can vary considerably in cancer cells. Ploidy also can be used to describe the copy number of individual chromosomes or other subsets of the genome.

Polyclonal samples: Samples derived from genetically diverse mixtures of cells, as opposed to samples derived from a single clone (monoclonal samples).

Precision: The precision of a measurement is the reproducibility of the measurement if performed independently multiple times. It is often expressed as a coefficient of variation (CV) or as a percent error. Precision expressed in this way can be reliably measured only for signals well above the background noise, for which the measurement error is proportional to the signal.

Rank consistent: A probe is considered to be rank consistent if it falls within the central tendency of the data observing consistent trends between the red and green channels. To compute this, one must rank-order the features in each channel by sorting them by their background subtracted signals. The highest rank is assigned to the most intense feature and the lowest rank, 1, is assigned to the least intense feature. Given a probe that has a red rank of n, if the corresponding green rank is also n, that feature is considered to be perfectly rank consistent. In general, it may not be possible to find probes perfectly rank consistent, so some variation is allowed. For example, if the rank in red is within 5% of its green rank, then the probe is considered rank consistent.

Resolution: In the context of aCGH measurements, resolution is the minimum size, in nucleotides, of an aberration that can be reliably detected in an assay. Depending on the goals of the experiment, resolution may also be defined as the uncertainty in the genomic locations of aberration boundaries.

Somatic: Generally taken to mean any cell forming the body of an organism, as distinguished from germ line cells. Somatic cells are distinct and significantly different from stem or germ line cells in that as DNA from somatic cells replicates over time it has the potential to change due to events during the DNA replication process. Daughter cells of those altered cells inherit the parents' altered DNA, thus propagating those changes

Surrogate: This is the poor man's way of dealing with the inconvenient explosion of logarithms as signals approach zero. If the ProcessedSignal is less than the ProcessedSignalError, it is replaced by the ProcessedSignalError (the "surrogate").

Two-color: A microarray assay in which a test and a reference sample are labeled with different fluorescent dyes and hybridized together to the array. Contrasted with a one-color assay, in which the test and reference samples are labeled with the same dye, and hybridized to different arrays. The decision to use a two-color or a one-color experimental design depends on many factors, including uniformity of array printing, hybridization temperature, mixing, and wash, and on the similarity between the test and reference samples. Maximum measurement precision is achieved by hybridizing two test-reference pairs, labeled in alternate colors, and averaging the results (a "dye-flip" experiment). However, this is expensive and, for many purposes, such high precision is unnecessary, especially when biological variation exceeds technical variation. A full discussion of these issues is beyond the scope of this chapter. In our experience, a two-color assay, with or without dye-flips, is the best choice for CGH assays.

REFERENCES

1. Barrett MA, Scheffer A, Ben-Dor A, Sampas N, Lipson D, Kincaid R, Tsang P, et al., Comparative genomic hybridization using oligonucleotide microarrays and total genomic DNA, *PNAS,* 101(51), 17765–17770, 2004.
2. Toruner GA, Streck DL, Schwalb MN, Dermody JJ, An oligonucleotide based array-CGH system for detection of genome-wide copy number changes including subtelomeric regions for genetic evaluation of mental retardation, *Am J Med Genet A,* 143(8), 824–829, 2007.
3. Agilent Feature Extraction Manual: http://www.chem.agilent.com/Scripts/PDS.asp?lPage=2547
4. Cleveland WS, Devlin SJ, Locally weighted regression: An approach to regression analysis by local fitting, *J Am Stat Assoc,* 83, 596–610, 1988; http://www.netlib.org/a/loess
5. Weng L, Dai H, Zhan Y, He Y, Stepaniants SB, Bassett DE, Rosetta error model for gene expression analysis, *Bioinformatics,* 22(9), 1111–1121, 2006.
6. Lipson D, Aumann Y, Ben-Dor A, Linial N, Yakhini Z, Efficient calculation of interval scores for DNA copy number data analysis, *Proc RECOMB '05,* LNCS 3500, p. 83, Springer–Verlag, 2005. Also in *J Computational Biol,* 13(2), 215–228, 2006.

7. Lipson D, Ph.D. thesis, http://www.cs.technion.ac.il/users/wwwb/cgi-bin/tr-get.cgi/2007/PHD/PHD-2007-05.pdf
8. Agilent CGH Analytics Manual: http://www.chem.agilent.com/Scripts/PDS.asp?lPage=29457
9. http://www.path.cam.ac.uk/~pawefish/ColonCellLineDescriptions/HT29.html
10. Redon R, Ishikawa S, Fitch KR, Feuk L, Perry GH, Andrews TD, Fiegler H, et al., Global variation in copy number in the human genome, *Nature*, 444, 444–454, 2006.

11 SNP Array-Based Analysis for Detection of Chromosomal Aberrations and Copy Number Variations

Ben Bolstad, Srinka Ghosh, and Yaron Turpaz

CONTENTS

Introduction 245
Algorithms 246
    General Algorithm Components 246
    CNAT4 246
        Copy Number Estimation 247
        Loss of Heterozygosity 256
        Other Algorithms for CN Analysis Using Affymetrix Mapping Arrays ... 258
Case Studies (Illustrate Workflow on 500K Data Using CNAT4) 260
Future Trends in CN Analysis 264
Acknowledgments 264
References 264

INTRODUCTION

Cytogeneticists are studying chromosomal aberrations and their related disease states; cancer researchers have validated correlations between carcinogenesis and genetic alterations in tumor cells [1]. These alterations involve allelic imbalances, manifested in the form of chromosomal copy number (CN) changes including amplifications, deletions, aneuploidy, loss of heterozygosity (LOH) and microsatellite instability. These events often indicate the activation of oncogenes and/or inactivation of tumor suppressor genes (anti-oncogenes). Variations in the form of copy number polymorphisms (CNP) can also occur in normal individuals [2,3]. Identification of the loci implicated in these aberrations can generate anchor points that facilitate oncogenomic and toxicogenomic studies. High-density oligonucleotide microarrays originally designed to facilitate the genotyping of a multitude of single nucleotide polymorphisms (SNPs) are being increasingly used for the accurate localization of genomic CN alterations.

Interpretation of these aberrations, however, is complicated by the following, among others: (a) CNPs in normal individuals, (b) CN aberrations possibly not accompanied by LOH and vice versa, and (c) preferential aberrations in one allele over the other. Motivated by the preceding rationale, we have developed a hidden Markov model-based framework to quantify total copy number (TCN), allele-specific copy number (ASCN) changes, and LOH. These analyses correlate the association and the extent of localization of ASCN and LOH changes using the Affymetrix GeneChip human mapping 500K array platform, which provides consistently high coverage with approximately half a million SNPs at a median inter-SNP distance of 2.5 kilobases.

ALGORITHMS

GENERAL ALGORITHM COMPONENTS

For a normal diploid genome a copy of a set of chromosome is acquired from each of the parents; the exceptions are chromosomes X and Y, in male, which are present as single copies. Due to abnormal cellular changes, for example with carcinogenesis, one or both chromosomal copies can be deleted or more copies acquired. Fundamentally, this results in CN changes. Loss of heterozygosity is a change most frequently associated with CN deletions. The inference of LOH is made when an allelic imbalance occurs in the tumor/test sample, provided those identical regions in the corresponding reference sample retain the normal heterozygosity. Loss of heterozygosity is also associated, albeit less frequently, with copy-neutral events; here, changes in allele-specific CN occur in a balanced manner such that the estimate of total CN remains diploid.

The following section discusses several currently available algorithms—namely, CNAT4, CNAG, dChip, GEMCA, CARAT, and PLASQ—and elucidates how the different algorithms implement the basic components to derive final estimates of CN and LOH. The fundamental components of the analyses are listed below.

Copy number
    Preprocessing of probe intensities
    Generation of SNP-level data summaries from probe intensity data
    Determination of test sample versus reference pairing
    Quantification of CN
    Segmentation of CN change regions
Loss of heterozygosity
    Determination of test sample versus reference pairing
    Quantification of LOH based on genotype calls
    Segmentation of LOH regions

CNAT4

This section provides an in-depth discussion of the CNAT4 (copy number analysis tool 4) algorithm for CN and LOH analysis of Affymetrix GeneChip human mapping arrays using the whole-genome sampling analysis (WGSA) [4]. The laboratory procedure requires that the DNA be fragmented separately by pairs of restriction enzymes (Nsp and Sty for the 500K, Xba and Hind for the 100K array, but only Xba for the 10K), and the respective SNP sequences carried on the enzyme fragments are hybridized separately onto two chips, each with about half of the total SNPs. The 500K set is therefore two "250K" arrays, one for each enzyme. Note that, on the newer versions of Affymetrix genotyping arrays, the fragments will be analyzed on a single chip. The CNAT algorithm was developed around the 500K set of arrays, but is equally applicable to the earlier 10K and 100K set of arrays.

The algorithmic workflows for both the CN and LOH estimation (summarized in the flowcharts shown in Figure 11.1) are categorized under the following:

Analysis of unpaired samples: In this case, samples potentially unrelated and spanning a diverse population are analyzed together.

Analysis of paired tumor–normal samples: In this case, the tumor and normal control are derived from tissues of the same individual with the normal tissue being obtained from the blood sample.

It should be noted that the TCN estimates are generated for both paired and unpaired experiments, whereas ASCN estimates are generated exclusively for paired experiments. Estimates of LOH are generated for both paired and unpaired experiments. The following discussion on CNAT4 is presented in two sections: CN estimation and LOH.

Copy Number Estimation

As with most microarray analysis algorithms, preprocessing makes up the initial stages of the CNAT4 algorithm where raw probe intensities are adjusted and combined to produce allele-specific summaries. In particular, CNAT4 uses only the "perfect match" (PM) probe intensities, ignoring the intensities of the "mismatch" (MM) probes on the array for all the supported chip types. In some situations it might also be additionally useful to prefilter SNPs based on certain criteria, such as length of the polymerase chain reaction (PCR) fragment the SNPs reside on, the GC content or sequence specificity, or other SNP quality control (QC) metrics. The fragment length and GC content distribution for the 10K, 100K (2 × 50K), and 500K (2 × 250K) arrays are summarized in Table 11.1.

Polymerase chain reaction fragment length-based filtration is particularly effective for estimation of CN in degraded DNA samples, such as formalin-fixed paraffin-embedded (FFPE) samples [5]. Generally for these samples, SNPs on smaller enzyme-digested fragments are less affected by degradation than those on the larger fragments. Of the three array types, the majority of the SNPs on the 500K arrays populate the lower end of the fragment-length spectrum, with a maximum length of 1143 base pairs. Therefore, the 500K array is potentially the most appropriate for FFPE analysis.

Because CN microarray experiments are typically performed using multiple chips, it is appropriate that any analysis method seek to reduce experimental noise due to interchip variation. CNAT4 uses a probe-level normalization for this purpose.

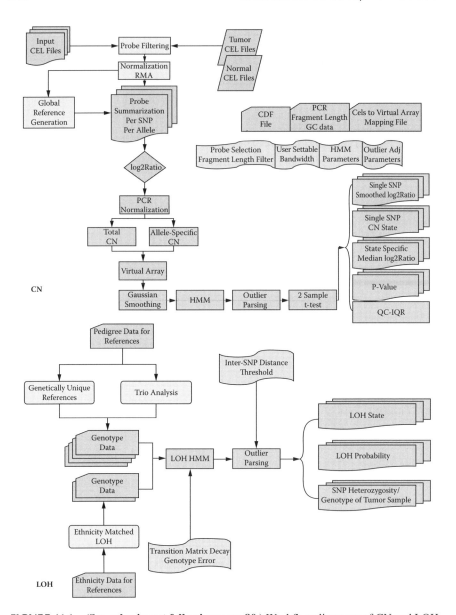

FIGURE 11.1 **(See color insert following page 80.)** Workflow diagrams of CN and LOH.

Specifically, it implements two different strategies: quantile normalization and median scaling.

Quantile normalization is particularly effective in normalizing nonlinearities in the data [6]. Probe-level quantile normalization is effective in the elimination of biases introduced by experimental covariates such as scanners, operators, and replicates. The sketch distribution against which all samples are quantile normalized

TABLE 11.1
Distribution of PCR Fragment Length and GC Content across the Different Arrays

Array Resolution	Restriction Enzyme	PCR Fragment Length (bp)		GC Content (%)	
		Min	Max	Min	Max
10K	Xba	202	1700	23.716	64.841
50K	Xba	200	2122	22.702	69.377
50K	Hind	245	2270	21.912	72.299
250K	Nsp	100	1143	17.108	68.762
250K	Sty	115	1093	17.162	71.556

is determined from a set of user-selected references. Since the quantile is a smooth and slow varying function, the quantile distribution for each chip is interpolated from approximately 50,000 PM probes sampled across each of the references. Sampling over 50,000 probes, as opposed to all available PM probes, facilitates computational efficiency without significantly affecting the outcome.

Median scaling constitutes a linear operation, where the scale factor is based on the median of all chip medians. A single chip median is computed based on the intensity of all probe pairs on the array. For paired-sample analysis, median scaling is often sufficient.

Next, the normalized individual PM probe intensities for each of the A and B alleles for each SNP are summarized to generate allele-specific summary values. The allele designation is based on the alphabetical order of the bases represented at the mutation. For example, for an SNP with a sequence AT {G/A} TTT, the allele A designates AT{A}TTT and allele B designates AT {G} TTT. A robust multichip summarization method is used to produce the summary values. This is done by default using the median polish algorithm because it takes into account both (a) sample-level effects arising from chip-to-chip (sample-to-sample) variation, and (b) feature-level effects arising from systemic differences in feature intensities due to the hybridization specificity of different probe sequences. The plier summarization (using Robust Multi-Chip Analysis [RMA]-like settings) algorithm can also be used for this purpose. Because every SNP has two alleles, each SNP has two summary values for each array in the data set.

In general, the summarized intensity of the A allele versus the B allele is governed by the quantity of the particular allele in the target genome. Hence, the ratio of the summarized A allele to B allele intensities in normal populations might exhibit an ethnicity bias. It should be noted that the A versus B allele distinction is different from the minimum versus maximum allele intensity distinction where the latter is analogous to allelic frequency. The minimum allele comprises the minimum intensity distribution of the allele-specific (A/B) SNP summaries. The output of the allele-specific analysis comprises the CN estimates for the min and max alleles.

Copy number estimates are generated from the SNP summary values. More importantly, these estimates are made relative to a reference sample or reference set.

In an unpaired analysis the reference set should be constituted using a pool of normal female reference samples. This is preferred because, among other reasons, pooling the data from multiple chips mitigates artifacts arising from outliers and random variation across samples. Three main factors govern reference set generation: number of samples, gender of samples, and ethnicity of samples.

The number of samples used in the reference set can have an effect on the CN estimation process. Too small a reference set will not adequately capture enough sample variation and could result in a lower quality reference array with increased noise. Very large reference set sizes become computationally slower. Around 25 normal samples have been determined to be a reasonable compromise.

It is recommended that only normal female samples are used in reference sets; otherwise, there could be interpretational difficulty in the estimation of changes observed in chromosome X. The aberration due to gender mixing is primarily evident in the form of bias offset and reduced discrimination of CN changes. If CN estimation in chromosome X is not a concern, a pooled set of male and female samples can be utilized to boost the overall reference sample size.

The references can either be ethnically matched to the test sample or not. Provided sample-size considerations are met, no ethnicity bias has been observed in overall CN estimation. However, it may be relevant, particularly in the determination of CN variants. In many cases, the ethnicity of a sample is unknown. However, if there is concern about dilution of the estimated CN from sampling across a diverse population set, the references can be restricted (if available) to the known ethnic origin of the sample. For a given ethnicity, more than 25 samples, independent of quality, did not enhance the convergence of the quantile sketch. If reference samples are unavailable in the lab, the HapMap references (see www.affymetrix.com or www.hapmap.org) can be downloaded and equivalent percentages can be pooled across ethnic diversity. Reference summary values for each SNP on each allele are computed by taking median values of the array-specific values.

In a paired analysis, the normal reference should be derived from a normal (blood) sample from the same individual that the tumor sample was obtained from. This will remove the possibility of interindividual variation and highlight the disease-related CN differences.

After establishing a reference set, TCN estimates are generated for each SNP by computing the $\log_2$ ratio of the sum of the two allele-specific summary values to the sum from the reference set, as shown in equation 11.1. Specifically, S_A and S_B represent the SNP summary values on the A and B alleles for the sample, and R_A and R_B represent these values as derived from the reference set. This estimate, referred to as the *logSum*, optimizes variance in the TCN at the cost of bias.

$$TCN_{\log Sum} = \log_2\left(\frac{S_A}{R_A + R_B} + \frac{S_B}{R_A + R_B}\right) \quad (11.1)$$

An alternative approach to estimating the TCN is referred to as *sumLog*, as shown in equation 11.2. This is an approximation to the *logSum* estimate and primarily optimizes the bias at the cost of variance. By default TCN is estimated by *logSum* in CNAT4.

SNP Array-Based Copy Number Analysis

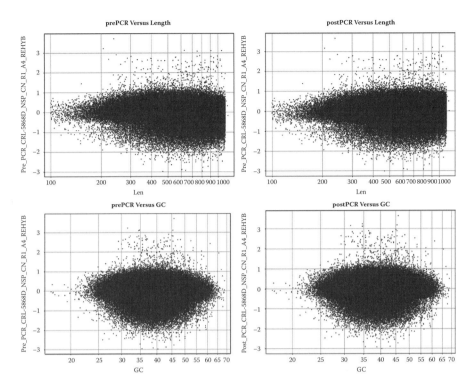

FIGURE 11.2 500K: Pre- (left) and post- (right) PCR correction versus PCR fragment length (top) and GC content (bottom).

$$TCN_{sumLog} = \log_2\left(\frac{S_A}{R_A}\right) + \log_2\left(\frac{S_B}{R_B}\right) \qquad (11.2)$$

Total CN constitutes the *rawest* form of the CN estimate. Next, CNAT4 performs a correction procedure, referred to here as "PCR normalization," to correct for potential artifacts introduced by the PCR process [7,8]. Equation 11.3 gives the relative CN for the ith SNP. The different PCR conditions are compensated for by equation 11.4, where $^c\Lambda_i^{1,2}$ represents the corrected CN, and $p(x)$ denotes the PCR amplification kinetics. In the quadratic PCR correction implemented as a linear regression, GC content and length of the PCR fragment that an SNP resides on are the covariates. $q(x,y)$ represents a variant on the implementation of $p(x)$. PCR correction has less of an impact on 500K compared to 100K data, as shown in the plots of fragment length and GC content versus CN for pre- and post-PCR normalization for 500K (Figure 11.2). Furthermore, in the case of the 500K arrays, length rather than GC content is the principal contributor to the PCR issue. A slight straightening of the curve is observed in the case of length; no change is observed for GC content (Figure 11.2).

$$\Lambda_i^{1,2} = \log_2\left(\frac{S_i^{sample1}}{S_i^{sample2}}\right) \text{ for } i\text{th SNP} \qquad (11.3)$$

$$\Lambda_i^{1,2} = {}^c\Lambda_i^{1,2} + p(x)$$

$$p(x) = \sum_{j=1}^{2}\left(a_j + b_j x_j + c_j x_j^2\right) \text{ where } j = GC, \text{ fragment length}$$

$$\Lambda_i^{1,2} = {}^cA_i^{1,2} + q(x,y)$$ (11.4)

$$q(x,y) = \left(A + bx + b'y + cx^2 + c'y^2\right) \text{ where } x = GC/y = \text{ fragment length}$$

In situations where both of the arrays in a set (100K, 500K) have been used for each sample, CNAT4 uses the concept of a virtual array. A virtual array is created by computationally combining data from the two arrays to create a single array. An interarray normalization is used to adjust for differences in the noise baseline and the variance in the two array types. Without the interarray normalization, there is a high probability of certain regions of the genome being erroneously labeled as CN aberrations. The interarray normalization parameters are estimated based on a subset of SNPs in the copy-neutral region (CNR). In an ideal case, the CNR is bounded by the hemizygous deletion state (CN = 1) at the low end, and a single copy gain state (CN = 3) at the high end. There are two main issues with the bounds defined as such:

1. Since CN states 1 and 3 are equivalent to

$$-1\left(\log_2\left(\frac{1}{2}\right)\right)$$

and

$$0.58\left(\log_2\left(\frac{3}{2}\right)\right)$$

respectively, the bounds are asymmetric with respect to 0.2.

2. A deletion or single copy gain state might actually represent a median value different (typically attenuated) from the archtypical values of −1 or 0.58. Hence, inclusion of all SNPs populating the region bounded by CN states 1 and 3 might include SNPs that are not copy neutral. Therefore, instead of fixed boundaries for the CNR, it is defined in terms of two bounds: upper and lower. In the data presented, all analysis is performed by considering symmetric bounds of ±0.2, the default in CNAT4.

The underlying normalization constitutes a *shift and scale* transformation, where the corrected $\log_2$ ratio for SNP_i in sample S ($\Lambda_i^{S'}$) is shifted by the mean of the SNPs in the CNR (equation 11.5) and is scaled by the variances of the arrays, as determined from the CNR (equation 11.6). These are combined in equation 11.7 to represent $\Lambda_i^{S'}$. Subsequent to interarray normalization, SNPs are merged across both arrays and sorted based on chromosomal and physical location.

$$\mu_S = average\left({}^c\Lambda\left[CNR[-], CNR[+]\right]\right)_s \tag{11.5}$$

where
    $CNR[+]$ = Copy – neutral region lower and upper bounds
    S = sample

$$\sigma_s = stdev\left({}^c\Lambda\left[CNR[-], CNR[+]\right]\right)_s \tag{11.6}$$

$$\Lambda_i^{1'} = \left(\frac{\sqrt{\sigma_2}}{\sqrt{\sigma_1}}\right)\left({}^c\Lambda_i^1 - \mu_1\right) \tag{11.7}$$

Depending on the inherent noise in the data there might be some advantage in performing smoothing prior to computing the final CN estimate and hidden Markov model (HMM) segmentation. Smoothing increases the signal to noise ratio (SNR) in the data and may enhance the demarcation between regions of CN change. When an explicit smoothing is required, CNAT 4 uses a Gaussian smoothing kernel. Smoothing is performed on PCR-normalized (and interarray-normalized in case of virtual arrays) CN data. It is performed for every index SNP_i by considering the CN contributions from all flanking SNPs encompassed in a window (W), as defined by $W \approx 2*\sigma_{mult}*\sigma$; where σ corresponds to the Gaussian bandwidth, and σ_{mult} is the sigma multiplier. Specifically, all flanking SNPs within the bounds defined by $\pm\sigma_{mult}*\sigma$ of SNP_i are considered. By default, 100 kb smoothing is used by CN4T4. Higher levels of smoothing are more likely to mask micro-aberrations and a smaller σ (instead of 100 kb) might be more appropriate in these cases.

After generating CN values for each SNP along the genome, the next thing of practical interest is to be able to clearly delineate regions of CN change. CNAT4 uses an HMM [9] to segment these regions. The underlying assumption in an HMM is that the experimental observation—CN and/or LOH value of an SNP—is generated by a hidden or unknown process. We are aided by the emitted or (experimentally) observed CN values, which are measured as continuous variables, because they hint at the true hidden CN states, which are discrete in value. In the (default) HMM used by CNAT4, there are five hidden states representing the following biological phenomena:

- Homozygous deletion or CN state = 0.
- Hemizygous deletion (haploid) or CN state = 1.
- Copy neutral (diploid) or CN state = 2.
- Single copy gain or CN state = 3.
- Amplification (multiple copy gain) or CN state ≥ 4.

According to the HMM, an SNP_i can exist in any one of the five hidden states (S). Based on the observed CN (O) of SNP_i and its neighboring SNPs, as well as the model λ (discussed later), the hidden state of the SNP_i needs to be determined. The same underlying HMM model is used for analysis of both paired and unpaired CN scenarios. A schematic of a five-state HMM is represented in Figure 11.3.

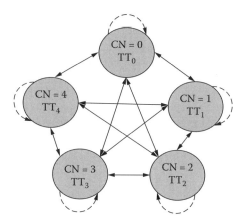

FIGURE 11.3 Schematic of a five-state ergodic HMM; self-transitions are denoted by dotted lines; non-self-transitions denoted by solid lines.

In a first-order HMM—as implemented here—it is assumed that the hidden state of any SNP is impacted by the status of the prior SNP in genomic space; this introduces the concept of the probability of sustaining/transitioning from one hidden state to another. In this implementation an ergodic model is supported. According to this model, at any point both same-state or self-transitions (denoted by dotted lines in Figure 11.3) as well as non-self-transitions (solid lines) are allowed. Hence, if SNP_i exists in hidden state 0, then the allowed transitions for SNP_{i+1} can be to any state from 0 through 4. If SNP_i and SNP_{i+1} are proximal, then there is a high probability that the latter will continue to exist in state 0 (based on the preceding example); conversely, if SNP_{i+1} were distal, then it might have increasing probability of transitioning to an amplification or copy-neutral state.

As mentioned, in the HMM approach it is assumed that SNP_i has a nonzero probability of existence in any one of the five states. The goal of the model is to determine the state that has the maximum likelihood of being its true underlying state. Therefore, the goal, stated succinctly, is to evaluate the probability of an observation sequence $\{O_1, O_2, O_3...O_n\}$ given the model λ. The three parameters of the model, λ, are prior probability, transition probability, and emission probability.

Prior probability or prior (π) refers to a priori knowledge of the chance that SNP_i exists in a particular state. Each state has a prior, with the sum across states being equal to 1. The priors are not assigned on a per-SNP basis, but instead to the population as a whole. Although CNAT4 allows selection of these priors for any experiment it may be difficult to do so. In general for all experiments, it suffices to assume that no a priori information is available about the states of the SNPs and begin with a uniform set of prior probabilities. In particular, the prior for each state is initialized to 0.2 (for a five-state model) after which the CNAT4 algorithm optimizes the priors for the individual CN states.

Transition probability refers to the probability of changing from one hidden state, S_i, to another S_j. In annotating CN aberrations, it is evident that a contiguous set of SNPs exists in a deletion state (micro-aberrations being an exception) and, in LOH, clusters of contiguous SNPs are shown to exhibit loss. Thus, a CN/LOH event

SNP Array-Based Copy Number Analysis

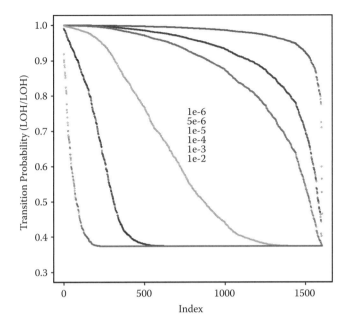

FIGURE 11.4 (See color insert following page 80.) Transition decay (β) as a function of inter-SNP distance.

is very seldom a singleton SNP effect. This indicates the existence of a correlation across the hidden states of neighboring SNPs. A factor affecting this is the significant variation in the SNP density across the 10K–500K SNP arrays. For 250K arrays the median inter-SNP distance is approximately 5 kb, whereas for the 10K arrays it is approximately 11 kb.

To eliminate dependencies on the inter-SNP distance, the transition matrix is modeled as a nonstationary one, as described in equation 11.8. In this equation, $\hat{\pi}$ refers to the prior matrix associated with the five states in CN (or two states in LOH); Δd refers to the inter-SNP distance, and β refers to the transition decay parameter. The decay parameter input into CNAT4 is β′ where β′ = 1/β and is represented in units of base pairs. A smaller β implies a slower transition from the current hidden state to a different hidden state and vice versa. This relationship is demonstrated in Figure 11.4.

$$\hat{I} \times \left(e^{-\beta \Delta d}\right) + \hat{\pi}\left(1 - e^{-\beta \Delta d}\right) \tag{11.8}$$

Emission probability reflects the probability with which the underlying state (*S*) is emitted to produce the observable (*O*). The two parameters that govern this probability are the mean (μ) and the allowed standard deviation (SD) of each hidden CN state. These parameters are used for the generation of the log-likelihood estimates. If the CN data are smoothed, then the expected SD should be tight (low); if the data are noisy, the SD should be high. As a rule, the lower the bandwidth is (implying that the underlying CN data are unsmoothed or noisy), the higher is the SD parameter.

**TABLE 11.2
HMM Initial SD as a Function
of Gaussian Bandwidth (σ)**

	logSum: Standard Deviation	
Bandwidth	States 0, 1, 3, 4	State 2
0	0.25	0.19
15K	0.15	0.13
30K	0.12	0.09
50K	0.11	0.08
100K	0.09	0.07
500K	0.06	0.03

Table 11.2 summarizes suggested SD as a function of bandwidth. These estimates are derived from a 1x–5x chromosome X titration study on replicate data sets where interlaboratory variability was also taken into consideration.

The HMM parameters that are user tunable relate to the preceding probabilities. It should be noted that the user input values constitute initial guesses or seeds for the parameters and are optimized by the underlying HMM algorithm. So while these are not expected to be precise by any means, a certain combination of values, as described later, can lead to difficulties with interpretation of data.

Finally, the discrete HMM or CN states are assigned based on the Viterbi algorithm, which is a global optimization method that is resilient to local CN perturbations. In addition to the smoothed ($\log_2$ ratio) CN estimate provided on a per-SNP basis, an HMM median CN estimate also computes the median CN for all contiguous SNPs belonging to a specific HMM state.

In order to generate statistical confidence of the estimates, a running t-test is performed on the PCR-normalized CN estimate on a per-chromosome/per-sample basis. The null hypothesis is that the mean CN in a given window (W) is equal to the baseline CN. The baseline is computed per sample, and the W corresponds to the Gaussian window size. In this regard, it is a windowed t-test, and therefore p-values cannot be computed for unsmoothed data.

In addition to the quantification and segmentation of TCN, ASCN is also estimated by CNAT4. This is computed exclusively for paired experiments. In this case, only the subset of SNPs that have heterogeneous, genotype {AB} calls in the reference are considered. On average this encompasses approximately 30% of the SNPs. For this subset, the PCR-normalized A and B allele data are transformed into min and max alleles, based on their intensity. The ASCN analysis pipeline, which comprises both quantification and segmentation, is analogous to the TCN pipeline (as described earlier) where the min and max allele data are processed independently.

Loss of Heterozygosity

Genotype data are critical for the estimation of loss of heterozygosity. Unlike the CN HMM, the LOH HMM implements two different models for the paired and unpaired analyses. The following outlines the implementation of the HMM for LOH analysis, which uses a two-state model:

- Retention of heterozygosity/normal or state = 0
- LOH or state = 1

Unlike the CN HMM, the prior probability in LOH is estimated empirically from the sample-specific genotype data. The paired LOH analysis is based only on the subset of SNPs that are heterozygous {AB} in the normal reference. The unpaired analysis includes all SNPs. Here, the prior is computed as a function of the SNP heterogeneity rate estimated on a per-SNP basis across all references and the genotyping error rate (ε), with a default of 0.02. The default is based on data derived from analysis of HapMap samples. SNP_i heterogeneity rate corresponds to the fraction of heterozygous calls for SNP_i in all pooled references, excluding the test sample. Hence, a small reference sample size might introduce errors in this estimation. For unpaired analysis, use of 40 references is considered optimal. This has been determined on the basis of the sensitivity and specificity estimated from 10 paired LOH data sets, performed in duplicate, which were subsequently analyzed in unpaired mode. The paired LOH data were considered as the ground truth in this experiment. The number of references was titrated in increments of five from 5 to 65. The relationship between the prior of LOH and retention is governed by the following property:

$$\pi_{LOH_{sample}} = 1 - \pi_{RET_{sample}} \tag{11.9}$$

The transition probability for LOH is computed in a manner analogous to that in CN. The only difference here is that the prior is a 2 × 1 as opposed to a 5 × 1 matrix.

In the case of LOH, the emission probabilities are computed empirically from the data. This estimation is significantly simpler for the paired case. In the paired case, since only the subset of SNPs with heterozygous calls in the reference is considered, the estimation of the emission probability follows this logic:

- If the hidden state is an LOH, then the probability of the observed state is:
  - Normal is ε.
  - LOH is $1 - \varepsilon$.
- If the hidden state is a retention, then the probability of the observed state is:
  - LOH is ε.
  - Normal is $1 - \varepsilon$.

The emission probability is controlled by the genotyping error rate, and hence if this value is set inappropriately, it will adversely affect the estimation of the emission probability.

For the unpaired case, the observations correspond to genotype calls while the hidden states correspond to LOH or retention. In this case, the emission probability for an LOH or retention event from homozygotes is computed as a function of the genotyping error rate and from heterozygotes is computed as a function of the SNP heterogeneity rate.

Loss of heterozygosity versus retention probability. These probabilities are analogous to the statistical confidence with which an LOH or retention event is observed. The LOH probability is estimated based on the marginal probability of the HMM. The marginal probability of an HMM state is the possibility that we will observe that state without the knowledge of any other events. The retention probability is estimated as 1-LOH probability. Hence, both these values are bounded by [1,0]. While the assignment of the final states in the LOH analysis is based on the Viterbi algorithm, the LOH probability is not computed from this; however, a correlation of 98% or higher has been established between the marginal and the Viterbi outcome. Loss of heterozygosity events have an LOH probability around 1.0, whereas the retention events encompass the entire spectrum from the min(LOH probability) to zero.

OTHER ALGORITHMS FOR CN ANALYSIS USING AFFYMETRIX MAPPING ARRAYS

Many other algorithms have been developed for CN quantification and segmentation on Affymetrix mapping arrays. The earliest CN algorithms were designed for the 10K mapping array. The first stage of the analysis is to combine the individual probe intensities into SNP summaries. In Bignell et al. [10], the algorithm proceeds by logging the average of the difference in PM and MM probe pairs. Huang et al. refine this to logging the average of the PM probes. Both then standardize these values using autosomal SNPs, and each centers the summary values around 0 by using the mean value of the autosomal SNPs. Huang et al. [11] extend this by using the sample standard deviation of all the autosomal SNPs to give variance 1.

For arrays in the reference set, genotype-specific mean and standard deviation values of these processed summary values are computed. An equation linking the summary value for an SNP and the mean intensity for that SNP in the reference set for SNPs of the same genotype is used to compute the CN estimate. The parameters of this model were estimated using a titration experiment. To segment regions, each SNP gets a test statistic value, and then regions along the genome with the same sign for this standardized test statistic are considered as candidate CN variant regions. A statistical test of the mean of these test statistic values is used to determine whether or not a region is exhibiting a CN change.

The next generation of human mapping arrays, the 100K set, introduced additional algorithms. Copy number analysis with regression and tree (CARAT) [12] provided a novel analysis technique using regression models and regression trees. In this algorithm, PM intensities are log transformed and then standardized using information from MM intensities from SNPs with the same genotype call. With CARAT, the dynamic model (DM) algorithm [13] is used for genotype determination. Next, using arrays in the reference set, probe selection is based on strong probe-dosage response to genotype calls. The selected probes are averaged for each allele to produce allele-specific SNP summaries. A regression adjustment is used to remove systematic fragment length and GC content effects. For test arrays, the allele-specific intensities are further adjusted based on their genotype call and the genotype-specific intensities of this SNP in the reference set. Allele-specific copy numbers are estimated using a model fitted to the reference set. For each SNP, total copy number is the sum of these two ASCN. To segment CN regions, CARAT first uses a 100-kb Gaussian smoother and then builds a regression tree using genomic physical position as the predictor and log-transformed CN estimate as the response.

dChip, a piece of software widely used for expression array analysis, provides functionality for CN analysis [14]. dChip uses an invariant set normalization method to remove systematic differences between multiple arrays. This method chooses a reference array, and for each subsequent array a rank-invariant set of probes is used to fit a normalizing relationship via a running median. This normalizing relationship is used to adjust the intensities on each array. For each probe, the A and B allele PM intensities are added, and similarly for the MM intensities. Then dChip uses the model-based expression index fit to the difference between these PM and MM intensities. A parameter from this model is then used as the SNP summary value. These are averaged across reference samples and the ratios are used for CN estimation. An HMM model provides segmentation of CN regions for dChip. The dChip software is available at www.dchip.org.

Another prominent algorithm is CNAG (copy number analyzer for GeneChip) [8]. In this algorithm, for each SNP allele, the PM probe intensities are added together, and the mean SNP value is equalized across all arrays. A regression model is used to correct for fragment length and GC content systematic effects. The CNAG supports both total CN and ASCN estimation. In the second case, it uses DM genotyping calls. To smooth CN estimates, it averages a fixed number of adjacent SNPs, irrespective of genomic spacing. Segmentation of CN regions is done using an HMM. The CNAG software is available at www.genome.umin.jp.

The genotyping microarry-based CNV analysis (GEMCA) algorithm [15] was designed specifically for the detection of small CN variations. It uses the preprocessing of the DM algorithm to summarize the probe intensities for each SNP allele and also the DM genotype calls. Gaussian mixture clustering is used to adjust for different allele affinities. The GIM (genomic imbalance map) algorithm, which accounts for fragment length and GC content of both the probe and fragment, is used to reduce systematic noise in the raw intensities. Because GEMCA is designed for detection of CNV within a population, it works in a pairwise manner. The SW-ARRAY algorithm, a modified form of the Smith–Waterman algorithm, is used to detect potential CNV regions. Then the maximum clique algorithm is used to identify the diploid group of samples for each particular CNV region. The frequency that each particular SNP in the potential CNV region appears as a CN change in each pairwise comparison with the diploid group is recorded. This allows GEMCA to give probabilities on the CNV region boundaries. In particular, it reports a 10 and 90% boundary for each end of the CNV region. GEMCA was used for the study of CNV in a HapMap population by Redon et al. [16]. The CN for each region is reported using the median ratio in that region. GEMCA is available at www2.genome.rcast.u-tokyo.ac.jp/CNV.

The probe-level allele-specific quantitation (PLASQ) algorithm [17,18] is another method for CN quantification of Affymetrix mapping arrays. To remove systematic differences between multiple arrays it uses probe-level quantile normalization. Based on a reference set data, it uses a generalized linear model, which relates the number of copies of each of the A and B alleles with normalized log-transformed probe intensities. The model accounts for whether the probe is centered or offset from the SNP site, and uses both PM and MM probe intensities. In general, the number of copies of each allele is not known a priori. However, for normal reference samples it can reasonably be assumed that the sum of these is 2, so the EM algorithm is used for fitting the generalized linear model. For test samples, the normalized probe intensities and the fitted model parameters are used to estimate the number of copies for each allele. For a total CN estimate, these are added. The PLASQ approach to segmentation of CN regions is to use the gain and loss analysis of DNA (GLAD) algorithm.

CASE STUDIES (ILLUSTRATE WORKFLOW ON 500K DATA USING CNAT4)

This section illustrates the usage of Affymetrix 500K mapping arrays for CN analysis with two different data sets using the CNAT4 algorithm. The data sets are the chromosome X titration and tumor–normal pairs, both of which are publicly available and may be downloaded directly from the Affymetrix Website (www.affymetrix.com).

Software implementing CNAT4 is available in two different forms. The first is as an extension of the GTYPE software, which provides a graphical user interface (GUI) tool, and the second is a command-line tool known as copynumber pipeline. Both are free and may be downloaded from www.affymetrix.com. The analysis results discussed in this chapter were generated using the command line tool, but could have been processed equivalently using the GUI tool.

The CNAT4 software tools provide four main workflows: unpaired CN, paired CN, unpaired LOH, and paired LOH (Figure 11.5). In other words, CNAT4 provides both CN and LOH values in two different modes. The paired modes are most useful for situations where we have normal and disease samples from the same individual, so that we can directly compare CN changes without dealing with interindividual CN variation. In most other cases, the unpaired modes are appropriate. In unpaired analysis each sample is compared to a reference set established using "normal" individuals.

The chromosome X titration data set consists of 12 data files from the Nsp and Sty 250K arrays. This data set is augmented with 48 HapMap normal male and female

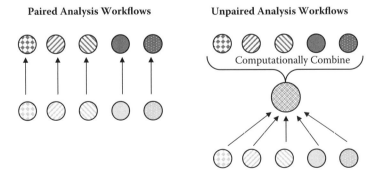

FIGURE 11.5 Paired and unpaired workflows. Arrows indicate how CN calls are made. The tails of the arrows indicate samples for which CN estimates are generated. The arrowheads point at the sample (or set) used as the reference.

SNP Array-Based Copy Number Analysis

samples on each of the two types of arrays, making for a total of 120 arrays. The 16 female HapMap samples are used for the reference set in this analysis since they can each reasonably be expected to have two copies of each chromosome, excluding smaller individual specific CN variations. As previously described, CNAT4 first preprocesses the arrays for each array type together and combines them to produce a virtual array, then carries out smoothing and segments regions of CN change using an HMM.

In CN workflows, CNAT4 produces an output of $\log_2$ CN ratios and HMM CN state calls. Figure 11.6 visualizes the analysis results for four different samples in this data set, showing the $\log_2$ CN estimates for each SNP and color coded by the HMM CN state for that SNP. Yellow represents a CN state call of 2 (corresponding to the normal diploid state). The prime difference between the samples is the number of X chromosomes present, while the majority of the other chromosomal regions are called as CN = 2 (with some individual-specific CNV). NA12264 (upper left) is a normal male where a single copy of the X chromosome should be present. In this case, the SNP markers on the X chromosome are colored blue (CN = 1), and the overall trend in the $\log_2$ CN ratios is lower than on the other chromosomes.

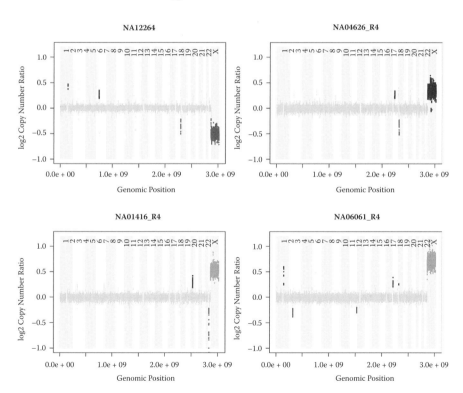

FIGURE 11.6 (See color insert following page 80.) CNAT4 CN estimates (smoothed over a 100K window), color coded by the HMM call for four samples with different X chromosome CNs: a normal HapMap male (NA12264; 1X), 3X (NA04626), 4X (NA01416), and 5X (NA06061) samples. Color coding: red: CN = 0, blue: CN = 1; yellow: CN = 2; black: CN = 3; green: CN = 4+.

NA04626 (upper right) is a sample containing three X chromosomes. The majority of points on the X chromosome are colored black (CN = 3), and the $\log_2$ ratios are elevated above the other chromosomes. Green designates a CN of 4 or more (recall that CNAT4 uses a five-state HMM model), which appears on samples NA01416 and NA06061, representing cases where there are four and five copies of the X chromosome, respectively. The X chromosome $\log_2$ ratios on the 5X samples are slightly higher than on the 4X sample.

An unpaired LOH analysis was also conducted on this data set. In an LOH analysis, CNAT4 returns a binary LOH call for each SNP location with 1 representing a loss of heterozygosity and 0 otherwise. The LOH calls are based on genotype calls—in this case, as determined by the Bayesian robust linear model with Mahalanobis distance classifier (BRLMM) algorithm [19]. LOH can be used, for example, to confirm a deletion-type event. In the context of the titration X data series, this means that, for normal males (only one X chromosome), we should expect to observe loss of heterozygosity across that chromosome.

In Figure 11.7, we see CNAT4 CN estimates for the X chromosome for a normal male sample, NA18605. Notice that almost all of the X chromosome has been correctly called as CN = 1, except on the far left-hand side where the CN = 2. The SNPs

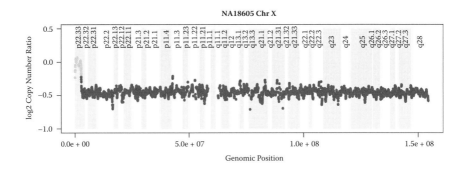

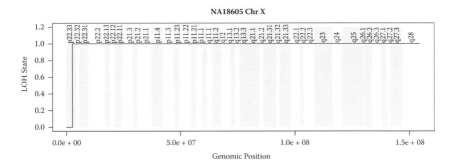

FIGURE 11.7 (See color insert following page 80.) CNAT4 CN estimates (top; with smoothing at 100K) and LOH calls (bottom) for the X chromosome of a HapMap male sample. The LOH state is represented by the gray line, which jumps from 0 to 1 at the end of the pseudo-autosomal region near the p-terminus.

SNP Array-Based Copy Number Analysis 263

in this region are part of the pseudo-autosomal region on the X chromosome, and the probes here detect both the X and Y chromosomes, which would be expected for a male. The second plot shows the LOH results for the X chromosome of this same sample. The pseudo-autosomal region is called as LOH = 0, and the rest of the X chromosome is correctly called as LOH = 1.

The tumor–normal pair data set consists of samples from nine different individuals. For each individual there is a normal peripheral blood sample and a tumor sample. These were each interrogated using both the Nsp and Sty 250K arrays, giving a total data set including 36 distinct arrays. The paired analysis workflows of CNAT4 are therefore applicable to this data set. In addition to total CN estimates, paired CN workflows also produce ASCN estimates.

Figure 11.8 shows CNAT4 output for tumor sample CRL-5868D. The first plot shows the total CN estimates as colored by HMM call, similar to the titration X data. Notice that there is a great deal of CN variation in this cancer tumor sample. The second plot shows the ASCN estimates, with the CN estimates for the min and max alleles coded in different colors. The final plot shows the LOH calls for this data set. The LOH regions line up with deletion regions shown in the two plots above.

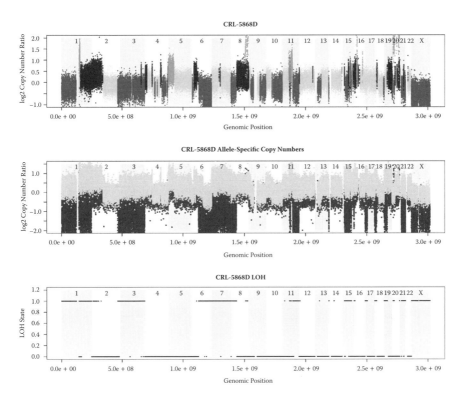

FIGURE 11.8 (See color insert following page 80.) Unsmoothed CNAT4 output produced for a tumor sample. From top to bottom: (a) Unsmoothed CNAT4 CN estimates colored by CN HMM state; (b) allele-specific CN estimates colored for the min and max alleles; and (c) LOH results for the data set.

FUTURE TRENDS IN CN ANALYSIS

Microarrays with increased SNP coverage in combination with a set of specific nonpolymorphic CN probes are becoming available. The Affymetrix SNP 5.0 array is a single array that contains all the SNPs on the two-array 500K set along with 420,000 additional nonpolymorphic probes. The Affymetrix SNP 6.0 array contains more that 900,000 SNPs and 946,000 nonpolymorphic probes. We expect these new lines of products to provide additional valuable information to cytogeneticists, cancer researchers, oncologists, and population geneticists. Algorithms will be developed to integrate the different types of data generated in a single experiment to estimate CN variations in greater accuracy with high-resolution boundaries determination. In the long term, we expect CN applications of microarrays to become a standard tool in clinical studies and molecular diagnostics solutions. Analysis methods and tools need to be developed to meet the requirements of clinical regulated environments.

ACKNOWLEDGMENTS

We would like to thank Paul Gardina for critical review of this manuscript.

REFERENCES

1. A. G. Knudson, Jr., Hereditary cancer, oncogenes, and antioncogenes, *Cancer Res,* 45, 1437–1443, 1985.
2. A. J. Iafrate, L. Feuk, M. N. Rivera, M. L. Listewnik, P. K. Donahoe, Y. Qi, S. W. Scherer, and C. Lee, Detection of large-scale variation in the human genome, *Nat Genet,* 36, 949–951, 2004.
3. J. Sebat, B. Lakshmi, J. Troge, J. Alexander, J. Young, P. Lundin, S. Maner, et al., Large-scale copy number polymorphism in the human genome, *Science,* 305, 525–528, 2004.
4. G. C. Kennedy, H. Matsuzaki, S. Dong, W. M. Liu, J. Huang, G. Liu, X. Su, et al., Large-scale genotyping of complex DNA, *Nat Biotechnol,* 21, 1233–1237, 2003.
5. S. Jacobs, E. R. Thompson, Y. Nannya, G. Yamamoto, R. Pillai, S. Ogawa, D. K. Bailey, and I. G. Campbell, Genome-wide, high-resolution detection of copy number, loss of heterozygosity, and genotypes from formalin-fixed, paraffin-embedded tumor tissue using microarrays, *Cancer Res,* 67, 2544–2551, 2007.
6. B. M. Bolstad, R. A. Irizarry, M. Astrand, and T. P. Speed, A comparison of normalization methods for high density oligonucleotide array data based on variance and bias, *Bioinformatics,* 19, 185–193, 2003.
7. S. Ishikawa, D. Komura, S. Tsuji, K. Nishimura, S. Yamamoto, B. Panda, J. Huang, et al., Allelic dosage analysis with genotyping microarrays, *Biochem Biophys Res Commun,* 333, 1309–1314, 2005.
8. Y. Nannya, M. Sanada, K. Nakazaki, N. Hosoya, L. Wang, A. Hangaishi, M. Kurokawa, et al., A robust algorithm for copy number detection using high-density oligonucleotide single nucleotide polymorphism genotyping arrays, *Cancer Res,* 65, 6071–6079, 2005.
9. L. Rabiner, A tutorial on hidden Markov models and selected applications in speech recognition, *Proc IEEE,* 77, 257–286, 1989.
10. G. R. Bignell, J. Huang, J. Greshock, S. Watt, A. Butler, S. West, M. Grigorova, et al., High-resolution analysis of DNA copy number using oligonucleotide microarrays, *Genome Res,* 14, 287–295, 2004.

11. J. Huang, W. Wei, J. Zhang, G. Liu, G. R. Bignell, M. R. Stratton, P. A. Futreal, et al., Whole genome DNA copy number changes identified by high-density oligonucleotide arrays, *Hum Genomics,* 1, 287–299, 2004.
12. J. Huang, W. Wei, J. Chen, J. Zhang, G. Liu, X. Di, R. Mei, et al., CARAT: A novel method for allelic detection of DNA copy number changes using high-density oligonucleotide arrays, *BMC Bioinformatics,* 7, 83, 2006.
13. X. Di, H. Matsuzaki, T. A. Webster, E. Hubbell, G. Liu, S. Dong, D. Bartell, et al., Dynamic model-based algorithms for screening and genotyping over 100K SNPs on oligonucleotide microarrays, *Bioinformatics,* 21, 1958–1963, 2005.
14. X. Zhao, C. Li, J. G. Paez, K. Chin, P. A. Janne, T. H. Chen, L. Girard, et al., An integrated view of copy number and allelic alterations in the cancer genome using single nucleotide polymorphism arrays, *Cancer Res,* 64, 3060–3071, 2004.
15. D. Komura, F. Shen, S. Ishikawa, K. R. Fitch, W. Chen, J. Zhang, G. Liu, et al., Genome-wide detection of human copy number variations using high-density DNA oligonucleotide arrays, *Genome Res,* 16, 1575–1584, 2006.
16. R. Redon, S. Ishikawa, K. R. Fitch, L. Feuk, G. H. Perry, T. D. Andrews, H. Fiegler, et al., Global variation in copy number in the human genome, *Nature,* 444, 444–454, 2006.
17. T. LaFramboise, D. Harrington, and B. A. Weir, PLASQ: A generalized liner model-based procedure to determine allelic dosage in cancer cells from SNP array data, *Biostatistics,* 8, 323–336, 2007.
18. T. LaFramboise, B. A. Weir, X. Zhao, R. Beroukhim, C. Li, D. Harrington, W. R. Sellers, and M. Meyerson, Allele-specific amplification in cancer revealed by SNP array analysis, *PLoS Comput Biol,* 1, e65, 2005.
19. Affymetrix, BRLMM: An improved genotype calling method for the GeneChip® human mapping 500K array Se, Santa Clara, CA: Affymetrix, 2006.

Index

A

Absolute rank deviation (ARD), 100
Accuracy, 52
AD, *see* Alzheimer's disease (AD)
AdditiveError term, 236
Affinity computation, 48
affycomp software and webtool, 2, 52
Affymetrix arrays, *see also* GeneChip Operating System (GCOS); Microarray Suite 5 (Mas5)
    cross-normalization comparison, 161
    exon array analysis, 208–209
    fundamentals, 154, 155–156
    future trends, 264
    MAQC measurements, 124
    normalization, 143, 160, 166
    probes usage, 160
    summarization, 7
    technical quality, 153–154
    workflow, with CNAT4 algorithm, 260–263
Affymetrix CEL files
    cross-normalization comparisons, 161
    exon data analysis workflow, 216
    filtering signal data, 219
    probe intensity files, 43
    signal intensities, 208
    SUB-SUB normalization, 27
Affymetrix GeneChip microarrays, *see also* Chip definition file (CDF)
    case study, 53–55
    data for study, 5
    differentiation detection, 29, 32
    expression measures comparison, 51–53
    focus on, 3
    fundamentals, xiii, 41–44
    GCRMA algorithm, 48–49
    intensity-dependent dye effect, 86
    joint expression variation, 136–137, 139
    MAS5.0 algorithm, 49–51
    number of probes, 20
    parameter selection, 27
    perfect match correction, 7
    preprocessing methods, 43–48
    research expansion, 2
    robust multichip analysis, 44–48
    variation retention, 100
    Website, 51
Affymetrix Integrated Genome Browser (IGB), 226
Affymetrix Power Tools, 214
Affymetrix spike-in data sets, 29, 32–33
affyQCReport, 4
Agilent Expression Microarray
    CXCR4-related pathways, 177
    fundamentals, 154–155
    ratio measurements, 157
    technical quality, 154
Algorithms
    Affymetrix mapping arrays, 258–260
    analyze networks algorithm, 182
    copy number analysis tool 4 algorithm, 247–258, 260–263
    dynamic model algorithm, 258
    GCRMA algorithm, 48–49
    genotyping microarray-based CNV analysis algorithm, 259
    Mahalanobis distance classifier algorithm, 262
    MAS5.0 algorithm, 49–51
    signal estimation, 213–214
Allele-specific copy number (ASCN), 246
Alternative expression technologies, 156–157
Alternative splicing
    biological interpretation, 229–230
    concepts and caveats, 214–216
    fundamentals, 205–207
    prediction, 214, 225–226
    splicing variants validation, 227–229

Index

Alzheimer's disease (AD), 4–5, 11, *see also* Differential expression (DE), data preprocessing methods
AN, *see* Analyze networks (AN) algorithm
Analysis of variance, *see* ANOVA (Analysis of Variance)
Analyze networks (AN) algorithm, 182
Annotation levels, 211
Anomalous probe sets and signals, 220, 223–224
ANOVA (Analysis of Variance)
  alternative splice predictions, 214, 225
  cDNA normalization, 101
  differentially expressed gene identification, 8
  modeling technical variation, 122
  normalization methods, 91–92
  quality control, 218
Apolipoprotein (APO) data set, 76–77, *see also* Spatial detrending and normalization methods
Applied Biosystems, 157
apt-midas command line program, 214, 225
apt-probeset-summarize program, 216
ARD, *see* Absolute rank deviation (ARD)
Area under the curve (AUC) value, 218
Array CGH (comparative genome hybridization) data, normalization
  background, 233–234
  centralization examples, 240
  data analysis, 234–239
  error model, 236–237
  fundamentals, xv, 233–234, 240–241
  normalization, 237–239
  polyclonal sample complications, 240
  workup, common processes, 234, 236
Array effect, ANOVA methods, 91
ASCN, *see* Allele-specific copy number (ASCN)
AUC, *see* Area under the curve (AUC) value
Autofluorescence changes, 153

B

Background correction
  Affymetrix GeneChip preprocessing steps, 3
  data preproccesing, 6
  defined, 3
  greatest effect, 14
  order of data preprocessing, 5
  preprocessing, 43
Background noise removal, 49
Baseline-array approach, 96
Bayes approach, 6
Bead Array, 156
BeadStudio software, 156
Between-microarray scale effect, 87
BIND, *see* Blind inversion problem (BIND)
Biocarta access, 152
Bioconductor software
  APO data set, loading, 76–77
  Bead Array, 156
  detecting spatial biases, 77
  installation and loading, 75–77
  normalization methods, 28
  open source implementations, 83
  spatial bias detection, 65, 77
  spatial normalization, 78
  workflow study, 53
Biological content, 168
Biological interpretation, 229–230
Biological interpretation, normalization selection
  alternative expression technologies, 156–157
  background, 153
  commercial arrays, 153–154, 159–160
  cross-normalization comparisons, 161, 163
  discussion, 166–168
  experimental design, 161, 163
  expression platform types, 154–156
  fundamentals, xiv, 151–152, 169–170
  history, 157–158
  imprecise measurements, causes and consequences, 160–161
  normalization methods, 157–160
  quality, 153–154
  results, 163
Biotique Local Integration System (BLIS), 226
Bisquare function, 51
Bland-Altman, *see* M-A plots
BLAT results, 227
Blind inversion problem (BIND), 21, 22, 37
BLIS (Biotique Local Integration System), 226
Block effect, 20
Boundary effects, 24
Brainarray, 7

Index

Breakdown point
 differentiation detection, 25–27, 36
 error model, aCGH specific, 237
 fundamentals, 241
BRLMM, *see* Mahalanobis distance classifier (BRLMM) algorithm
Bscale
 k-NN evaluation, 113
 k-NN LOOCV error functional assessment, 112
 scale normalization strategies comparison, 111

C

Cancer Genome array, 156
Candidate splicing event selection, 226–227
Canonical pathways map distribution, 176–177
CARAT, *see* Copy number analysis with regression and tree (CARAT)
Case studies
 Affymetrix GeneChips, 53–55
 CNAT4, 260–263
CDF, *see* Chip definition file (CDF)
cDNA (complementary DNA) arrays
 external controls, 36
 intensity-dependent dye effect, 86
 k-NN classification comparison, 101–103
 mRNA detection, 207
cDNA (complementary DNA) arrays, normalization survey
 ANOVA methods, 91–92
 baseline-array approach, 96
 comparison criteria, 99–101
 data transformation, 89–90
 dye-swap normalization, 96–97
 evaluation, 101
 external control strategies, 95–96
 fundamentals, 83–84
 global techniques, 90–91
 housekeeping-gene-related approaches, 94–95
 iGloess method, 92–93
 iLloess method, 93–94
 intensity- and spatial-dependent dye effects minimization, 98–99
 intensity-dependent dye effect, 85–86, 92–95
 intensity ratios and values, 89–90
 k-NN classification comparison, 101–103
 logarithmic transformation, 89
 minimizing strategies, 89–99
 normalization methods, 90–99
 Qspline, 96
 reducing unwanted variations, 100
 retaining biological variations, 100–101
 scale effect, 87–89
 semilinear models, 95–96
 spatial-dependent dye effect, 86–87, 97–98
 total gene approaches, 92–95
 unwanted data variations, 84–89
Celera, 157
CEL file, *see* Affymetrix CEL files
Centralization, 237, 241
Centralization curve, 237, 241
Centralization examples, 240
CGH, *see* Comparative genomic hybridization (CGH)
ChemTree, 174
Chimpanzees, *see* Primate brain expression data set
Chip definition file (CDF), 7, 22, *see also* Affymetrix GeneChip microarrays
Chi-square distribution, 132
Chromosomal aberrations, SNP array-based analysis
 Affymetrix mapping arrays, 258–260
 algorithms, 246–260
 case studies, 260–263
 CNAT4, 246–260
 copy number estimation, 247–256
 fundamentals, xv–xvi, 245–247
 future trends, 264
 general components, 246
 heterozygosity loss, 256–258
Classification, *k*-NN
 cDNA microarray data, 101–103
 diagnostic plot dye bias removal, 112
 diagnostic plot dye effect removal, 104, 106
 double-bias-removal methods, 109–110
 evaluation summary, 112, 115–116
 intensity- and spatial-dependent strategies, 104–110

LOOVC classification errors, 106–110, 112
normalization methods, 101–103
Qspline-related approaches, 110
scale normalization strategies, 111–116
single-bias-removal methods, 106, 109
Clone library, 158
CMOS-like technology, 156
CN, *see* Copy number (CN)
CNAG, *see* Copy number analysis for GeneChip (CNAG)
CNAT4 (copy number analysis tool 4) algorithm, *see also* Copy number (CN)
case studies, 260–263
copy number estimation, 247–256
fundamentals, 246–247
heterozygosity loss, 256–258
CNP, *see* Copy number polymorphisms (CNP)
CodeLink
fundamentals, 126
global normalization techniques, 91
intensity-dependent dye effect, 86
interlaboratory variation, 132–134, 136
linearity, 127–128
MAQC measurements, 124, 126
measurement error, 128, 130–132
normalization, 126–127, 143
Combimatrix, 156–157
Commercial array quality, 153–154
Comparative functional genomics, 32–34
Comparative genomic hybridization (CGH), 82, 233, *see also* Array CGH (comparative genome hybridization) data, normalization
Comparison criteria, 99–101
Complementary DNA, *see* cDNA (complementary DNA) arrays
Composite normalization, 69
Conditional residual analysis for microarrays (CRAM), 64
Copy-neutral region, 252
Copy number analysis for GeneChip (CNAG), 259
Copy number analysis tool 4 (CNAT4) algorithm, *see also* Copy number (CN)
case studies, 260–263
copy number estimation, 247–256

fundamentals, 246–247
heterozygosity loss, 256–258
Copy number analysis with regression and tree (CARAT), 258
Copy number (CN), *see also* Single nucleotide polymorphism (SNP), array-based analysis
estimation, 247–256
fundamentals, 241
ratios, 237
Copy number polymorphisms (CNP), 245
Core annotation levels, 211
Core metaprobe set, 217
Covariance matrix, 131, 139
CRAM, *see* Conditional residual analysis for microarrays (CRAM)
Criss-cross regression, 143
cRNA, 42
Cross-normalization comparisons, 161, 163
Cross-platform, 2
CXCR4-related pathways, 176–177
Cyclic loess normalization, 160

D

DABG, *see* Detection above background (DABG)
Data, 5
Data analysis, aCGH data
background, 233–234
error model, 236–237
fundamentals, xv, 233–234
normalization, 237–239
workup, common processes, 234, 236
Data integration software, 174
Data preprocessing
background correction, 6
data, 5
differentially expressed gene identification, 8–9
discussion, 13–15
fundamentals, xiii, 1–8
materials and methods, 5–9
motivation, 1–2
normalization, 6–7
perfect match correction, 7
results, 2, 9–12

Index

summarization, 7–8
when performed, 1
Data transformation, 89–90
Data variations, unwanted
    data transformation, 89–90
    intensity-dependent dye effect, 85–86
    intensity ratios and values, 89–90
    logarithmic transformation, 89
    minimizing strategies, 89–99
    scale effect, 87–89
    spatial-dependent dye effect, 86–87
dChip/dChip PM/dChip PM-MM
    Affymetrix arrays, 259
    classifier error, effect on, 168
    commonality between gene lists, 163
    cross-normalization comparisons, 161
    CXCR4-related pathways, 177
    disease distribution, 180
    image processing components, 168
    Microarray Suite 5 comparison, 168
    network generation and analysis, 183
    normalization methods, 159–160
    reproducibility, 163
    sample-to-sample variance, 163
DE, *see* Differential expression (DE), data preprocessing methods
Degree of differentiation, 23
Detecting spatial bisases, 77
Detection above background (DABG)
    filtering strategies, 223
    signal estimation, 213–214, 217
detectSpatialBias method, 67
Diagnosis, differentiation detection, 37
Diagnostic plot dye bias removal, 112
Diagnostic plot dye effect removal, 104, 106
Differential expression (DE), data preprocessing methods
    background correction, 6
    data, 5
    data preproccesing, 5–8
    differentially expressed gene identification, 8–9
    discussion, 13–15
    fundamentals, xiii, 1–5
    materials and methods, 5–9
    motivation, 1–2

    normalization, 6–7
    perfect match correction, 7
    results, 2, 9–12
    summarization, 7–8
Differentially expressed (DE) gene identification, 8–9, *see also* Data preprocessing
Differentiated fraction, 24, 36
Differentiation detection
    Affymetrix spike-in data set, 29, 32
    breakdown, 25–27, 36
    comparative functional genomics, 32–34
    degree of differentiation, 23
    diagnosis, 37
    differentiated fraction, 24, 36
    discussion, 36–37
    examples and results, 27–35
    external controls, 36
    fundamentals, xiii, 20–21
    high breakdown estimation, 25–26
    least trimmed squares, 25–26
    linear approximation, 24–25
    linear subarray transformation, 37
    lung cancer data set, 35
    methods, 21–27
    mismatched probes, 28
    multiple arrays, 26
    nonlinear array transformation, 37
    parameter selection, 27
    perturbed spike-in data set, 32
    primate brain expression data set, 32–34
    reference, 26
    simple linear regression, 26
    spatial pattern, 24
    statistical breakdown, 26–27
    statistical principle, 21–23
    subarray normalization, 27–28
    subarrays, 24
    SUB-SUB normalization, 27
    symmetric solutions, 26
    transformation, 37
    trimmed solutions, 26
    undifferentiated probe set, 24
    variation reduction, 27–28
Diploid peak, 239
Discovery Gate, 174
Discriminant coordinates, 133–134, 148–149

Discussions
    data preprocessing methods, differential expression, 13–15
    differentiation detection, 36–37
    interpretation of normalization selection, 166–168
    joint expression variation, modeling, 141–142
    normalization selection, biological interpretation, 166–168
Disease distribution, 179–181
dLRsd, 242
DM, see Dynamic model (DM) algorithm
Double-bias-removal methods, 109–110
Down-regulated mice, 72
Dye bias, 63, 112
Dye effect
    ANOVA methods, 91
    intensity- and spatial-dependent, 98–99
    intensity-dependent, 85–86, 92–95
    removal, 104, 106
    spatial-dependent, 86–87, 97–98
Dye normalization, 236
Dye swap
    mRNA detection, 208
    normalization methods, 96–97
Dynamic model (DM) algorithm, 258

E

Emission probability, 255–256
Error model, aCGH data analysis, 236–237
Escherichia coli genes
    Affymetrix arrays, 208
    number of genes, 205
    performance of procedures, 42
Euclidean distance, 50
Evaluation
    k-NN classification comparison, 112, 115–116
    normalization methods, 101
Examples and results
    Affymetrix spike-in data set, 29, 32
    comparative functional genomics, 32–34
    interpretation of normalization selection, 163
    lung cancer data set, 35
    mismatched probes, 28
    parameter selection, 27
    perturbed spike-in data set, 32
    primate brain expression data set, 32–34
    subarray normalization, 27–28
    SUB-SUB normalization, 27
    variation reduction, 27–28
Exon array analysis
    Affymetrix arrays, 208–209
    alternative splicing, 205–207, 214, 225–226, 229–230
    anomalous probe sets and signals, 220, 223–224
    biological interpretation, alternative splicing, 229–230
    candidate splicing event selection, 226–227
    design, 209–213
    exon arrays, 209–216
    filtering signal data, 219–225
    filtering strategies, 222–225
    fundamentals, xv, 216–217, 230
    ideal splicing event, 219
    identification of splicing events, 214–216
    low confidence predictions, 224–225
    low signal, 222–223
    mRNA detection, 207–208
    prediction, alternative splicing, 214, 225–226
    quality control, 218
    signal estimation, 213–214, 217
    splicing variants validation, 227–229
    workflow, 216–229
Experimental design, 161, 163
Expression Console, 218
Expression measures comparison, 51–53
Expression platform types, 154–156
Extended annotation levels, 211
External controls
    differentiation detection, 36
    strategies, normalization methods, 95–96
External RNA Controls Consortium (ERCC), 158
Extraction, genes, 8

F

FACS, see Fluorescently activated cell sorting (FACS)
False discovery rate (FDR), 9, 100
False negatives/positives, 159

Index

FDR, *see* False discovery rate (FDR)
Feature Extraction software
  common workup processes, 234, 236
  fundamentals, 155
FFPE, *see* Formalin-fixed paraffin-embedded (FFPE) samples
Filtering
  anomalous probe sets and signals, 220, 223–224
  filtering strategies, 222–225
  ideal splicing event, 219
  low confidence predictions, 224–225
  low signal, 222–223
  methods, 8
  strategies, 222–225
FISH data, 238
Fitting, measurement error, 130
Fluorescently activated cell sorting (FACS), 157
Fluorescently labeled RNA, binding, 42
Formalin-fixed paraffin-embedded (FFPE) samples, 247
Fraction, differentiated, 24, 36
Full annotation levels, 211
Functional analysis steps
  canonical pathways map distribution, 176–177
  disease distribution, 179–181
  fundamentals, 175–176
  GeneGo process distribution, 177, 179
  GO process distribution, 179
Functional assessment, k-NN LOOCV error, 112
Future trends
  direction of efforts, 15
  SNP array-based analysis, 264

G

GAPD/ACTB ratios, 167
GAPDH (glyceraldehyde-3-phosphate dehydrogenase), 94
GCOS, *see* GeneChip Operating System (GCOS)
GCRMA, *see* GeneChip (GG)-robust multichip average (GCRMA)
GEMCA, *see* Genotyping microarray-based CNV analysis (GEMCO) algorithm
Gene-array interaction effect, 92

GeneChip (GG)-robust multichip average (GCRMA)
  algorithm, preprocessing methods, 48–49
  background correction, 4, 6, 12
  comparing expression measures, 52
  differentially expressed genes, 163
  Microarray Suite 5 comparison, 51
  normalization methods, 159
  ratio compression, 163
  reproducibility, 163
  sample-to-sample variance, 163
GeneChip Human Exon 1.0 ST array, 210–212
GeneChip Operating System (GCOS), 18, *see also* Microarray Suite 5 (Mas5)
Gene-dye interaction effect, 92
Gene effect, ANOVA methods, 92
Gene expression omnibus (GEO), 53
Gene extraction, 8
GeneGo
  disease distribution, 179
  enrichment analysis, 175
  functional analysis steps, 177, 179
  MAQC project, 185
  network generation and analysis, 181
Gene onotology (GO) and processes
  followup step, 173
  functional analysis steps, 179
  housekeeping gene changes, 168
  Metadiscovery Suite, 174
Gene regulatory network, *see* Neural network-based spatial and intensity normalization
GeneSpring, 174
GeneSpring GX 7.2, 152
GenMapp access, 152
Genomic DNA, fluorescence measurements, 158
Genotyping microarray-based CNV analysis (GEMCO) algorithm, 259
GEO, *see* Gene expression omnibus (GEO)
Global least trimmed squares (LTS), 37, *see also* Least trimmed squares (LTS)
Global Lowess, 37, *see also* Lowess normalization
Global median (Gmedian) normalization method
  double-bias-removal methods, 109
  global normalization techniques, 91, 101

274　　　Index

　　k-NN evaluation, 112
　　removing dye effects, 109
Global normalization, 6–7, 90–91, *see also*
　　　　dChip/dChip PM/dChip PM-MM;
　　　　Dye swap; Loess normalization
GO, *see* Gene onotology (GO) and processes
Gridding, 42–43

H

Heterozygosity, *see* Loss of heterozygosity (LOH)
Hidden Markov model (HMM)
　　Affymetrix arrays, 259
　　case study, 261–263
　　copy number estimation, 253–254, 256
Hierarchical clustering, 12
High breakdown estimation, 25–26
High-level analysis, preprocessing, 43
High-throughput characteristics, 15
HMM, *see* Hidden Markov model (HMM)
Housekeeping-gene-related approaches, 94–95
HT-29 cell line, centralization, 240
Humans, number of genes, 205

I

Ideal mismatch (IM), 50
Ideal splicing event, 219
iGloess method
　　cDNA normalization, 101
　　intensity-dependent dye effect minimization,
　　　　92–93
　　removing dye effects, 106
　　scale normalization strategies comparison,
　　　　112
iGsGloess
　　double-bias-removal methods, 109
　　intensity- and spatial-dependent dye effects,
　　　　99
　　removing dye effects, 104, 106
iLloess method
　　intensity-dependent dye effect minimization,
　　　　93–94
　　overcoming limitations, 95
　　removing dye effects, 106
　　scale normalization strategies comparison, 112

Illumina platform
　　bead-based technology, 156
　　fundamentals, 154, 156
　　technical quality, 154
IM, *see* Ideal mismatch (IM)
Image processing components, 168
Imprecise measurements, causes and
　　　　consequences, 160–161
Inferential literacy, 15
Inflammation array, 156
Inforsense, 174
Installation and loading R and Bioconductor
　　　　software, 75–77
Intensity- and spatial-dependent dye effects
　　　　minimization, 98–99
Intensity- and spatial-dependent strategies
　　diagnostic plot dye effect removal, 104, 106
　　double-bias-removal methods, 109–110
　　LOOVC classification errors, 106–110
　　Qspline-related approaches, 110
　　single-bias-removal methods, 106, 109
Intensity bias
　　detection, 67, 69–72
　　fundamentals, 63
Intensity-dependent dye effect, 85–86
Intensity ratios and values, 89–90
Interlaboratory variation
　　CodeLink platform results, 132–134, 136
　　single-platform MAQC measurement
　　　　decomposition, 147–149
International Vocabulary of Metrology, 124
Interpretation, normalization selection
　　alternative expression technologies, 156–157
　　background, 153
　　commercial array quality, 153–154
　　cross-normalization comparisons, 161, 163
　　discussion, 166–168
　　experimental design, 161, 163
　　expression platform types, 154–156
　　fundamentals, xiv, 151–152, 169–170
　　imprecise measurements, causes and
　　　　consequences, 160–161
　　normalization methods, commercial arrays,
　　　　159–160
　　normalization methods, history, 157–158
　　results, 163

Index

Introns and intronic regions, 206
iSTspline, 106, 108
Iterative reweighted least squares, 143
IterPLIER, 213–214, 217

J

Joint expression modeling variation
 Affymetrix platform results, 136–137, 139
 CodeLink platform results, 126–136
 discussion, 141–142
 fundamentals, xiv, 121–124
 interlaboratory variation, 132–134, 136, 147–149
 linearity, 127–128, 144–146
 measurement error, 124–126, 128, 130–132, 146–147
 normalization, 126–127, 142–144
 statistical analysis, 142–149
Joint loess (iGsGloess), 99

K

KeGG, 152, 173
k-NN classification comparison
 cDNA microarray data, 101–103
 diagnostic plot dye bias removal, 112
 diagnostic plot dye effect removal, 104, 106
 double-bias-removal methods, 109–110
 evaluation summary, 112, 115–116
 intensity- and spatial-dependent strategies, 104–110
 LOOVC classification errors, 106–110, 112
 normalization methods, 101–103
 Qspline-related approaches, 110
 scale normalization strategies, 111–116
 single-bias-removal methods, 106, 109
Knock-out mice, 72
Knowledge databases, 174

L

Langmuir isotherm model, 126, 128
Latin square design, 29
Least trimmed squares (LTS)
 Affymetrix spike-in data sets, 29, 32
 comparing two arrays, 21

 diagnosis, 37
 differentiated fractions, 36
 differentiation detection, 25–26
 nonlinear array transformation, 37
 perturbed spike-in data sets, 32
 SUB-SUB normalization, 27
 usage of mismatched probes, 28
Leave one out cross-validation (LOOCV), classification errors
 double-bias-removal methods, 109–110
 future directions, 116
 intensity- and spatial-dependent strategies, 106–110
 k-NN classification, 101, 103
 Qspline-related approaches, 110
 scale normalization strategies, 112
 single-bias-removal methods, 106, 109
 variation retention, 101
Library of clones, 158
Linear approximation, 24–25
Linearity
 CodeLink platform results, 127–128
 single-platform MAQC measurement decomposition, 144–146
Linear subarray transformation, 37
Local median filter, 98
Loess normalization, *see also* Lowess normalization
 Agilent arrays, 155
 method, 160
 print-tip, 69, 70
Logarithmic transformation, 89
Logarithmic values comparison, 132
Log ratio
 bias normalization, 67
 common workup processes, 236
 fundamentals, 242
 normalization, aCGH specific, 238–239
 spatial bias detection, 64–65
LOH, *see* Loss of heterozygosity (LOH)
LOOCV, *see* Leave one out cross-validation (LOOCV), classification errors
Loss of heterozygosity (LOH), 266–268, 274–276, 278–279, 281, 283–284, 304, 306
 case study, 262
 CNAT4, 246, 256–258

hidden Markov model, 256–258
paired and unpaired experiments, 247
transition probability, 254
Low confidence predictions, 224–225
Lowess normalization, *see also* Loess normalization
nonlinear array transformation, 37
oligonucleotides arrays, 20
variation reduction, 28
Low-level analysis, preprocessing, 43
Low signal filtering, 222–223
Lung cancer data set
differentiation detection, 35
dye bias removal, 112
scale effect, 87–88
Lymphoma data set, 112

M

Machine learn approaches, 168
MAD, *see* Median absolute deviation (MAD)
Mahalanobis distance classifier (BRLMM) algorithm, 262
MapEditor, 174
M-A plots
comparing expression measures, 52
intensity-dependent dye effect, 86
nonlinear array transformation, 37
oligonucleotides, 20
removing dye effects, 106
variation reduction, 28
MAQC, 185, *see also* Joint expression modeling variation
Marray package, 78
MAS, *see* Maskless array synthesis (MAS) technology
Mas5, *see* Microarray Suite 5 (Mas5)
Maskless array synthesis (MAS) technology, 156
Massively parallel signature sequencing (MPSS), 157
Materials and methods, data preprocessing, 5–9
background correction, 6
data, 5
data proccesing, 5–8
differentially expressed gene identification, 8–9

fundamentals, 5
normalization, 6–7
perfect match correction, 7
summarization, 7–8
Material-to-material differences, 134
MBEI, *see* Model-Based Expression Indexes (MBEI)
Mean square errors (MSE), 101
Measurement error
CodeLink platform results, 128, 130–132
joint expression variation model, 124–126
single-platform MAQC measurement decomposition, 146–147
Measurement error model, 123
Median absolute deviation (MAD), 99, 218
Median filter, 70
Median polish procedure
robust microarray analysis, 46
summarization, 4, 7–8
Median scaling, 249
Megaclone, 157
M-estimation regression, 46–47
MetaCore
disease distribution, 180
fundamentals, 174
network generation and analysis, 181, 183, 185
sorting of similar genes, 179
MetaDiscovery platform, 174
MetaDrug, 174
MetaLink, 174
Microarray Quality Control (MAQC) project, *see* Joint expression modeling variation
Microarray Suite 5 (Mas5), *see also* GeneChip Operating System (GCOS)
algorithm, preprocessing methods, 49–51
background correction, 4, 6
commonality between gene lists, 163
comparing expression measures, 52
cross-normalization comparisons, 161
CXCR4-related pathways, 177
dChip comparison, 168
differentially expressed genes, 163
normalization methods, 159
perfect match correction, 7
perfect match minus mismatch method, 156

Index

ratio expansion, 163
reproducibility, 163
sample-to-sample variance, 163
summarization, 7–8
usage, 3
MIDAS (multi-instrument data analysis system), 214, 225
Minimizing strategies
   data transformation, 89–90
   intensity ratios and values, 89–90
   logarithmic transformation, 89
   unwanted data variations, 89–99
Mismatched (MM) probes
   Affymetrix arrays, 208–209, 259
   background correction, 6
   compared to perfect match probes, 20
   copy number estimation, 247
   differentiation detection, 28
   fundamentals, 42
   intensities, 258
   interlaboratory variation, 139
   Microarray Suite 5, 49–50
   perfect match correction, 7
MM, *see* Mismatched (MM) probes
Model-Based Expression Indexes (MBEI), 37
Modeling, joint expression variation
   Affymetrix platform results, 136–137, 139
   CodeLink platform results, 126–136
   discussion, 141–142
   fundamentals, xiv, 121–124
   interlaboratory variation, 132–134, 136, 147–149
   linearity, 127–128, 144–146
   measurement error, 124–126, 128, 130–132, 146–147
   normalization, 126–127, 142–144
   statistical analysis, 142–149
Modeling software, Metadiscovery Suite, 174
mRNA
   comparing expression levels, 3
   differentiation of levels, 20
   exon array analysis, 207–208
   global normalization techniques, 91
   levels, measurement error model, 124
   lung cancer data, 35
   nonlinear array transformation, 37

normalization impact, 25
primate brain expression data set, 33
technical quality, 153
MSE, *see* Mean square errors (MSE)
Multi-instrument data analysis system (MIDAS), 214, 225
Multiple arrays, 26
Muscle contraction, 179

N

National Center for Biotechnology Information (NCBI)
   exons in human genome, 206
   Gene Expression Omnibus, 53
   microarray technology background, 2
NCBI, *see* National Center for Biotechnology Information (NCBI)
Negative control probes, 242
NetAffx Data Analysis Center, 227
Neural network-based spatial and intensity normalization, 70–72
NI, *see* Normalized intensities (NI)
NimbleGen maskless array synthesis (MAS) technology, 156
nnNorm package
   normalization, 70, 78
   spatial bias detection, 65, 67
   spatial bias removal benefits, 72–73
Noncontrol probes, 242
Nonlinear array transformation, 37
Non-normalized M values (Nonorm)
   double-bias-removal methods, 109, 110
   intensity-dependent dye effect, 86
   k-NN evaluation, 112
   removing dye effects, 104, 109
   scale normalization strategies comparison, 111
   spatial-dependent dye effect, 87
Nonorm, *see* Non-normalized M values (Nonorm)
Normalization
   aCGH data analysis, 237–239
   Affymetrix GeneChip preprocessing steps, 3
   ANOVA methods, 91–92
   baseline-array approach, 96

CodeLink platform results, 126–127
commercial arrays, 159–160
comparison criteria, 99–101
composite normalization, 69
data preproccesing, 6–7
defined, 3
dye-swap normalization, 96–97
evaluation, 101
external control strategies, 95–96
fundamentals, 44, 67
global normalization, 6–7
global techniques, 90–91
greatest effect, 14
history, 157–158
housekeeping-gene-related approaches, 94–95
iGloess method, 92–93
iLloess method, 93–94
intensity- and spatial-dependent dye effects minimization, 98–99
intensity-dependent dye effect minimization, 92–95
k-NN classification comparison, 101–103
median filter, 70
Microarray Suite 5 comparison, 51
neural network-based spatial and intensity normalization, 70–72
order of data preprocessing, 5
print-tip Loess normalization, 69, 70
Qspline, 96
quantiles normalization, 7
reducing unwanted variations, 100
retaining biological variations, 100–101
scale effect, 99
semilinear models, 95–96
single-platform MAQC measurement decomposition, 142–144
spatial-dependent dye effect minimization, 97–98
total gene approaches, 92–95
within-print-tip-group type, 24
Normalization, array CGH data
background, 233–234
centralization examples, 240
data analysis, 234–239
error model, 236–237
fundamentals, xv, 233–234, 240–241
normalization, 237–239
polyclonal sample complications, 240
workup, common processes, 234, 236
Normalization selection, biological interpretation
alternative expression technologies, 156–157
background, 153
commercial arrays, 153–154, 159–160
cross-normalization comparisons, 161, 163
discussion, 166–168
experimental design, 161, 163
expression platform types, 154–156
fundamentals, xiv, 151–152, 169–170
history, 157–158
imprecise measurements, causes and consequences, 160–161
normalization methods, 157–160
quality, 153–154
results, 163
Normalized intensities (NI), 214
Normalized unscaled standard errors (NUSE)
lower quality array identification, 55
probe level model, 47

O

Oligonucleotide arrays, 207, 234
Omics data normalization
canonical pathways map distribution, 176–177
disease distribution, 179–181
functional analysis steps, 175–181
fundamentals, xiv–xv, 173–176
GeneGo process distribution, 177, 179
GO process distribution, 179
metadiscovery data analysis suite, 174
network generation and analysis, 181–183, 185
Optical noise parameters, 48
Orangutan, *see* Primate brain expression data set
Order of values, 168

P

Papillary thyroid cancer (PTC), *see also* Differential expression (DE), data preprocessing methods
data for study, 5
overlapping transcripts, 12

Index 279

preprocessing technique, 10–11
study fundamentals, 4
Parameter selection, 27
Parametric normalization, 124
PCOS, see Polycystic ovary syndrome (PCOS)
PCR, see Polymerase chain reaction (PCR)
Penalized discriminant coordinates, 148
Perfect match minus mismatch (PM-MM) method, 156
Perfect match (PM) correction
    Affymetrix GeneChip preprocessing steps, 3
    data preproccesing, 7
    order of data preprocessing, 5
    summarization, 4
Perfect match (PM) probes
    Affymetrix arrays, 208–209, 259
    combined with summarization, 5
    copy number estimation, 247
    fundamentals, 42
    intensities, 258
    Microarray Suite 5, 49–50
    number of probes, 20
    usage of mismatched probes, 28
Perturbed spike-in data set, 32
PGK1/LDHA, 167
Phylosopher, 174
PipelinePilot, 174
PLASQ, see Probe-level allele-specific quantitation (PLASQ) algorithm
PLIER, see Probe logarithmic intensity error (PLIER)
PLM, see Probe level model (PLM)
Ploidy, 237–238, 242
Plotted values comparison, 132
PM, see Perfect Match (PM) correction; Perfect Match (PM) probes
PM-MM, see Perfect match minus mismatch (PM-MM) method
Point-tip (PT) group
    intensity-dependent dye effect, 86, 94
    scale effect, 89
Polyclonal samples, 240, 242
Polycystic ovary syndrome (PCOS), 4–5, 10, see also Differential expression (DE), data preprocessing methods
Polymerase chain reaction (PCR)

cDNA microarrays, 83–84
copy number estimation, 247
validation, 228–229
Populations, 211–212
Precision
    array CGH, 233
    comparing expression measures, 52
    fundamentals, 242
Prediction, alternative splicing, 214
Preprocessing
    fundamentals, 43–44
    GCRMA algorithm, 48–49
    lack of consensus, 3
    MAS5.0 algorithm, 49–51
    robust multichip analysis, 44–48
    steps, 3
Preprocessing and normalization, Affymetrix GeneChips
    case study, 53–55
    expression measures comparison, 51–53
    fundamentals, xiii, 41–44
    GCRMA algorithm, 48–49
    MAS5.0 algorithm, 49–51
    preprocessing methods, 43–48
    robust multichip analysis, 44–48
    Website, 51
Primate brain expression data set, 32–34
Principal components analysis (PCA), 218
Print-tips
    Loess normalization, 69, 70
    spatial bias detection, 65, 67
Prior probability, 254
Probe-level allele-specific quantitation (PLASQ) algorithm, 259
Probe level model (PLM)
    GCRMA, 49
    robust microarray analysis, 46–47
Probe-level quantile normalization, 248, see also Quantile normalization
Probe logarithmic intensity error (PLIER)
    commonality between gene lists, 163
    cross-normalization comparison, 161
    quality control, 218
    sample-to-sample variance, 163
    signal estimation, 213, 217
Probe selection region (PSR), 211

Prostate cancer, 4
PSR, *see* Probe selection region (PSR)
PT, *see* Point-tip (PT) group
PTC, *see* Papillary thyroid cancer (PTC)

Q

Q-Q plots, 21, 37
Qspline
    intensity- and spatial-dependent strategies, 110
    LOOVC classification errors, 110
    nonlinear array transformation, 37
    normalization methods, 96
QsplineG, 106
QsplineR, 106
Quality, commercial arrays, 153–154
Quality control, workflow, 218
Quantile normalization, *see also* Normalization
    copy number estimation, 248
    method, 159
    nonlinear array transformation, 37
    robust microarray analysis, 45
    statistical principles, 23
Quantile sketch normalization, 217
Quantitative (Q-)PCR, validation, 228–229
Quantitative real-time PCR (qRT-PCR), 152

R

Rank consistency, 242
Rank-consistent probes, 237
RankGene
    differentially expressed gene identification, 8–9
    differentially expressed gene selection, 5
    gene-ranking tool only, 9
    preprocessing techniques, 11
    study results, 9
    usage, 4
Real-time polymerase chain reaction (RT-PCR)
    filtering signal data, 219
    validation, 228–229
Reducing unwanted variations, *see* Variation reduction
Reference, differentiation detection, 26

Regression normalization, 23, *see also* Normalization
Relative log expression (RLE)
    lower quality array identification, 55
    probe level model, 48
Relative rank deviation (RRD), 100
Repeatability and reproducibility
    differential expression measurement, 139
    interlaboratory variation, 133
    MAQC-like study, 141
    modeling technical variation, 122
    standards, 158
Resolution, 233, 243
Resolver, 174
Results and examples
    Affymetrix spike-in data set, 29, 32
    comparative functional genomics, 32–34
    interpretation of normalization selection, 163
    lung cancer data set, 35
    mismatched probes, 28
    parameter selection, 27
    perturbed spike-in data set, 32
    primate brain expression data set, 32–34
    subarray normalization, 27–28
    SUB-SUB normalization, 27
    variation reduction, 27–28
Retaining biological variations, 100–101
Retention probability, 258
RLE, *see* Relative log expression (RLE)
RMA, *see* Robust microarray analysis (RMA); Robust multiarray averaging (RMA); Robust multichip analysis (RMA)
RMAExpress software, 53–55
RMS, *see* Root mean square (RMS)
RNA, 36, 42
Robust microarray analysis (RMA)
    alleles, 249
    background correction, 4, 6, 12
    comparing expression measures, 52
    copy number estimation, 249
    GCRMA algorithm, 48
    Microarray Suite 5 comparison, 51
    multiple arrays and reference, 26
    normalization methods, 159
    perfect match correction, 7
    quality control, 218

Index   **281**

    signal estimation algorithms, 213
    summarization, 4, 7–8
    usage, 3
    workflow study, 53
Robust multiarray averaging (RMA), 163
Robust multichip analysis (RMA), 44–48
Robust PAC, 214
Root mean square (RMS), 236
Rosetta error model, 236
Roundworms, number of genes, 205
RPSL19/ILF2, 167
RRD, *see* Relative rank deviation (RRD)
R software package
    affyQCReport, 4
    APO data set, loading, 76–77
    implementation, 75–78
    installation and loading, 75–77
    open source implementations, 83
    spatial bias detection, 77
    spatial normalization, 78
RT-PCR, *see* Real-time polymerase chain reaction (RT-PCR)

S

SAM, *see* Significance analysis of microarrays (SAM)
Sample size importance, 15
SAPE, *see* Streptavidin-phycoerythrin (SAPE)
SB, *see* Specific background (SB)
Scaling
    copy number estimation, 249
    shift-and-scale transformation, 252
    unwanted data variations, 87–89
Selection
    candidate splicing event, 226–227
    parameters, differentiation detection, 27
Selection, biological interpretation
    alternative expression technologies, 156–157
    background, 153
    commercial arrays, 153–154, 159–160
    cross-normalization comparisons, 161, 163
    discussion, 166–168
    experimental design, 161, 163
    expression platform types, 154–156
    fundamentals, xiv, 151–152, 169–170

    history, 157–158
    imprecise measurements, causes and consequences, 160–161
    normalization methods, 157–160
    quality, 153–154
    results, 163
Selection, data preprocessing methods
    background correction, 6
    data, 5
    data preproccesing, 5–8
    differentially expressed gene identification, 8–9
    discussion, 13–15
    fundamentals, xiii, 1–5
    materials and methods, 5–9
    motivation, 1–2
    normalization, 6–7
    perfect match correction, 7
    results, 2, 9–12
    summarization, 7–8
Semilinear models (SLMs), 95–96
Sentrix Human-6 Expression BeadChip, 156
Shadow-masking technology, 155
Shift-and-scale transformation, 252
Signal adjustment, 43
Signal estimation, 217
Significance analysis of microarrays (SAM)
    differentially expressed gene identification, 8–9
    differentially expressed gene selection, 5
    gene-selection/gene-ranking tool, 5
    hierarchical clustering, 12
    preprocessing techniques, 11
    study results, 9–10
    usage, 4
Simple linear regression, 26
Single-bias-removal methods, 106, 109
Single-dye-removal strategies, 98
Single nucleotide polymorphism (SNP), 82
Single nucleotide polymorphism (SNP), array-based analysis
    Affymetrix mapping arrays, 258–260
    algorithms, 246–260
    case studies, 260–263
    CNAT4, 246–260
    copy number estimation, 247–256

fundamentals, xv–xvi, 245–247
future trends, 264
general components, 246
heterozygosity loss, 256–258
Site-to-site differences, 134, 142
SKY data, 238–240
sLfilterW3
   removing dye effects, 104, 106, 109
   scale normalization strategies comparison, 112
sLfilterW7
   removing dye effects, 104, 106, 109
   scale normalization strategies comparison, 112
sLloess
   double-bias-removal, 109
   removing dye effects, 104, 106
   scale normalization strategies comparison, 112
SLMs, see Semilinear models (SLMs)
SNP, see Single nucleotide polymorphism (SNP)
Solexa, 157
Somatic cells, 234, 243
Spatial biases, detection
   composite normalization, 69
   fundamentals, 64–67
   intensity biases, 67, 69–72
   median filter, 70
   neural network-based spatial and intensity normalization, 70–72
   normalization methods, 67, 69–72
   print-tip Loess normalization, 69
Spatial bias removal benefits, 72–73, 75
Spatial dependence, 63
Spatial-dependent dye effect, 86–87, 97–98
Spatial detrending and normalization methods
   APO data set, 76–77
   Bioconductor software, 75–78
   composite normalization, 69
   detecting spatial bisases, 77
   fundamentals, xiii–xiv, 61–67, 78–79
   installation and loading software, 75–77
   intensity biases, 67, 69–72
   median filter, 70
   neural network-based spatial and intensity normalization, 70–72

normalization methods, 67, 69–72
print-tip Loess normalization, 69
removal of spatial biases, benefits, 72–73, 75
R implementation, 75–78
spatial biases detection, 64–72
spatial normalization, 78
Spatial lowess (SGloess), 97–98
Spatial normalization, Bioconductor software, 78
Spatial pattern
   Affymetrix spike-in data sets, 29, 32
   differentiation detection, 24
   lung cancer data, 35
Spatial plots, 87
Specific background (SB), 50
Spike-ins, see also Affymetrix spike-in data sets
   choice of, 158
   comparing expression measures, 51
   data preprocessing, 3
   issues, 14–15
   semilinear models, 95
Splicing, see Alternative splicing
Spot effect, 90
Spotfire, 174
Spotted arrays, 207–208
Standard deviations, 146
Stanford Microarray Database (SMD), 83
Statistical analysis
   joint expression variation
      interlaboratory variation, 147–149
      linearity, 144–146
      measurement error, 146–147
      normalization, 142–144
Statistical breakdown, normalization, 26–27
Statistical principle, normalization, 21–23
Statistics-centered data analysis tools, 174
Strand hybridization, 207
Stratification, 27–28
Streptavidin-phycoerythrin (SAPE), 42
Stress array, 156
Subarrays
   differentiation detection, 24
   normalization, 27–28
SUB-SUB normalization
   Affymetrix spike-in data sets, 32
   differentiation detection, 27

Index   283

nonlinear array transformation, 37
perturbed spike-in data sets, 32
primate brain expression data set, 33–34
transformation, 37
Summarization
  Affymetrix GeneChip preprocessing steps, 3
  data preproccesing, 7–8
  defined, 3
  fundamentals, 44
  order of data preprocessing, 5
  robust microarray analysis, 46
Surrogate, 243
Survey, normalization methods and assessment
  ANOVA methods, 91–92
  baseline-array approach, 96
  biological variations, retaining, 100–101
  cDNA microarrays and data, 83–84, 101–103
  comparison criteria, 99–101
  data transformation, 89–90
  diagnostic plot dye bias removal, 112
  diagnostic plot dye effect removal, 104, 106
  double-bias-removal methods, 109–110
  dye bias removal, 112
  dye effect removal, 104, 106
  dye-swap normalization, 96–97
  evaluation, 101, 112, 115–116
  external control strategies, 95–96
  fundamentals, 82–83, 116
  future directions, 116
  global techniques, 90–91
  housekeeping-gene-related approaches, 94–95
  iGloess method, 92–93
  iLloess method, 93–94
  intensity- and spatial-dependent dye effects minimization, 98–99
  intensity- and spatial-dependent strategies, 104–110
  intensity-dependent dye effect, 85–86, 92–95
  intensity ratios and values, 89–90
  k-NN classification comparison, 101–103
  logarithmic transformation, 89
  LOOVC classification errors, 106–110, 112
  minimizing strategies, 89–99
  normalization methods, 90–99
  Qspline, 96, 110

scale effect, 87–89
scale normalization strategies, 111–116
semilinear models, 95–96
single-bias-removal methods, 106, 109
spatial-dependent dye effect, 86–87, 97–98
total gene approaches, 92–95
unwanted variations, 84–89, 100
Swiss-Prot, 174
Syllago, 174
Symmetric solutions, 26

T

TaqMan technology, 157
Target distribution, 159
Taylor expansion, 25
Total copy number (TCN)
  copy number estimation, 250
  fundamentals, 246
  paired and unpaired experiments, 247
Training neural networks, 72
Transformation, differentiation detection, 37
Transition probability, 254–255
Translational medicine packages, 174
Trimmed solutions, 26
Tukey biweight
  Microarray Suite 5, 50
  summarization, 4, 7–8
Two-channel DNA and protein data
  APO data set, 76–77
  Bioconductor software, 75–78
  composite normalization, 69
  detecting spatial bisases, 77
  fundamentals, xiii–xiv, 61–67, 78–79
  installation and loading, 75–77
  intensity biases, 67, 69–72
  median filter, 70
  neural network-based spatial and intensity normalization, 70–72
  normalization methods, 67, 69–72
  print-tip Loess normalization, 69, 72
  removal of spatial biases, benefits, 72–73, 75
  R implementation, 75–78
  spatial biases detection, 64–72
  spatial normalization, 78
Two-color assay, 243

U

UCSC genome browser, 225
Undifferentiated probe set, 24
Universal RNA, 161
Unwanted data variations
  data transformation, 89–90
  intensity-dependent dye effect, 85–86
  intensity ratios and values, 89–90
  logarithmic transformation, 89
  minimizing strategies, 89–99
  scale effect, 87–89
  spatial-dependent dye effect, 86–87

V

Validation, splicing variants, 227–229
Values, order of, 168
Variance stabilization
  achieving, 3
  transformation, 37
Variation, joint expression modeling
  Affymetrix platform results, 136–137, 139
  CodeLink platform results, 126–136
  discussion, 141–142
  fundamentals, xiv, 121–124
  interlaboratory variation, 132–134, 136, 147–149
  linearity, 127–128, 144–146
  measurement error, 124–126, 128, 130–132, 146–147
  normalization, 126–127, 142–144
  statistical analysis, 142–149
Variation reduction
  comparison criteria, 100
  differentiation detection, 27–28
Variety effect, ANOVA methods, 92
Viterbi outcome, 258
Volcano plots, 137

W

WBscale
  k-NN evaluation, 113
  k-NN LOOCV error functional assessment, 112
  scale normalization strategies comparison, 111
Weighted residuals, 130–131
Weighted sum of squares, 145
WGSA, see Whole-genome sampling analysis (WGSA)
Whole-genome sampling analysis (WGSA), 247
Within-group variability, 73
Within-microarray scale effect, 89
Within-print-tip-group normalization, 24
Workflow, exon array analysis, 216–229
  alternative splice prediction, 225–226
  anomalous probe sets and signals, 220, 223–224
  candidate splicing event selection, 226–227
  filtering signal data, 219–225
  filtering strategies, 222–225
  fundamentals, 216–217
  ideal splicing event, 219
  low confidence predictions, 224–225
  low signal, 222–223
  quality control, 218
  signal estimation, 217
  splicing variants validation, 227–229
Workflow, 500K mapping arrays with CNAT4 algorithm, 260–263
Workup, common processes, 234, 236
Wrapper methods, 8
Wscale
  k-NN LOOCV error functional assessment, 112
  scale normalization strategies comparison, 111

X

Xenobase, 174

Y

Yeasts
  number of genes, 205
  subarray normalization, 28